Physical Oceanography of Estuaries
(and Associated Coastal Waters)

Physical
Oceanography
of Estuaries
(and Associated Coastal Waters)

Charles B. Officer

Partner, Marine Environmental Services,
Hanover, New Hampshire

Adjunct Professor, Earth Sciences Department,
Dartmouth College, Hanover, New Hampshire

Guest Investigator, Department of Geology and Geophysics,
Woods Hole Oceanographic Institution, Woods Hole, Massachusetts

A Wiley-Interscience Publication

JOHN WILEY & SONS
New York · London · Sydney · Toronto

Copyright © 1976 by John Wiley & Sons, Inc.

All rights reserved. Published simultaneously in Canada.

No part of this book may be reproduced by any means,
nor transmitted, nor translated into a machine language
without the written permission of the publisher.

Library of Congress Cataloging in Publication Data:

Officer, Charles B
   Physical oceanography of estuaries (and associated coastal waters)

   "A Wiley-Interscience publication."
   Bibliography: p.
   Includes index.
   1. Estuarine oceanography. 2. Coasts. I. Title.

GC97.033   551.4′609   75-45330
ISBN 0-471-65278-4

Printed in the United States of America

10 9 8 7 6 5 4 3 2 1

Preface

It has been my intention in this book to give a coordinated treatment of the physical oceanography of estuaries and related bodies of water, such as tidal rivers, straits, bays, and lagoons. The emphasis has been, first, on the mathematical theory and, second, on applications of the theory by various investigators to particular estuarine problems. I have attempted to present the theory in such a manner and to such a degree that following this a reader might feel at ease with the published literature in the field.

I hope that the book may be of interest and use not only to those investigating various aspects of estuarine oceanography but also to those investigating related ecological and engineering problems and to those who, for various reasons, wish to become acquainted with this subject. There is increased interest in existing and potential future pollution problems of estuaries; were it not for such interest it is doubtful that this book would have been written.

The book assumes a mathematical background through calculus and differential equations. It also assumes a reasonable background in physics.

At the outset it might be helpful to give an abbreviated history of the manner in which this theory and its applications have evolved. During the latter half of the last century and the early part of this century contributions have come principally from physical oceanographers and geophysicists in western Europe and the United Kingdom. Over the past 50 years contributions have continued to come from these sources as well as from investigators in the United States and Japan. During this same time period, however, the emphasis of the physical oceanographer has changed from coastal oceanography to deep sea oceanography because of the availability of research funds, ships, and techniques, and because deep sea oceanography had in the past received only modest attention. In addition, over the past 20 years numerous contributions have come from the civil engineering profession. And still more recently there has been a multitude of investigations, surveys, and the like from various governmental, academic, and industrial groups related in one way or another to the environmental concerns of estuaries. There has not been a continued, concerted investigation. Contributions have come from individuals with differing professional backgrounds, and there are necessarily differing results and conclusions among them. Part of the concern in this book has been to resolve these differences as far as possible and to present a coordinated synthesis of the whole.

The approach in this book has been toward an analytic description of the physical processes involved. Little mention is made of computer or physical models and modeling techniques. It is appreciated that such models are often useful, and even essential, particularly for inclusion of the actual configuration of a given estuary. However, a word of caution should be made in the interpretation of results from any model, namely, these model results depend directly on the analytic equations used as their base; and a careful examination of the simplifying assumptions in the defining equations as well as any empirical and scaling factor inputs is essential.

There are several general references which I should like to bring to the reader's attention here. These books and articles can be considered the basic reference points for much of what follows but may not be referred to specifically in any given chapter section. For elementary physical and geophysical theory the reader may refer to Page (1935) and Officer (1974). In hydrodynamics the text by Lamb (1932) is recommended, and in engineering hydraulics the books by Streeter (1948, 1958) and that edited by Rouse (1950) are suggested. In physical oceanography the books by Proudman (1953) and by Defant (1960, 1961) are suggested. For the physical oceanography of estuaries the review articles by Pritchard (1952) and Cameron and Pritchard (1965) and by Bowden (1962, 1967) and the introductory text by Dyer (1973) are useful. For the engineering approach to estuary water movements and the modeling of estuaries the book by Thomann (1972) and the compilations edited by Ippen (1966) and by Ward and Espey (1971) are suggested. For details on the process of diffusion from a point source the book by Csanady (1973) and that edited by Frenkiel and Sheppard (1959) are useful. For a general overview of the biology, chemistry, and physics of estuaries the symposium edited by Lauff (1967) is informative. Still further are the more recent symposia on sedimentation in estuaries edited by Nelson (1973) and on the physical processes responsible for pollutant dispersal in the sea edited by Kullenberg and Talbot (1974). Special reference must also be made to the series of reports, of limited distribution, from the Woods Hole Oceanographic Institution by Stommel and co-workers (Stommel, 1951; Stommel and Farmer, 1952a–c; Francis et al., 1953), portions of which have been published (Arons and Stommel, 1951; Stommel 1953; and Stommel and Farmer, 1952, 1953).

I express my thanks for useful discussions and helpful suggestions during the writing of this book to Dean F. Bumpus, Bostwick H. Ketchum, John H. Ryther, and Henry Stommel, all of the Woods Hole

Oceanographic Institution, to Stuart L. Kupferman and Dennis F. Polis of the University of Delaware, to Keith R. Dyer, R. Kirby, and W. R. Parker of the Institute of Oceanographic Sciences, and to K. F. Bowden of the University of Liverpool. I also express my thanks to Dartmouth College for the use of the facilities at their various libraries and to the Woods Hole Oceanographic Institution for the use of their reference library of unpublished and limited distribution reports.

CHARLES B. OFFICER

Hanover, New Hampshire
December 1975

Contents

1 Introduction

    1-1. Description of estuaries 1
    1-2. Classification of estuaries 3

PART 1 THEORY

2 Hydrodynamics

    2-1. Hydrodynamic preliminaries 9
    2-2. Equations of motion 12
    2-3. Equations of continuity 14
    2-4. Continuity, mixing, and diffusion 18
    2-5. Turbulence and diffusion 26
    2-6. Diffusion scale effects 31
    2-7. Forces 36
    2-8. Turbulence and internal friction 45
    2-9. Boundary layer and bottom friction. 53
    2-10. Scaling and integrated equations 58
    2-11. Hydraulics of open channel flow 61
    2-12. Hydrodynamic numbers and coefficients 67

3 Tidal phenomena

    3-1. Introduction 70
    3-2. Tidal waves 70
    3-3. Tidal cooscillation 80
    3-4. Tidal waves with friction 84
    3-5. Tidal energies and energy losses 93
    3-6. Bores 96

4 Circulation

    4-1. Controls, transitions, and jumps 97
    4-2. Fjord type entrainment flow 102
    4-3. Arrested salt wedge flow 105

4-4. Stratified flow in straits 112
4-5. Horizontal density gradient flow 116
4-6. Two dimensional density gradient flow 125
4-7. Tidal motion and near bottom effects 129
4-8. Fronts 132
4-9. Transverse effects 139
4-10. Some lateral effects on circulation 144

5 Mixing

5-1. Simplified mixing concepts 155
5-2. Overmixing 162
5-3. Entrainment mixing 165
5-4. Longitudinal dispersion across a vertical section . . 174
5-5. Horizontal, one dimensional tidal mixing . . . 178
5-6. Vertical velocity shear, circulation, and mixing effects . 191
5-7. Lateral velocity shear, circulation, and mixing effects . 202
5-8. Diffusion induced circulation 203
5-9. Vertical stability and vertical mixing . . . 209
5-10. Seasonal thermocline and heat effects 218

6 Pollutant dispersion

6-1. Introduction 226
6-2. Longitudinal conservative dispersion 227
6-3. Longitudinal nonconservative dispersion 230
6-4. Vertical dispersion effects 235
6-5. Geometry and source considerations 237
6-6. Velocity shear and point source dispersion . . . 240
6-7. Coupled nonconservative systems 245

PART II APPLICATIONS

7 Great Britain

7-1. Introduction 253
7-2. Mersey estuary and Liverpool bay 254
7-3. Severn estuary and Bristol channel 266
7-4. Southampton estuary and the Solent 274
7-5. Thames estuary 278
7-6. Tay estuary 285
7-7. Red wharf bay, Anglesey 287
7-8. Irish sea 294

8 Europe

8-1. Hardangerfjord and Nordfjord 298
8-2. Frierfjord and Oslofjord 303
8-3. Randers fjord and Schultz's grund 307
8-4. Elbe and Ems estuaries 309
8-5. Bosporus, Dardanelles, Gibraltar, and Bab el Mandeb 310
8-6. European coastal waters 314

9 Americas, East Coast

9-1. St. John estuary 322
9-2. Miramichi, Penobscot, and Kennebec estuaries . . 324
9-3. Connecticut and Merrimack rivers 326
9-4. Hudson river and associated waters 335
9-5. Raritan estuary 341
9-6. Delaware estuary 344
9-7. Chesapeake bay and associated waters. . . . 349
9-8. James estuary 360
9-9. Savannah and Charleston harbors. 366
9-10. Florida peninsula waters 367
9-11. Mississippi river 374
9-12. Galveston bay 378
9-13. Amazon river 379
9-14. Bay of Fundy 380
9-15. Long Island sound. 382

10 Americas, West Coast

10-1. Alaska inlets. 385
10-2. British Columbia inlets. 389
10-3. Vancouver island inlets. 397
10-4. Alberni inlet. 402
10-5. Duwamish and Snohomish rivers. 406
10-6. Columbia river estuary. 408
10-7. Grays harbor and Yaquina estuaries 411
10-8. San Francisco bay and associated waters . . . 412
10-9. Inlets of Chile 418
10-10. Straits of Juan de Fuca and Georgia 418

11 Asia, Australia, and Japan

11-1. Vellar estuary 422
11-2. Hooghly estuary 428

11-3. Chao Phya estuary   428
11-4. Australian estuaries   429
11-5. Rivers of Japan   433
11-6. Osaka and Ariake bays   437

Bibliography   441

Author Index   455

Subject Index   **459**

Physical Oceanography of Estuaries
(and Associated Coastal Waters)

CHAPTER 1

Introduction

1-1. DESCRIPTION OF ESTUARIES

This book is divided into two parts. The first part deals with the physical theory of tidal motion, circulation, and mixing that may occur under various estuarine conditions. The theory is developed in detail and from first principles, so that the reader may ascertain to his own satisfaction how well the various assumptions and approximations in the theory apply to the physical processes being described. The second part presents applications of this theory and provides general descriptions of various estuarine conditions that occur throughout the world. The emphasis in the second part is worldwide, rather than on estuaries in a given few localities, which may have certain similarities in observable conditions, in order to point out the wide variety of physical oceanographic conditions that can and indeed do occur.

The purpose of this introductory chapter is mainly to define several terms used continuously throughout the rest of the book. To start, the definition of an *estuary* given by Pritchard (1952) and Cameron and Pritchard (1965) is used here. Thus an estuary is defined as a semi-enclosed body of water having a free connection with the open sea and within which sea water is measurably diluted with fresh water derived from land drainage. Traditionally the term estuary has been applied to the lower reaches of a river into which sea water intrudes and mixes with fresh water draining seaward from the land. The term has been extended to include bays, inlets, gulfs, and sounds into which several rivers empty and in which the mixing of fresh and salt water occurs.

Various attempts have been made to classify estuaries by type. Such a classification scheme for an estuary as a whole is difficult to devise, for there are hardly any two estuaries that are exactly the same as regards geometric and bathymetric configuration, physical oceanographic characteristics of circulation and mixing, or both. Further, any given estuary may show well mixed or stratified conditions, for example, as a function of longitudinal distance along the estuary, season of the year, or even in some cases phase of the tidal cycle.

1

One can nevertheless distinguish for each estuary or estuary system various geometric conditions. One can have an estuary with a single principal river at its head gradually opening into the sea at its mouth, such as the Delaware estuary, or one that opens in a more discontinuous manner into the sea, such as the Raritan estuary. One can also have an estuary with several tributary rivers which may form estuaries in themselves, such as Chesapeake bay or even a more complex waterway system such as New York or San Francisco harbor. Alternatively, one can have a tidal river that opens with little change in breadth into the sea, such as the Connecticut river, or one with a distributary system, such as the Mississippi river. One can have a fjord with a restricted sill at its entrance, such as most of the Norwegian fjords and to a lesser extent the inlets of British Columbia. One can also distinguish associated water bodies such as straits, canals, bays, lagoons, and seas with restricted access.

Distinctions can also be made as to the physical oceanographic conditions, in terms of the salinity distribution, that exist for any given estuary or portion of an estuary. These conditions are governed principally, as the reader might presume, by the geometry of the estuary, the magnitude of river flow into the estuary, and the magnitude and extent of the tidal motion. The usual distinctions made are that of a well mixed condition in which there is essentially no variation in the salinity in a vertical column (e.g., the Thames estuary), one that is stratified with a halocline between the upper and lower portions of the water column of nearly constant salinity (e.g., the James and Mersey estuaries), and an arrested flow estuary in which there is an interface between two different water types. In the last-mentioned category a further subdivision can be made into a salt wedge estuary with a stable salt wedge underlying a strong, freshwater river flow (e.g., the Mississippi river and some of the rivers in Japan), and a fjord type entrainment flow in which there is a relatively stagnant deep water mass overlain by a thin river runoff flow (e.g., summer conditions for the Norwegian fjords). Further, there can be extreme conditions in which an estuary is entirely fresh or entirely of sea water. An example of an estuary that varies from one of these extremes to the other is the Vellar estuary in India, which consists entirely of fresh water during the monsoon season and entirely of oceanic water during the drought season.

The descriptive terms used here are those mentioned above, namely, *well mixed, stratified, arrested salt wedge,* and *fjord entrainment* type flow conditions. The stratified condition is sometimes further subdivided into a *weakly or partially stratified* condition in which there is a change in salinity of only a few parts per thousand from surface to bottom, and a *strongly or highly stratified* condition in which there is a change in salinity of several parts per thousand from surface to bottom.

To continue, Simmons (1955) makes an interesting observation, based on his field experience, as to the distinctions that can be made in any given estuary. He notes that the physical oceanographic conditions of an estuary can normally be ascertained from the ratio of the river runoff over a tidal cycle into an estuary to the tidal prism of the estuary. When this ratio is of the order of 10^0, an arrested flow condition normally exists; when it is of the order of 10^{-1}, a partially stratified condition normally exists; and when it is of the order of 10^{-2}, a well mixed condition normally exists.

Rochford (1951) makes a useful longitudinal division of conditions that can exist in an estuary, based on his observations of longitudinal salinity profiles. The estuary is divided into four zones—freshwater, longitudinal salinity gradient, tidal, and marine—as measured from the estuary head to its mouth. The terms freshwater zone and marine zone are self explanatory. The tidal zone is the region in which the full effect of tidal mixing can be seen. The longitudinal salinity gradient zone is the region in which there is a fairly rapid change in salinity from nearly freshwater values to nearly oceanic values, presumably related to the change in circulation conditions from that dominated by river flow to that dominated by estuarine density gradient flow. For any given estuary any one or more of these zones may be curtailed or may be nonexistent.

In the above discussion vertical salinity variations, and secondarily longitudinal salinity variations, were used to define the physical oceanographic conditions that may exist in any given section of an estuary. We should logically include lateral variations as well. Unfortunately not enough is known of such lateral variations at this time to permit such categorical distinctions to be made.

1-2. CLASSIFICATION OF ESTUARIES

The best classification system developed so far, at least in our opinion, is the one presented by Hansen and Rattray (1966). It is based on theoretical derivations; it includes only simple, observable quantities, rather than derived quantities; and it describes a continuum, which appears to be the proper method of approach.

This system is summarized in the stratification-circulation diagrams of Figs. 1-1 and 1-2. In both figures the ordinate is the ratio of the tidal averaged salinity difference from the bottom to the surface, $\bar{s}_b - \bar{s}_s$, to the depth and tidal averaged salinity value $\langle \bar{s} \rangle$ at a given location in an estuary; and the abscissa is the ratio of the net circulation velocity at the surface averaged over a tidal cycle $\bar{v}_s$ to the cross section, averaged, net river runoff flow velocity $\langle \bar{v}_x \rangle$. The quantity ν plotted in Fig. 1-1 is the

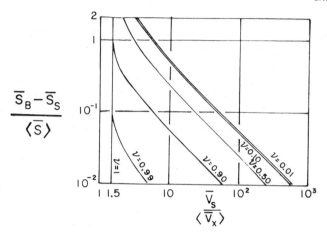

Fig. 1-1   Reproduced with permission from D. V. Hansen and M. Rattray, 1966, *Limnology and Oceanography*, **11,** 319–326, Fig. 1.

ratio of the tidal diffusion salt flux to the total upestuary longitudinal dispersion salt flux. When $\nu = 1$, the upestuary longitudinal dispersion flux is due entirely to tidal diffusion. As $\nu \to 0$, the upestuary longitudinal dispersion flux is due almost entirely to the net circulation flux contribution related to the combined effects of the vertical variation of the tidal averaged salinity and net circulation velocity profiles.

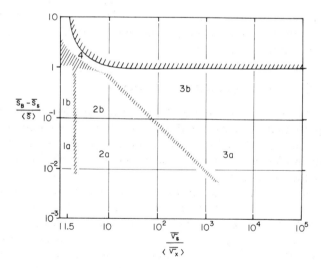

Fig. 1-2   Reproduced with permission from D. V. Hansen and M. Rattray, 1966, *Limnology and Oceanography*, **11,** 319–326, Fig. 2.

Figure 1-2, with the division of the diagram into various characteristic regions, is obtained principally from the considerations shown in Fig. 1-1. In type 1 the net flow is seaward at all depths, and the upestuary salt transfer is entirely due to tidal diffusion. For type 2 the net flow reverses at depth, and both tidal diffusion and net circulation contribute to the upestuary salt flux. The distinction between category *a* and *b* in both cases is that between a well mixed and a stratified condition. Fjord entrainment flow is in category 3, with essentially no tidal diffusion flux contribution, and type 4 is characteristic of an arrested salt wedge flow condition.

PART I

Theory

PART I

Theory

CHAPTER 2

Hydrodynamics

2-1. HYDRODYNAMIC PRELIMINARIES

In the theoretical developments in this and the following chapters it is assumed that the reader has a mathematical background through calculus and differential equations. It is also assumed that he has a reasonable background in physics and is familiar with the manipulations of elementary vector analysis. Although the various theoretical derivations are given in some detail, the physical principles involved in these derivations are emphasized; steps involving simply algebraic or other types of rationalization are often omitted, although the procedures involved in such steps are clearly stated so that the reader may follow them through if he chooses.

In hydrodynamics we deal with a continuum, i.e., a fluid in motion in which there is relative motion of one part with respect to the next. This is to be contrasted with problems in dynamics, such as the motion of a particle or a small finite number of particles under the influence of a given force field and interaction forces among the particles for which there is a discrete number of equations, the solutions of which define the motion of the particles; or such as the motion of a rigid body where distances between fixed points in the body remain constant; or such as elastic wave propagation for which the motions are small and are about an equilibrium position. We must develop a separate formulation to describe hydrodynamic motion before we can develop the equations of motion and their solutions.

We may seek a solution that gives a description of the velocity of the fluid, its pressure, and its density at all points in space it occupies at all instants in time, or seek a solution that gives a description of the motion of each particle. The equations obtained by the two methods are referred to, respectively, as being in *Eulerian* or *Lagrangian* form. We generally seek a solution in terms of the first method; however, it is necessary for the derivation of the equations of motion (i.e., the application of Newton's second law for the forces acting on and the acceleration of a given particle) to use the concepts of the second method. For the solutions our attention is fixed on a particular point in the observer's inertial system; and the velocity of the fluid, its pressure, and its density at this point and all other points are determined as functions of time.

9

Now, let the vector **v** denote the linear velocity of the fluid at a point $P(x, y, z)$ at time t, and v_x, v_y, v_z denote the components of **v** along the three coordinate directions. To calculate the rate at which any function $F(x, y, z, t)$ varies for a moving particle of the fluid, we note that at time $t + \delta t$ the particle originally at $P(x, y, z)$ is now at $Q(x + \delta x, y + \delta y, z + \delta z)$, where Q is given in terms of the components of the linear velocity as $Q(x + v_x \, \delta t, \ y + v_y \, \delta t, \ z + v_z \, \delta t)$. Then, the value of F at Q is

$$F(x + v_x \, \delta t, \ y + v_y \, \delta t, \ z + v_z \, \delta t, \ t + \delta t)$$

$$= F + \frac{\partial F}{\partial x} v_x \, \delta t + \frac{\partial F}{\partial y} v_y \, \delta t + \frac{\partial F}{\partial z} v_z \, \delta t + \frac{\partial F}{\partial t} \delta t \quad (2\text{-}1)$$

If we now introduce the symbol D/Dt to denote a differentiation following the motion of the fluid, the new value of F may also be expressed, by definition, as $F + (DF/Dt) \, \delta t$, so that from (2-1) we obtain

$$\frac{DF}{Dt} = \frac{\partial F}{\partial t} + v_x \frac{\partial F}{\partial x} + v_y \frac{\partial F}{\partial y} + v_z \frac{\partial F}{\partial z} \quad (2\text{-}2)$$

or

$$\frac{DF}{Dt} = \frac{\partial F}{\partial t} + \mathbf{v} \cdot \nabla F \quad (2\text{-}3)$$

where the last expression of (2-3) is the scalar product of the vector **v** and the vector ∇F, ∇ (del) is the vector gradient operator, and ∇F is the

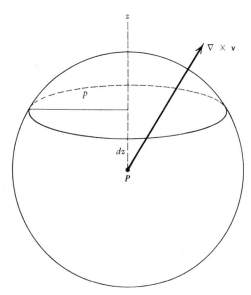

Fig. 2-1

gradient of the scalar *F*. For the acceleration of a particle of the fluid we then have, substituting **v** for *F*,

$$\frac{D\mathbf{v}}{Dt} = \frac{\partial \mathbf{v}}{\partial t} + \mathbf{v} \cdot \nabla \mathbf{v} \tag{2-4}$$

Let us next consider the rotation of neighboring fluid particles about $P(x, y, z)$ at the instant *t*. Since the medium is not rigid, different neighboring particles may have different angular velocities about *P*; and we cannot avail ourselves of the usual relation between the angular velocity **ω** and the curl of the linear velocity $\nabla \times \mathbf{v}$ for a rigid body that $2\mathbf{\omega} = \nabla \times \mathbf{v}$. Nevertheless, we can show that $\nabla \times \mathbf{v}$ at *P* represents both in magnitude and direction twice the *average* angular velocity about *P* of the fluid particles. Consider a small sphere with *P* as center as shown in Fig. 2-1. Then, for an annular ring of radius *p* about the *Z* axis at a distance of *dz* from *P*, we have, applying Stokes' theorem of vector analysis to the area bounded by the ring,

$$\int_{\sigma} \nabla \times \mathbf{v} \cdot d\mathbf{\sigma} = \oint \mathbf{v} \cdot d\mathbf{\lambda}$$

$$\left[|\nabla \times \mathbf{v}|_z + \left. \frac{\partial}{\partial z}(\nabla \times \mathbf{v}) \right|_z dz \right] \pi p^2 = 2\pi p \bar{v}_\lambda$$

$$|\nabla \times \mathbf{v}|_z + \left. \frac{\partial}{\partial z}(\nabla \times \mathbf{v}) \right|_z dz = 2\bar{\omega}_z$$

where **σ** is the area of the ring, **λ** is its periphery, $|\nabla \times \mathbf{v}|_z$ is the *z* component of $\nabla \times \mathbf{v}$, $\bar{v}_\lambda$ is the average value of the tangential component of **v** around the circumference of the annular ring, and $\bar{\omega}_z$ is the corresponding average angular velocity, given by $\bar{\omega}_z = \bar{v}_\lambda/p$. If we average this over two annular rings equidistant from *P* on opposite sides, the term in *dz* will necessarily disappear. Hence, as the volume of the sphere can be divided into such pairs of annular rings we have

$$|\nabla \times \mathbf{v}|_z = 2\bar{\omega}_z$$

for the entire sphere. Since similar expressions hold for the other components, the mean angular velocity of the particles contained in a small sphere with *P* as center is

$$\bar{\mathbf{\omega}} = \tfrac{1}{2}\nabla \times \mathbf{v} \tag{2-5}$$

If the mean angular velocity is everywhere zero, **v** by definition is an irrotational vector. In this case the motion is said to be *irrotational*. Now it is shown in elementary vector analysis that any irrotational vector may be expressed as the gradient of a scalar function of position in space. So if the motion is irrotational, we may write

$$\mathbf{v} = -\nabla\Phi \tag{2-6}$$

The scalar function of position Φ is known as the *velocity potential*, since the velocity is obtained from it by the same mathematical operation as that used to obtain the force from the potential energy in the case of a conservative dynamic system. For such a situation we may seek a solution, generally more easily, in terms of the scalar function Φ rather than in terms of the vector function $\mathbf{v}$.

A surface over which Φ is a constant is known, similar to that in gravitational potential theory, as an *equipotential surface*. The *stream lines* in the fluid are curves having everywhere the direction of the velocity $\mathbf{v}$. It is a fundamental property of the gradient $\nabla\Phi$ that it is perpendicular to the surfaces $\Phi(x, y, z) = \text{constant}$. Therefore the stream lines are perpendicular to the equipotential surfaces, the differential equation of these lines being

$$\frac{dx}{v_x} = \frac{dy}{v_y} = \frac{dz}{v_z} \tag{2-7}$$

2-2. EQUATIONS OF MOTION

In the first instance we deal with a *perfect fluid*, i.e., a fluid that cannot support a tangential stress. The equilibrium stress relations at a point then reduce down simply to a hydrostatic pressure. Let us then examine what effect such an internal hydrostatic pressure distribution has on the equation of motion. Consider a rectangular parallelepiped as shown in Fig. 2-2 of volume $dx\,dy\,dz$ fixed relative to the coordinate axes. If we denote by

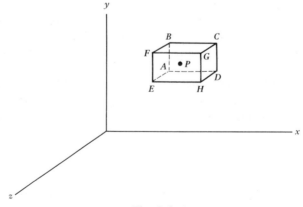

Fig. 2-2

p the pressure at the center P of the parallelepiped, the force on the fluid inside the parallelepiped due to the hydrostatic pressure on the face *ABFE* will be the pressure multiplied by the area of the face, or

$$\left(p - \frac{\partial p}{\partial x}\frac{dx}{2}\right) dy\, dz$$

acting in the positive x direction, and that on the face *DCGH* will be

$$\left(p + \frac{\partial p}{\partial x}\frac{dx}{2}\right) dy\, dz$$

acting in the negative x direction. The net force in the positive x direction due to the pressure on these two faces will then be simply

$$-\frac{\partial p}{\partial x}\, dx\, dy\, dz$$

Similar expressions hold for the forces in the y and z directions for the faces perpendicular to these two directions. Dividing by the volume $dx\, dy\, dz$ and then reducing the parallelepiped to zero, we obtain for the net force per unit volume at P,

$$-\mathbf{i}\frac{\partial p}{\partial x} - \mathbf{j}\frac{\partial p}{\partial y} - \mathbf{k}\frac{\partial p}{\partial z} = -\nabla p \tag{2-8}$$

where $\mathbf{i}$, $\mathbf{j}$, and $\mathbf{k}$ are the unit vectors along the x, y, and z axes, respectively.

In addition to the internal stress due to the pressure there may be external or body forces acting on the elements of the fluid, such as the force of gravity. If $\mathbf{F}$ denotes the external force per unit mass and ρ the density of the fluid, the total force per unit volume will be $\rho\mathbf{F} - \nabla p$; and, applying Newton's second law, the equation of motion for a particle of the fluid will be

$$\rho\frac{D\mathbf{v}}{Dt} = \rho\mathbf{F} - \nabla p \tag{2-9}$$

or, using (2-4),

$$\frac{\partial \mathbf{v}}{\partial t} + \mathbf{v}\cdot\nabla\mathbf{v} = \mathbf{F} - \frac{1}{\rho}\nabla p \tag{2-10}$$

This is clearly not a very simple differential equation, and it is not amenable to some of the general solutions for partial differential equations of physics. We may, however, obtain a useful integration under the simplifying assumptions given below. Often the external force may be derivable from a potential, such as in the case for the force of gravity. In this case we can integrate (2-10) provided further that (*a*) the motion is irrotational and (*b*) the density ρ is a function of the pressure p only. Putting $-\nabla\Omega$ for $\mathbf{F}$ in (2-10), we have

$$\frac{\partial\mathbf{v}}{\partial t}+\mathbf{v}\cdot\nabla\mathbf{v}=-\nabla\Omega-\frac{1}{\rho}\nabla p \qquad (2\text{-}11)$$

As the motion is assumed irrotational, $\nabla\times\mathbf{v}$ is zero, so that from vector analysis we have for the expansion of the triple vector product:

$$(\nabla\times\mathbf{v})\times\mathbf{v}=\mathbf{v}\cdot\nabla\mathbf{v}-(\nabla\mathbf{v})\cdot\mathbf{v}=0$$

or

$$\mathbf{v}\cdot\nabla\mathbf{v}=(\nabla\mathbf{v})\cdot\mathbf{v}=\nabla(\tfrac{1}{2}\mathbf{v}\cdot\mathbf{v})=\nabla(\tfrac{1}{2}v^2) \qquad (2\text{-}12)$$

Substituting (2-12) into (2-11) and putting $-\nabla\Phi$ for $\mathbf{v}$, we obtain

$$-\nabla\frac{\partial\Phi}{\partial t}+\nabla(\tfrac{1}{2}v^2)=-\nabla\Omega-\frac{1}{\rho}\nabla p \qquad (2\text{-}13)$$

Taking the scalar product of (2-13) with an arbitrary position vector $d\boldsymbol{\lambda}$ and remembering the definition of a gradient, we obtain

$$-d\left(\frac{\partial\Phi}{\partial t}\right)+d(\tfrac{1}{2}v^2)=-d\Omega-\frac{1}{\rho}\,dp \qquad (2\text{-}14)$$

Integrating we obtain

$$\int\frac{dp}{\rho}=\frac{\partial\Phi}{\partial t}-\Omega-\tfrac{1}{2}v^2+G(t) \qquad (2\text{-}15)$$

where the constant of integration may be a function of time since our integration is with respect to the space coordinates only, and by hypothesis ρ is a function of p only. If the motion is steady, Φ is not a function of time, and equation (2-15) reduces to

$$\int\frac{dp}{\rho}+\Omega+\tfrac{1}{2}v^2=C \qquad (2\text{-}16)$$

where C is a constant. Equation (2-16) is known as *Bernoulli's equation*. If in addition the fluid is incompressible, ρ will be a constant; and (2-16) reduces further to

$$\frac{p}{\rho}+\Omega+\tfrac{1}{2}v^2=D \qquad (2\text{-}17)$$

where D is a constant.

2-3. EQUATIONS OF CONTINUITY

The equation of continuity is simply a mathematical statement of the conservation of mass. We may derive the hydrodynamic equation of continuity by either an Eulerian or a Lagrangian approach, and it is instructive to do both.

Following the Eulerian approach let us consider the continuity of mass flow into and out of a small volume fixed in our inertial reference system, taken for convenience as a small rectangular parallelepiped as shown in Fig. 2-3. Our derivation then simply equates the net mass flow into the

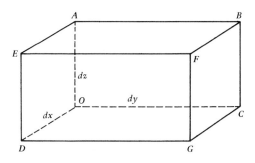

Fig. 2-3

volume to the increase in mass per unit time within the volume. Let the vector **V** represent the mass of fluid flowing through a unit cross section normal to the flow per unit time. Then, if ρ is the density of the fluid and **v** its velocity,

$$\mathbf{V} = \rho\mathbf{v} \tag{2-18}$$

since the mass of fluid flowing through a unit cross section per unit time is that contained in a right prism having a unit area base and an altitude **v**. Then, with reference to Fig. 2-3, the mass of fluid flowing in through *OABC* per unit time is that contained in a prism of base $dy\,dz$ and altitude v_x, or

$$\rho v_x\,dy\,dz = V_x\,dy\,dz$$

and that flowing out through *DEFG* is

$$\left[\rho v_x + \frac{\partial}{\partial x}(\rho v_x)\,dx\right]dy\,dz = \left[V_x + \frac{\partial V_x}{\partial x}\,dx\right]dy\,dz$$

So the net increase in mass of the fluid inside the parallelepiped per unit time due to the flow through these two faces is

$$V_x\,dy\,dz - \left[V_x + \frac{\partial V_x}{\partial x}\,dx\right]dy\,dz = -\frac{\partial V_x}{\partial x}\,dx\,dy\,dz$$

Similarly the net increase in mass per unit time due to the flow through

OAED and *CBFG* is

$$-\frac{\partial V_y}{\partial y}\, dx\, dy\, dz$$

and to that through *OCGH* and *ABFE* is

$$-\frac{\partial V_z}{\partial z}\, dx\, dy\, dz$$

By adding these three expressions and dividing by the volume $dx\, dy\, dz$ of the parallelepiped, the total increase in mass per unit volume per unit time due to the excess of the flow inward over the flow outward is then

$$-\frac{\partial V_x}{\partial x}-\frac{\partial V_y}{\partial y}-\frac{\partial V_z}{\partial z}=-\nabla \cdot \mathbf{V}$$

where $\nabla \cdot \mathbf{V}$ is, from vector analysis, the divergence of the vector $\mathbf{V}$. But this is just the time rate of increase in density; hence

$$\frac{\partial \rho}{\partial t}=-\nabla \cdot \mathbf{V} \tag{2-19}$$

or

$$\frac{\partial \rho}{\partial t}+\nabla \cdot \rho\mathbf{v}=0 \tag{2-20}$$

This is the *equation of continuity*.

Following the Lagrangian approach, let Q be the volume of a small moving element of the fluid. We then have on account of the constancy of mass, following the fluid motion,

$$\frac{D}{Dt}\,\rho Q=0$$

or

$$\frac{1}{Q}\frac{DQ}{Dt}+\frac{1}{\rho}\frac{D\rho}{Dt}=0 \tag{2-21}$$

To calculate the value of the first term in (2-21) let the volume element in question at time t be a rectangular space dx, dy, dz having one corner at O and edges parallel to the three coordinate axes. Then at a later time the same volume element will form an oblique parallelepiped. The side of original length dx will now be oriented so that its projections on the three coordinate axes are

$$\left(1+\frac{\partial v_x}{\partial x}\,dt\right)dx \qquad \frac{\partial v_y}{\partial x}\,dt\,dx \qquad \frac{\partial v_z}{\partial x}\,dt\,dx$$

A similar orientation holds for the other two sides. We then have, from vector analysis, that the volume of the parallelepiped is the triple scalar product of its three sides. Neglecting product terms of the small quantities $\partial v_n/\partial m$ we then have for the change in volume per unit volume per unit time,

$$\frac{1}{Q}\frac{DQ}{Dt}=\frac{\partial v_x}{\partial x}+\frac{\partial v_y}{\partial y}+\frac{\partial v_z}{\partial z} \tag{2-22}$$

As we see, expression (2-22) measures the rate of dilatation of the fluid and is sometimes referred to as the *expansion*. Substituting (2-22) into (2-21), we obtain

$$\frac{D\rho}{Dt}+\rho\left(\frac{\partial v_x}{\partial x}+\frac{\partial v_y}{\partial y}+\frac{\partial v_z}{\partial z}\right)=0 \tag{2-23}$$

or

$$\frac{D\rho}{Dt}+\rho\nabla\cdot\mathbf{v}=0 \tag{2-24}$$

which is the form of the equation of continuity in terms of the derivative D/Dt following the motion. Substituting (2-3) into (2-24) we obtain

$$\frac{\partial\rho}{\partial t}+\mathbf{v}\cdot\nabla\rho+\rho\nabla\cdot\mathbf{v}=0$$

or

$$\frac{\partial\rho}{\partial t}+\nabla\cdot\rho\mathbf{v}=0 \tag{2-25}$$

which is the same as (2-20).

If the fluid is incompressible, which is often the assumption in hydrodynamics, ρ is a constant and equation (2-20) or (2-25) becomes

$$\nabla\cdot\mathbf{v}=0 \tag{2-26}$$

or

$$\frac{\partial v_x}{\partial x}+\frac{\partial v_y}{\partial y}+\frac{\partial v_z}{\partial z}=0 \tag{2-27}$$

If, in addition, the motion is irrotational, we obtain, substituting (2-6) into (2-26),

$$\nabla\cdot\nabla\Phi=0 \tag{2-28}$$

This differential equation is simply Laplace's equation.

The equation of continuity is of fundamental importance in hydrodynamics. For many hydrodynamic problems the method of attack consists simply in finding a solution for this differential equation that satisfies the boundary conditions.

We have so far examined application of the principle of continuity to the derivation of an equation of continuity for mass transport. Let us now look at its application to other quantities occurring in hydrodynamics. *Salinity s* is usually defined as a ratio expressing the number of grams of salt per kilogram of sea water. With reference to expression (2-18) the flux of salt **S** expressed as mass of salt flow per unit time is then

$$\mathbf{S} = \rho s \mathbf{v} \tag{2-29}$$

Following through the same derivation as given before, or simply using ρs for ρ in (2-20), we obtain

$$\frac{\partial}{\partial t}(\rho s) + \nabla \cdot \rho s \mathbf{v} = 0 \tag{2-30}$$

which can be reduced, using (2-20), to

$$\rho\left(\frac{\partial s}{\partial t} + \mathbf{v} \cdot \nabla s\right) + s\left(\frac{\partial \rho}{\partial t} + \nabla \cdot \rho \mathbf{v}\right) = 0$$

$$\frac{\partial s}{\partial t} + \mathbf{v} \cdot \nabla s = 0 \tag{2-31}$$

or, in terms of D/Dt,

$$\frac{Ds}{Dt} = 0 \tag{2-32}$$

We could have written (2-32) down directly. It is simply the direct statement that any particular quantity of water moves without change in total salt content. If the conditions are stationary, (2-31) reduces to

$$\mathbf{v} \cdot \nabla s = 0 \tag{2-33}$$

or

$$v_x \frac{\partial s}{\partial x} + v_y \frac{\partial s}{\partial y} + v_z \frac{\partial s}{\partial z} = 0 \tag{2-34}$$

We may extend this to any conservative quantity c associated with fluid motion, i.e., one that remains unchanged for any particular quantity of sea water, for which we have the same equation as (2-32),

$$\frac{Dc}{Dt} = 0 \tag{2-35}$$

or

$$\frac{\partial c}{\partial t} + \mathbf{v} \cdot \nabla c = 0 \tag{2-36}$$

2-4. CONTINUITY, MIXING, AND DIFFUSION

As mentioned in the previous section, many physical oceanographic problems can be solved simply by the application of the considerations of continuity along with the associated boundary conditions. We start in this section with a few simple examples and then consider somewhat more complex examples in the latter part of the section.

Consider a body of water having one major inflow section and one major outflow section, as shown in Fig. 2-4. Let A_1 and A_2 be the cross

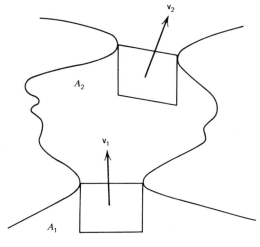

Fig. 2-4

sectional areas at the entrance and exit, and $\mathbf{v}_1$ and $\mathbf{v}_2$ the average water velocities normal to these two sections. Let r denote the mean rainfall plus freshwater river flow over evaporation per unit surface area between the two cross sections, and let B denote the total surface area between the two sections. Then, assuming that the density ρ is constant, that there is complete mixing of the salt water entering A_1 and the fresh water represented by rB, and that the conditions are stationary, we have for continuity of water mass,

$$A_1 v_1 + rB = A_2 v_2 \qquad (2\text{-}37)$$

and for continuity of the salt mass

$$A_1 v_1 s_1 = A_2 v_2 s_2 \qquad (2\text{-}38)$$

where s_1 and s_2 are the average salinities of the sea water at sections A_1 and A_2, respectively. Solving (2-37) and (2-38) for $A_1 v_1$ and $A_2 v_2$, we obtain

$$A_1 v_1 = \frac{rB}{(s_1/s_2) - 1} \qquad (2\text{-}39)$$

and

$$A_2 v_2 = \frac{rB}{1 - (s_2/s_1)} \tag{2-40}$$

From (2-39) and (2-40) we see that, when r is positive, i.e., when the influx of fresh water exceeds evaporation, the current is in the direction of decreasing salinity; and when r is negative, the current is in the direction of increasing salinity, as expected. Now, if we denote by h the mean depth of water beneath the surface B, we see from (2-40) that the time necessary to empty the volume between the two cross sections, or the *flushing time*, is given by

$$t = \frac{hB}{A_2 v_2} = \left(1 - \frac{s_2}{s_1}\right)\frac{h}{r} \tag{2-41}$$

If A, v, and s can be regarded as functions of a single space coordinate under the same conditions as above, and if r can be regarded as a constant, e.g., no influx of fresh water by river flow except through the section A_1, (2-37) and (2-38) can be generalized to

$$\frac{\partial}{\partial x}(Av) = br \tag{2-42}$$

and

$$\frac{\partial}{\partial x}(Avs) = 0 \tag{2-43}$$

where $b = b(x)$ is the width of the area B normal to x. Combining these two equations we obtain

$$Av\frac{\partial s}{\partial x} + brs = 0 \tag{2-44}$$

Let us consider next the similar problem of a partially enclosed body with inflow to and outlet from the body through one principal opening, channel, or the like. Let us presume that there is a stratified flow of the incoming and outgoing currents as shown in Fig. 2-5. Then, (2-37) to (2-41) apply. We must also have a net vertical mass movement between the two stratified flows to complete the flow circuit with a mean vertical velocity w across the stratification boundary such that wB is equal to $A_1 v_1$.

So far we have assumed that there is complete mixing and that the transfer of any particular property of the fluid is by mass transfer alone. We are now interested in examining the mixing processes themselves, and we first look at a phenomenon known as diffusion. For our purpose *diffusion* is defined as the dispersion of a property of the fluid, such as salinity, without any net mass transfer of the fluid itself. In physical

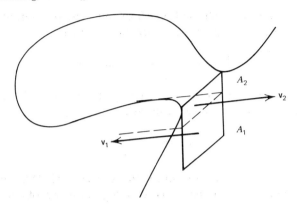

Fig. 2-5

oceanography we recognize diffusive transfer due to molecular motion of the water particles, in which case it is referred to as *molecular diffusion,* and due to turbulence of the water masses, in which case it is referred to as *eddy diffusion.* The corresponding dispersion of the fluid due to the mass transfer of the fluid is known as *advection.*

Consider the diffusion across a unit cross section normal to the z axis as shown in Fig. 2-6. As a result of turbulence or molecular motion we have a mass of fluid $m_d(z)$ passing down through the cross section per unit time for fluid above the plane $z = 0$ and a corresponding mass of fluid $m_u(z)$ passing up through the cross section per unit time for fluid below the plane $z = 0$. Since there is no net mass transfer of fluid, we can write

$$\int_- m_u(z) \, dz = \int_+ m_d(z) \, dz \tag{2-45}$$

where the first integral is taken for negative values of z and the second

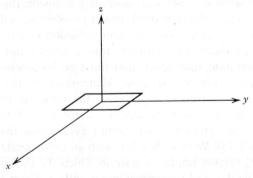

Fig. 2-6

for positive values of z. The salt flux S or salt mass per unit area per unit time progressing in the positive z direction is then

$$S = \int_- m_u(z)s(z)\,dz - \int_+ m_d(z)s(z)\,dz \qquad (2\text{-}46)$$

Since these integrals generally have appreciable values only in the neighborhood of the plane $z = 0$, we can write for $s(z)$,

$$s(z) = s_0 + \frac{\partial s}{\partial z}\,dz \qquad (2\text{-}47)$$

Substituting (2-47) into (2-46) and with the use of (2-45), and assuming $\partial s/\partial z$ to be constant over the range of integration, we obtain

$$S = -\left[\int_- zm_u(z)\,dz + \int_+ zm_d(z)\,dz\right]\frac{\partial s}{\partial z}$$

$$= -\eta\frac{\partial s}{\partial z} \qquad (2\text{-}48)$$

where we have made the substitution η for the expression in the brackets. The quantity η is known as the *coefficient of diffusion*. Given a particular $m_u(z)$ and $m_d(z)$ for molecular or turbulent motion, the coefficient of diffusion can be evaluated in terms of other physical quantities. Equation (2-48) states that the diffusive salt flux in a given direction is proportional to the salinity gradient in that direction. In general the flux of any conservative quantity is proportional to the concentration gradient in the same direction.

The derivation given above for (2-48) illustrates both the strengths and the weaknesses of this method of approach to diffusion. It clearly shows the dependence of the diffusive salt flux on the salinity gradient. However, it also shows in the form of the integrals the dependence on the scale of the diffusion process. For molecular diffusion this scale is much smaller than the scale of any hydrodynamic problem we will consider. For turbulent diffusion, which is the dominant diffusion process in all physical oceanographic problems we consider, this is usually not the case. The scale of the maximum turbulence that must be considered in any given problem is comparable to the space dimensions of the problem. For example, in the dispersion of a pollutant the diffusion process depends also on the spatial and time history of the distribution of the pollutant; in effect, in such scale problems, we cannot even make the simplification from (2-46) to (2-48). We are then left with an unfortunate dilemma. The turbulence itself usually implies a state in which the instantaneous velocities exhibit irregular and sometimes apparently random fluctuations, so that in practice usually only the statistical properties of such motion can be recognized and subjected to analysis; however, even in such a statistical analysis of the diffusive effects of such turbulence we now see that we often cannot describe such turbulent diffusion as a physical property of a given hydrodynamic situation but that it also depends on the distribution of the quantity being diffused as well. We discuss this further later in this section, in Sections 2-5 and 2-6, and in Chapters 5 and 6.

In the present analysis, if the coefficient of diffusion η is a constant, which is generally the case for molecular diffusion but not for eddy diffusion, (2-48) may be generalized to

$$\mathbf{S} = -\eta \, \nabla s \qquad (2\text{-}49)$$

If η is not a constant, but a function of the spatial coordinate directions, it is generally more convenient to use the three scalar relations

$$S_x = -\eta_x \frac{\partial s}{\partial x} \qquad S_y = -\eta_y \frac{\partial s}{\partial y} \qquad S_z = -\eta_z \frac{\partial s}{\partial z} \qquad (2\text{-}50)$$

rather than to consider η in its tensor form, since usually for such cases the eddy coefficient of diffusion is appreciable for only one coordinate direction and can be neglected for the other two.

As in Section 2-3 we can express the diffusive flux of salt in terms of a diffusion velocity $\mathbf{w}$ as

$$\mathbf{S} = \rho s \mathbf{w} \qquad (2\text{-}51)$$

similar to (2-29). The equation of continuity applies equally to salt diffusion so that, if ρ is a constant, we have, from (2-30),

$$\frac{\partial s}{\partial t} = -\nabla \cdot (s \mathbf{w}) \qquad (2\text{-}52)$$

or

$$\rho \frac{\partial s}{\partial t} = -\nabla \cdot \mathbf{S} \qquad (2\text{-}53)$$

If (2-49) applies, we will then have, from (2-53),

$$\frac{\partial s}{\partial t} = \frac{\eta}{\rho} \nabla^2 s \qquad (2\text{-}54)$$

or

$$\frac{\partial s}{\partial t} = K \nabla^2 s \qquad (2\text{-}55)$$

where we have made the substitution

$$K = \frac{\eta}{\rho} \qquad (2\text{-}56)$$

The quantity K is referred to as the *kinematic coefficient of diffusion*. We generally use the diffusion coefficient K rather than η; for physical oceanographic problems, using cgs units, the numerical values of K and η are essentially the same. The partial differential equation (2-55) is a familiar form encountered in heat conduction, electromagnetism, and diffusion problems. It is sometimes referred to as the *Fickian diffusion equation* to distinguish it from other methods of describing the oceanographic diffusion process. If (2-49) does not apply, we will have, from (2-50),

$$\frac{\partial s}{\partial t} = \frac{\partial}{\partial x}\left(K_x \frac{\partial s}{\partial x}\right) + \frac{\partial}{\partial y}\left(K_y \frac{\partial s}{\partial y}\right) + \frac{\partial}{\partial z}\left(K_z \frac{\partial s}{\partial z}\right) \qquad (2\text{-}57)$$

For essentially all problems in physical oceanography the eddy diffusion coefficient is much larger than the molecular diffusion coefficient, so that we are usually concerned with an equation of the form (2-57), which as stated previously may often be reduced to a one dimensional form. If we are dealing with a nonconservative quantity, instead of the salinity, or with a problem involving sources and sinks for either a conservative or nonconservative quantity, we will have to alter the equation of continuity accordingly, and consequently (2-55) and (2-57). The eddy diffusion coefficients η_x, η_y, and η_z are sometimes referred to, in physical oceanographic literature, as the *austausch* or *exchange coefficients*.

We now derive the equation for eddy diffusion through an alternative method which provides us with a different relation for the eddy diffusion coefficients. Consider a medium in which there is turbulence. The instantaneous velocity $\mathbf{v}$ may be taken as equal to a mean velocity $\bar{\mathbf{v}}$ and a velocity deviation $\mathbf{v}'$. The limits taken for the mean immediately become a problem. For example, in oceanographic problems, if the mean is taken over several days, it will include tidal variations; if it is taken over several years, it will include seasonal variations. The same reasoning also applies to the spatial means that determine the scale of turbulence being considered. Nevertheless, we assume that the velocity deviation is sufficiently well behaved across its zero position that mean values can be defined. In taking averages the following principles are adopted. If A and B are dependent variables being averaged with respect to any one of the independent variables x, y, z, or t,

$$\overline{\bar{A}B} = \bar{A}\bar{B}$$

since $\bar{A}$ is already an averaged value and is invariable over the second averaging and

$$\overline{\frac{\partial A}{\partial t}} = \frac{1}{2\tau}\int_{t-\tau}^{t+\tau}\frac{\partial A}{\partial t}\,dt = \frac{1}{2\tau}[A(t+\tau)-A(t-\tau)]$$

$$= \frac{1}{2\tau}\frac{\partial}{\partial t}\int_{t-\tau}^{t+\tau}A\,dt = \frac{\partial}{\partial t}\left[\frac{1}{2\tau}\int_{t-\tau}^{t+\tau}A\,dt\right]$$

$$= \frac{\partial \overline{A}}{\partial t}$$

where the overbar denotes a mean value. We may then write

$$v_x = \bar{v}_x + v_x' \qquad v_y = \bar{v}_y + v_y' \qquad v_z = \bar{v}_z + v_z' \tag{2-58}$$

where

$$\overline{v_x'} = \overline{v_y'} = \overline{v_z'} = 0$$

and $$\tag{2-59}$$

$$\frac{\overline{\partial v_x'}}{\partial t} = \frac{\overline{\partial v_x'}}{\partial x} = \frac{\overline{\partial v_x'}}{\partial y} = \frac{\overline{\partial v_x'}}{\partial z} = 0$$

and similarly for the partial derivatives with respect to v_y' and v_z'. Substituting relations (2-58) into the equation of continuity (2-27) for ρ a constant, we have, taking the mean value,

$$\frac{\partial \bar{v}_x}{\partial x} + \frac{\partial \bar{v}_y}{\partial y} + \frac{\partial \bar{v}_z}{\partial z} = 0$$

or $$\tag{2-60}$$

$$\nabla \cdot \bar{\mathbf{v}} = 0$$

From (2-60) and (2-27) we then also have

$$\frac{\partial v_x'}{\partial x} + \frac{\partial v_y'}{\partial y} + \frac{\partial v_z'}{\partial z} = 0$$

or $$\tag{2-61}$$

$$\nabla \cdot \mathbf{v}' = 0$$

Equations (2-60) and (2-61) may be considered the equations of continuity for the mean velocity and for the velocity deviation, respectively. We may also take in the same way the instantaneous salinity s to be equal to a mean salinity $\bar{s}$ and a salinity deviation s' with relations similar to (2-58) and (2-59). Substituting into the equation of continuity for salt mass (2-31) and taking the mean value, we have, using the relation (2-61),

$$\frac{\partial \bar{s}}{\partial t} + \bar{\mathbf{v}} \cdot \nabla \bar{s} + \overline{\mathbf{v}' \cdot \nabla s'} = 0$$

$$\frac{\partial \bar{s}}{\partial t} + \bar{\mathbf{v}} \cdot \nabla \bar{s} + \nabla \cdot \overline{(s'\mathbf{v}')} = 0 \tag{2-62}$$

or

$$\frac{\partial \bar{s}}{\partial t} + \bar{v}_x \frac{\partial \bar{s}}{\partial x} + \bar{v}_y \frac{\partial \bar{s}}{\partial y} + \bar{v}_z \frac{\partial \bar{s}}{\partial z} + \frac{\partial}{\partial x} \overline{(s'v'_x)} + \frac{\partial}{\partial y} \overline{(s'v'_y)} + \frac{\partial}{\partial z} \overline{(s'v'_z)} = 0 \qquad (2\text{-}63)$$

If we now define the functions K_x, K_y, and K_z by

$$K_x = -\frac{\overline{s'v'_x}}{\partial \bar{s}/\partial x} \qquad K_y = -\frac{\overline{s'v'_y}}{\partial \bar{s}/\partial y} \qquad K_z = -\frac{\overline{s'v'_z}}{\partial \bar{s}/\partial z} \qquad (2\text{-}64)$$

assuming of course that such definitions have some physical significance (2-63) becomes

$$\frac{\partial \bar{s}}{\partial t} = \frac{\partial}{\partial x}\left(K_x \frac{\partial \bar{s}}{\partial x}\right) + \frac{\partial}{\partial y}\left(K_y \frac{\partial \bar{s}}{\partial y}\right) + \frac{\partial}{\partial z}\left(K_z \frac{\partial \bar{s}}{\partial z}\right) - \bar{v}_x \frac{\partial \bar{s}}{\partial x} - \bar{v}_y \frac{\partial \bar{s}}{\partial y} - \bar{v}_z \frac{\partial \bar{s}}{\partial z} \qquad (2\text{-}65)$$

which is the same as (2-57) with the addition of the advection term $\bar{\mathbf{v}} \cdot \nabla \bar{s}$. Equation (2-65) states that the local time rate of change in salinity $\partial \bar{s}/\partial t$ within a small volume is equal to the diffusion into the volume minus the advection out of the volume.

2-5. TURBULENCE AND DIFFUSION

We now look at, in somewhat more detail, the process of diffusion by turbulent water motion in order to gain a better understanding of the process itself, the complications one has to consider, and the simplifications that sometimes can be made. To do so we examine in three different ways one of the simplest diffusion problems, that of one dimensional diffusion from an instantaneous source of finite quantity at $x = 0$ and $t = 0$.

First, let us look at the solution of the Fickian diffusion equation (2-55), corresponding to this source condition. This is one of the elementary problems in heat conduction concerning which the reader may refer to any standard text, such as Carslaw and Jaeger (1959). We restate it briefly here. By taking the appropriate partial derivatives we see that

$$c = At^{-1/2}e^{-x^2/4Kt} \qquad (2\text{-}66)$$

is a particular solution of (2-55). We further see that $c \to 0$ as $t \to 0$ for all $x \neq 0$, and that $c \to \infty$ as $t \to 0$ for $x = 0$ in such a way that its integral over the x axis

$$\int_{-\infty}^{\infty} c\, dx = \int_{-\infty}^{\infty} At^{-1/2}e^{-x^2/4Kt}\, dx = 2A\sqrt{\pi K} \qquad (2\text{-}67)$$

is finite for all t, thus meeting our given source condition. We must normalize our source to unity in order to be able to compare directly with the probability distributions of the following discussions. From (2-67) we have then, for our final solution,

$$c = \frac{1}{\sqrt{4\pi Kt}} e^{-x^2/4Kt} \tag{2-68}$$

This is a familiar Gaussian distribution in x, which spreads out in the x direction as t increases. From the symmetry of (2-68) about the x axis we see that the mean value of the distance of the concentration away from the distribution center is zero, or $\bar{x} = 0$. We shall, however, be interested in the mean square value $\overline{x^2}$ which from (2-68) is given by

$$\overline{x^2} = \int_0^\infty \frac{2x^2}{\sqrt{4\pi Kt}} e^{-x^2/4Kt} \, dx = 2Kt \tag{2-69}$$

We note then that $\overline{x^2}$ is directly proportional to the time t and to the eddy diffusion coefficient K, remembering of course that this result is based on an assumption of the validity of (2-55).

To examine the mechanics of the turbulent diffusion process let us consider what occurs following instantaneous release of a finite quantity of marked water particles at the origin at $t = 0$. We take a somewhat fictitious mechanism and presume that the turbulence is such that each particle moves through a distance l with a constant velocity v and that another path also of length l then starts but that its direction is independent of the initial direction. We thus have a discontinuous but random diffusion process. The probability, then, of finding a given particle at fixed multiple distances l away from the origin after four such movements is as shown in Fig. 2-7. As stated this is a usual beginning problem in physical statistics, concerning which the reader may refer to an appropriate text such as that of Lindsay (1941). The solution for the probability

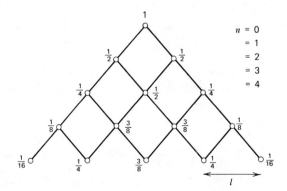

Fig. 2-7

of finding a given particle at any fixed multiple of l away from the origin is given in terms of the ratios of factorials, which for large values of n may be approximated by

$$P = \frac{2}{\sqrt{2\pi n}} e^{-x^2/2nl^2} \tag{2-70}$$

Again the distribution is Gaussian. Equating the coefficients of the exponentials in (2-68) and (2-70), we have

$$4Kt = 2nl^2$$

$$2Kt = vlt$$

$$K = \tfrac{1}{2}vl = \tfrac{1}{2}v^2 t_0 = \tfrac{1}{2}\frac{l^2}{t_0} \tag{2-71}$$

since

$$t = nt_0 = n\frac{l}{v}$$

where t_0 is the time required for each movement. [In the normalization the probability P refers to finding a particle over a distance $\pm l$ on each side of a given location so that in taking the integral similar to (2-67) the integrand of (2-70) must be divided by $2l$, which then gives the same value for K as (2-71) when the amplitude coefficients of (2-68) and (2-70) are compared.] This analysis provides us with an interesting result, namely, that when the turbulence is considered as a discontinuous process the eddy diffusion coefficient is proportional to the square of the turbulence dimensions and inversely proportional to the time scale of the turbulence. From (2-69) and (2-71) we have further that

$$\overline{x^2} = 2Kt = nl^2 \tag{2-72}$$

which we could also have obtained directly from our original assumptions in that

$$\overline{x^2} = \overline{(y_1 + y_2 + y_3 + \cdots + y_n)^2} = nl^2$$

where $y_1, y_2, y_3, \ldots, y_n$ are each numerically equal to l and where, since there is no correlation between the directions of successive jumps, $y_m y_n = 0$ for $m \neq n$.

To continue further in an examination of the mechanics of the turbulent diffusion process we consider it as a continuous rather than a discontinuous movement of water particles. In addition we must consider that there is some correlation between the velocity of a particle at a given instant in time and that of the same particle at some later time, or equivalently, that there is a correlation of the velocities between adjacent particles at the same instant of time. We may correctly assume that this correlation value approaches zero in some manner depending on the scale and details of the turbulence as the time or spacing interval is increased. For simplicity we consider a field of turbulent flow that is statistically uniform, so that the value of $\overline{v^2}$, where v is here the turbulent water velocity, and the mean square of all the derivatives of v with respect to time are the same at all points. This development is due to Taylor (1921); the reader may also refer to Goldstein (1938), Sutton (1949), or Hinze (1959). If x is the coordinate of a marked particle at time t, taking the same source condition as before,

$$x = \int_0^t v \, dt_1$$

and

$$\frac{1}{2}\frac{d}{dt}\overline{x^2} = \frac{1}{2}\frac{d}{dt}\overline{x}^2 = \overline{x\frac{dx}{dt}} = \overline{xv_t} = \overline{v_t \int_0^t v \, dt_1} \qquad (2\text{-}73)$$

where v_t is the velocity of the given particle at time t. Now to find the average value of $v_t \int_0^t v \, dt_1$, we may imagine the whole time from 0 to t to be divided into n intervals, as shown in Fig. 2-8. In each interval we make a summation, or averaging, over all particles. Thus in the sth interval we have a contribution,

$$\overline{v_t v_{t-\tau}}\left(\frac{t}{n}\right) \qquad (2\text{-}74)$$

Now the correlation coefficient between two velocities separated by a time interval τ is defined as

$$R_\tau = \frac{\overline{v_t v_{t+\tau}}}{\overline{v^2}} = \frac{\overline{v_t v_{t-\tau}}}{\overline{v^2}} \qquad (2\text{-}75)$$

so that if $d\tau$ is taken as the time interval T/n and we let $n \to \infty$, we will have for (2-74),

$$R_\tau \overline{v^2} \, d\tau \qquad (2\text{-}76)$$

Substituting (2-76) into (2-73) and remembering that $\overline{v^2}$ is a constant, we

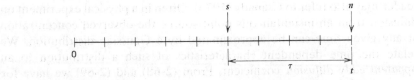

Fig. 2-8

have

$$\frac{1}{2}\frac{d}{dt}\overline{x^2} = \overline{xv_t} = \overline{v^2}\int_0^t R_\tau \, dt \qquad (2\text{-}77)$$

or

$$\overline{x^2} = 2\overline{v^2}\int_0^t \int_0^{t_1} R \, d\tau \, dt_1 \qquad (2\text{-}78)$$

Now, for small values of t we expect that R_τ would not differ appreciably from unity over the range of the integration, so that (2-78) becomes

$$\overline{x^2} = 2\overline{v^2}\int_0^t t_1 \, dt_1 = \overline{v^2}t^2 \qquad (2\text{-}79)$$

The motion is entirely correlatable, and we have a simple advection effect with the root mean square distance traveled being given, as expected, by the product of the root mean square velocity and the time. For large values of t we expect that the first integral will approach some limiting value I as R_τ approaches zero and remain constant for all values of t_1 greater, so that (2-78) then becomes

$$\overline{x^2} = 2\overline{v^2}\int_0^t I \, dt_1 = 2\overline{v^2}It \qquad (2\text{-}80)$$

This has the same dependence as the solution (2-69) for the Fickian diffusion equation, so that the effective eddy diffusion coefficient is then, in terms of the present analysis,

$$K = \overline{v^2}I = \overline{v^2}\int_0^\infty \frac{\overline{v_t v_{t+\tau}}}{\overline{v^2}} \, d\tau \qquad (2\text{-}81)$$

Both (2-79) and (2-80) give an expression for the diffusive transport in terms of the turbulent velocities. We begin to see quantitatively how the scale of the turbulence is involved in determining the diffusive transport, in that for times small compared to the scale of the turbulence the transport is by simple advection and for time scales comparable to the scale of the turbulence the transport is as described by the Fickian diffusion equation.

We may extend this type of analysis a further step through the following consideration, given by Batchelor (1949), concerning which the reader may also refer to Csanady (1973). Often in a physical experiment on diffusion from an instantaneous point source the observed concentration at any given time can be approximated by a Gaussian distribution. We relate the time dependent characteristics of such a distribution to an *apparent eddy diffusion* coefficient. From (2-68) and (2-69) we have for the one dimensional form of a Gaussian distribution in terms of the observable quantity $\overline{x^2}$ and with the inclusion of a mean advection velocity $\overline{v_x}$, which we assume to be constant,

$$\bar{c} = \frac{1}{\sqrt{2\pi\overline{x^2}}} e^{-(x-\bar{v}_x t)^2/2\overline{x^2}} \tag{2-82}$$

where we now consider that $\overline{x^2} = \overline{x^2}(t)$ but is not necessarily of the form (2-69). To compare with a defining equation of the form (2-65) we then have from (2-82) that

$$\frac{\partial \bar{c}}{\partial t} + \bar{v}_x \frac{\partial \bar{c}}{\partial x} = \frac{1}{2}\left[-\frac{1}{\overline{x^2}} + \frac{(x-\bar{v}_x t)^2}{(\overline{x^2})^2} \right] \bar{c} \frac{d\overline{x^2}}{dt} \tag{2-83}$$

$$= \frac{1}{2} \frac{d\overline{x^2}}{dt} \frac{\partial^2 \bar{c}}{\partial x^2}$$

which on comparison with (2-65) and relation (2-77), considering K_x to be a function of t only, gives

$$K_x = \frac{1}{2} \frac{d\overline{x^2}}{dt} = \overline{v_x'^2} \int_0^t R_\tau \, d\tau \tag{2-84}$$

The same analysis for the three dimensional case gives similar relations for K_y and K_z in terms of $\overline{y^2}$ and $\overline{z^2}$, respectively.

2-6. DIFFUSION SCALE EFFECTS

We have seen in the previous section for the phenomenon of turbulent diffusion with particular regard to the results of relations (2-79) and (2-80) that when the time scale is large, approaching steady state, we can use the Fickian diffusion equation, and that when the time scale is very small we have a simple advection effect. Here we look at some of the effects within the intermediate range between these two extremes. It will be clear from the discussion that we can do this only in a relatively imperfect manner.

To start let us look at the consequences, and inconsistencies with the observed facts, of a direct application of the Fickian diffusion equation in this intermediate time range. We have from (2-69) that for an instantaneous source that the time rate of change of $\overline{x^2}$ is

$$\frac{d\overline{x^2}}{dt} = 2K \tag{2-85}$$

This relation asserts that $d\overline{x^2}/dt$ is independent of $\overline{x^2}$. It is not in accordance with observations. We find in an unbounded ocean or atmosphere that the rate of spreading increases enormously with $\overline{x^2}$, showing the effects of the larger scale turbulence as the spreading increases. For small values of $\overline{x^2}$ the large scale turbulence is seen only as an advection effect, having much the same effect on all portions of the cluster, whereas for larger values of $\overline{x^2}$ the larger scale turbulence increases the spreading of the cluster.

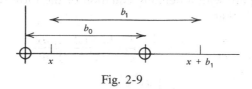

Fig. 2-9

We may look at this same consequence in another manner. If two particles at $t = 0$ are placed at $x = 0$ and at $x = b_0$, respectively, as shown in Fig. 2-9, at time t the probability of the first being at x and the second at $x + b_1$ is, respectively, from (2-68),

$$c_1 = \frac{1}{\sqrt{4\pi Kt}}\, e^{-x^2/4Kt} \tag{2-86}$$

and

$$c_2 = \frac{1}{\sqrt{4\pi Kt}}\, e^{-(x+b_1-b_0)^2/4Kt} \tag{2-87}$$

The probability $P(b_0, b_1)$ that these two particles will be a distance b_1 apart at time t is simply the product of these two expressions integrated over all x, or

$$
\begin{aligned}
P(b_0, b_1) &= \frac{1}{4\pi Kt} \int_{-\infty}^{\infty} e^{-x^2/4Kt}\, e^{-(x+b_1-b_0)^2/4Kt}\, dx \\
&= \frac{1}{4\pi Kt}\, e^{-(b_1-b_0)^2/8Kt} \int_{-\infty}^{\infty} e^{-[\sqrt{2}x+(b_1-b_0)/\sqrt{2}]^2/4Kt}\, dx \\
&= \frac{1}{2\sqrt{2\pi Kt}}\, e^{-(b_1-b_0)^2/8Kt}
\end{aligned}
\tag{2-88}
$$

Therefore we see that the Fickian equation of diffusion leads to the result that the probability of a pair of particles originally a distance b_0 apart being a distance b_1 apart after a time interval t depends solely on the quantity $(b_1 - b_0)^2$ and not on the original separation b_0. Again this is contrary to observation. In the ocean the probability of large values of $(b_1 - b_0)^2$ increases greatly with the initial separation of the two particles.

The logical resolution of these sorts of difficulties appears to be, at least to us, in the approach given by Taylor (1921), the essentials of which were discussed in the previous section. Another somewhat more intuitive approach, which is very interesting and instructive in itself, is given by Richardson (1926). As the diffusion phenomena are scale dependent we might consider changing the independent variable from a coordinate distance x to a separation variable l between any two particles of the fluid, presuming that such a definition can physically be made. The variable l is referred to as the *neighbor separation*. Correspondingly, rather than using as the dependent variable the concentration $c(x)$ of a diffusing substance as a function of position, we need a *neighbor concentration* $q(l)$, expressing the concentration as a function of the mutual separation of particles rather than as a function of position in space. The mathematical transformation from $c(x)$ to $q(l)$ is then simply the probability integral

$$q(l) = \int_{-\infty}^{\infty} c(x)c(x+l)\, dx \qquad (2\text{-}89)$$

We then postulate that there is a Fickian equation relation between this new independent and new dependent variable, which we write in the form

$$\frac{\partial q}{\partial t} = \frac{\partial}{\partial t}\left[F(l)\frac{\partial q}{\partial l}\right] \qquad (2\text{-}90)$$

The quantity $F(l)$ is analogous to the ordinary diffusivity K and may be referred to as the *neighbor diffusivity*. Because $F(l)$ is a function of l, it is possible to reconcile it with some of the observed facts without demonstrating that (2-90) has physical validity. Indeed, this is the whole reason for changing the independent variable from x to l.

The form of (2-90) provides us with a simple means for obtaining values of $F(l)$, where $F(l)$ is presumably an increasing function of l. If identifiable particles, such as floats, small dye patches, or the like, are released in pairs at a fixed initial separation l_0 apart, and if the separation l_1 is measured again after a time interval t which is small enough that $l_1 - l_0$ averages only a small fraction of l_0, we can consider $F(l)$ in (2-90) to be a constant $F(l_0)$. Then the solution to (2-90) will be of the same form as (2-68) with x replaced by $l_1 - l_0$, so that the relation corresponding to (2-69) may now be written as

$$F(l_0) = \frac{\overline{(l_1 - l_0)^2}}{2t} \qquad (2\text{-}91)$$

Relation (2-91) then provides us with a means of determining $F(l_0)$ from observations of $l_1 - l_0$ for various initial separations l_0. Although no direct comparison can be made between $F(l)$ and K, which is important to emphasize, we see from the form of (2-88) that, for the special case of $F(l)$ being a constant, $F = 2K$.

From such separation measurements by Richardson and Stommel (1948), Stommel (1949), Ichiye (1951), Olson (1952), and Olson and Ichiye (1959), as summarized by Ichiye and Olson (1960), it is found that $F(l)$ follows a relation

$$F(l) = \epsilon l^{4/3} \qquad (2\text{-}92)$$

where in cgs units ε is given by $\epsilon = 0.0246$.

We can obtain the same four thirds power law in this case for N, the kinematic coefficient of viscosity defined in (2-137), rather than F through the following theoretical considerations of turbulence. These theoretical considerations were originally presented by von Weizsäcker (1948) and by Heisenberg (1948) and have been elaborated and extended by others. The theory itself is extraordinarily complex, and we are concerned here only with its essentials and only its simplified form. Let us consider a continuous scale of eddies of all sizes which are horizontally isotropic. We also presume that there is a constant supply of kinetic energy available to the large eddies, from atmospheric wind circulation in the case of the deep ocean, and from tidal motion in the case of estuaries. This atmospheric wind circulation or estuarine tidal motion then effectively defines the upper limit of the scale of the turbulent motion. We also presume that in the turbulent region, or what is the same for motion with large Reynolds numbers, the energy sifts down in a regular manner without loss from the larger to the smaller eddies, eventually being dissipated as heat energy for the smallest scale eddies. We further presume that the turbulent motion is kinematically similar, i.e., all statistical characteristics of the turbulence are independent of the absolute linear dimensions being considered. Now we know that, when turbulent motion is defined by equations of the form (2-134), the kinetic energy flux per unit volume is of the form (2-139), or

$$D \propto v'v'\left(\frac{\partial \bar{v}}{\partial x}\right) \qquad (2\text{-}93)$$

where v' is a generalized turbulent velocity component, $\bar{v}$ is a generalized mean velocity, and x is a generalized coordinate. We have presumed a continuous scale of eddies; thus the actual values of $\bar{v}$ and v' and the turbulent scale being considered depend on the averaging time used in defining $\bar{v}$. We also have presumed that, above the scale size for which molecular motions and dissipation become important, this kinetic energy flux is constant and independent of the arbitrary division into mean and turbulent flows. For any given averaging time we may write for the mean velocity in terms of its spatial derivative an equational relation of the form

$$\bar{v} = a \frac{\partial \bar{v}}{\partial x} l$$

where l is a characteristic scale distance being considered and a is an undetermined coefficient. Now as we have presumed that the motion is kinematically similar, the coefficient remains unchanged as regards scale, and

$$\frac{\partial \bar{v}}{\partial x} \propto \frac{\bar{v}}{l} \qquad (2\text{-}94)$$

regardless of the scale or averaging time being considered. From the same kinematic similarity condition we can also write

$$v' \propto \bar{v} \qquad (2\text{-}95)$$

as a function of the scale. We then have, substituting (2-94) and (2-95) into (2-93),

$$v' \propto l^{1/3} \qquad (2\text{-}96)$$

and from (2-137),

$$N \propto l^{4/3} \qquad (2\text{-}97)$$

which is the desired result.

We can carry this type of analysis an additional step with the following consideration. From (3-34) or (3-36) the energy density E of the turbulent fluctuations for any given averaging time or scale dimension is proportional to v'^2 or, from (2-96),

$$E \propto l^{2/3} \qquad (2\text{-}98)$$

We can also think of the energy characteristics of the turbulence in terms of a turbulence spectrum $F(k)$, where k is the wave number of (3-72), the reciprocal of the wavelength, or what is the same in this case the scale dimension l. We then note that E for any given scale being considered is the integral of $F(k)$ over all k from the defining scale out to infinity, or

$$E = \int_{k}^{\infty} F(k)\, dk \propto l^{2/3} \propto k^{-2/3} \qquad (2\text{-}99)$$

so that

$$F(k) \propto k^{-5/3} \qquad (2\text{-}100)$$

We could alternatively think of this in terms of a characteristic oscillation frequency Γ, much the same as in (2-71), so that

$$v' \propto \Gamma l \qquad (2\text{-}101)$$

and, following the above line of reasoning,

$$E \propto \Gamma^{-1} \qquad (2\text{-}102)$$

and

$$F(\Gamma) \propto \Gamma^{-2} \qquad (2\text{-}103)$$

Clearly in some of the concepts in this and the previous section we are approaching the subject of turbulence spectra, which we may consider mathematically as being defined as the Fourier transform of the Eulerian counterpart to the correlation function (2-75), or

$$R(\tau) = \int_0^\infty F(f) \cos 2\pi f\tau \, df$$
$$F(f) = 4 \int_0^\infty R(\tau) \cos 2\pi f\tau \, d\tau \qquad (2\text{-}104)$$

where f is the spectral frequency. (For a derivation of this relation the reader may refer to Taylor, 1938.) Although there has been considerable investigation into turbulence spectra and into the energy transformation processes from larger generating eddies down to smaller scale eddies in fluid mechanics and in physical oceanography, we have little direct concern with these investigations here. We are principally interested in their gross effects. In this sense there is usually an upper limit for most estuarine or coastal oceanographic problems in the turbulent motions, being specifically the scale of the turbulence-generating, tidal motions. We are generally interested in averaging times over a tidal cycle. And, most important, the scale magnitude of the resultant turbulent diffusion and internal frictional effects is determined by the scale of the tidal motions. Effectively we might consider that the magnitude term of any diffusion or viscosity coefficient is determined by the scale of the tidal, turbulence generating motions, and the multiplying coefficient by the details of the internal spectral distribution.

For further information on the material covered in this section the reader may refer to Richardson (1926), Stommel (1949), Ichiye and Olson (1960), Taylor (1938), von Weizsäcker (1948), and Heisenberg (1948).

2-7. FORCES

In this and the following two sections we discuss the external and internal forces that are important in the determination of estuary circulation.

These include forces due to the water surface slope and to the slopes of internal discontinuity surfaces, forces associated with Coriolis acceleration, wind stresses, internal frictional stresses, and bottom frictional stresses. Generally the horizontal circulation is the dominant motion in an estuary, and the emphasis here is on the forces affecting such horizontal motion.

One of the simpler and more important driving forces for horizontal water motion in an estuary is that caused by the horizontal pressure gradient associated with the slope of the water surface. Referring to Fig. 2-10, we have a right handed coordinate system with the x axis pointing downestuary, i.e., toward the ocean, the z axis vertical downward with origin at the ocean surface, and the y axis at right angles to x and z. The surface slope is $\partial \zeta_1 / \partial x$ where ζ_1 is the vertical displacement downward of the water surface away from an equilibrium (level) surface. We have also included an internal interface at $z = h_1$, with slope $\partial \zeta_2 / \partial x$ and with fluids of constant density ρ_1 and ρ_2, respectively, above and below the interface. Then the pressure in the upper medium at a point $P(x, z)$ is simply

$$p_1 = p_a + g\rho_1(z - \zeta_1) \qquad (2\text{-}105)$$

where p_a is the atmospheric pressure. If we assume that the atmospheric pressure is constant along the water surface, the horizontal pressure gradient will be

$$\frac{\partial p_1}{\partial x} = -g\rho_1 \frac{\partial \zeta_1}{\partial x} \qquad (2\text{-}106)$$

For the lower medium we have similarly at $Q(x, z)$,

$$p_2 = p_a + g\rho_1(\zeta_2 - \zeta_1) + g\rho_2(z - \zeta_2) \qquad (2\text{-}107)$$

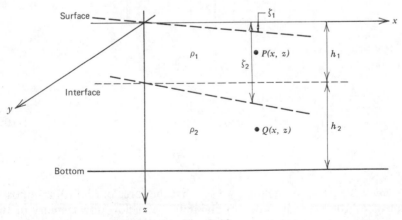

Fig. 2-10

and

$$\frac{\partial p_2}{\partial x} = -g(\rho_2 - \rho_1)\frac{\partial \zeta_2}{\partial x} - g\rho_1 \frac{\partial \zeta_1}{\partial x} \qquad (2\text{-}108)$$

In the simplest case expression (2-106) is the driving horizontal pressure gradient force of (2-10) in the direction of flow and in the steady state is balanced by a bottom frictional resistance term. For the lower medium in an estuary with a stratified flow the flow is upestuary, so that $\partial \zeta_2/\partial x$ is negative and of such magnitude that numerically the first term of (2-108) is greater than the second term. It should also be noted that, in most estuaries that exhibit stratified flow, both (2-106) and (2-108) are usually considerably larger than their difference, of equivalently that the two components of the stratified flow are usually much larger than their difference, i.e., the net river flow.

Another force of comparable importance in describing the net horizontal circulation in an estuary is that caused by any internal density gradient. In this instance let us start with a vertically mixed estuary so that the density varies in some manner from river water at the head of the estuary to sea water at the mouth of the estuary as a function of the x coordinate only, $\rho = \rho(x)$, as shown in Fig. 2-11. Then we have, for

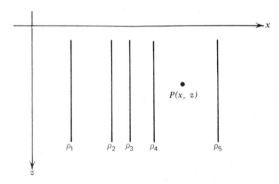

Fig. 2-11

a level estuary surface,

$$p = p_a + g\rho(x)z \qquad (2\text{-}109)$$

and

$$\frac{\partial p}{\partial x} = gz\frac{\partial \rho}{\partial x} \qquad (2\text{-}110)$$

As $\partial \rho/\partial x$ is a positive quantity, we see that the force of (2-110) is opposite that of (2-106); and in order to maintain a net flow downestuary of the entering river water its vertically integrated value must be less than

(2-106). In the case of a stratified estuary, $\rho = \rho(x, z)$, we have

$$p = p_a + g \int_0^z \rho(x, z)\, dz \tag{2-111}$$

and

$$\frac{\partial p}{\partial x} = g \int_0^z \frac{\partial \rho}{\partial x}\, dz \tag{2-112}$$

where, since we have assumed a level surface, there are no contributions from the partial differentiation with respect to the limits of integration. If the constant density surfaces can be approximated by parallel planes, as shown in Fig. 2-12, we will have

$$\rho = \rho(z') = \rho(z \cos \theta + x \sin \theta) \tag{2-113}$$

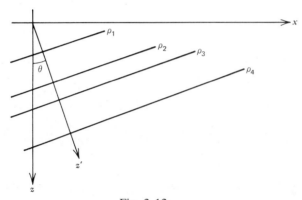

Fig. 2-12

from which we have

$$\frac{\partial \rho}{\partial z} = \rho' \cos \theta \qquad \frac{\partial \rho}{\partial x} = \rho' \sin \theta \tag{2-114}$$

so that expression (2-112) becomes

$$\frac{\partial p}{\partial x} = g \tan \theta \int_0^z \frac{\partial \rho}{\partial z}\, dz$$
$$= g \tan \theta [\rho(z) - \rho(0)]$$
$$= g \tan \theta\, h(z) \tag{2-115}$$

where $h(z)$ is the vertical density excess over that at the surface. Expression (2-115) may be considered intermediate between the two extremes of complete vertical mixing and complete vertical stratification. We see that (2-115) reduces to (2-108) in the latter extreme as $h(z) \rightarrow \rho_2 - \rho_1$ and $\tan \theta \rightarrow -\partial \zeta_2 / \partial x$.

Another force that should be considered is that of the gravitational attraction of the moon and the sun. As the moon and the sun move relative to the earth, a variable gravitational force acts on the waters of an estuary. In nature tidal phenomena are extremely important in estuaries, but not directly in regard to the effect of this force on the estuary. Rather the tidal motion in an estuary is that due to the ocean tides entering the restricted inlet of an estuary; in a sense one can think of an oceanic driving tidal force at the mouth of the estuary. Furthermore, a common occurrence in estuaries is that, in addition to the bulk of the oceanic tidal energy being expended in bottom friction, a certain amount is also expended in doing work on the moon rather than the more normal case, which one might expect, of the moon doing work in creating estuary tides. Tidal phenomena in estuaries are the subject of the next chapter and enter in various ways into the succeeding two chapters.

One of the more important forces in deep sea oceanography is *Coriolis acceleration.* It is usually not important in estuaries, because of their restricted lateral dimensions and shallower water depths. It can become important when one considers wide bodies of coastal waters such as lagoons. In general, we consider a coordinate system fixed in a given estuary to be an inertial reference. This is of course not the case; for this coordinate system rotates about the earth with the earth's rotational velocity and revolves around the sun in the earth's planetary orbit. To use Newton's second law of motion we must employ an inertial reference system. Carrying through the vector analysis to refer this earth rotating coordinate system to an inertial system, concerning which the reader may refer to any standard text such as Page (1935), introduces several acceleration and angular velocity terms related to the moving coordinate system, the most important of which for the earth is the Coriolis acceleration. For a particle on the earth subject to the force of earth gravity and the effects due to the rotation of the earth, we obtain from this analysis the acceleration, or force per unit mass, relative to the earth as given by

$$\mathbf{f} = \mathbf{g} - 2\boldsymbol{\omega}_0 \times \mathbf{v} \qquad (2\text{-}116)$$

where $\mathbf{g}$ is the gravitational force per unit mass, $\boldsymbol{\omega}_0$ is the angular velocity of the earth's rotation, and $\mathbf{v}$ is the water particle velocity. The expression $-2\boldsymbol{\omega}_0 \times \mathbf{v}$ is the *Coriolis force* per unit mass. Equation (2-10) then becomes

$$\frac{\partial \mathbf{v}}{\partial t} + \mathbf{v} \cdot \nabla \mathbf{v} = \mathbf{g} - 2\boldsymbol{\omega}_0 \times \mathbf{v} - \frac{1}{\rho} \nabla p \qquad (2\text{-}117)$$

Using a right handed coordinate system as shown in Fig. 2-13, the

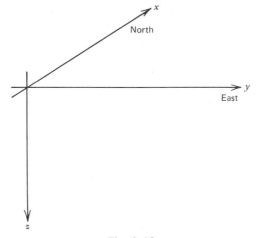

Fig. 2-13

components of $\boldsymbol{\omega}_0$ are, for the northern hemisphere,

$$\omega_x = \omega_0 \cos \varphi \qquad \omega_y = 0 \qquad \omega_z = -\omega_0 \sin \varphi \qquad (2\text{-}118)$$

where φ is the latitude. The scalar form of (2-117) then becomes

$$\frac{\partial v_x}{\partial t} + v_x \frac{\partial v_x}{\partial x} + v_y \frac{\partial v_x}{\partial y} + v_z \frac{\partial v_x}{\partial z} = -2\,\omega_0 v_y \sin \varphi - \frac{1}{\rho}\frac{\partial p}{\partial x}$$

$$\frac{\partial v_y}{\partial t} + v_x \frac{\partial v_y}{\partial x} + v_y \frac{\partial v_y}{\partial y} + v_z \frac{\partial v_y}{\partial z} = 2\,\omega_0 v_x \sin \varphi + 2\,\omega_0 v_z \cos \varphi - \frac{1}{\rho}\frac{\partial p}{\partial y} \qquad (2\text{-}119)$$

$$\frac{\partial v_z}{\partial t} + v_x \frac{\partial v_z}{\partial x} + v_y \frac{\partial v_z}{\partial y} + v_z \frac{\partial v_z}{\partial z} = g - 2\,\omega_0 v_y \cos \varphi - \frac{1}{\rho}\frac{\partial p}{\partial z}$$

For most problems we are concerned only with horizontal motion. Further, the Coriolis term in the third equation in (2-119) is always much smaller than the gravitational term. Equations (2-119) then reduce to

$$\frac{\partial v_x}{\partial t} + v_x \frac{\partial v_x}{\partial x} + v_y \frac{\partial v_x}{\partial y} = -2\,\omega_0 v_y \sin \varphi - \frac{1}{\rho}\frac{\partial p}{\partial x}$$

$$\frac{\partial v_y}{\partial t} + v_x \frac{\partial v_y}{\partial x} + v_y \frac{\partial v_y}{\partial y} = 2\,\omega_0 v_x \sin \varphi - \frac{1}{\rho}\frac{\partial p}{\partial y} \qquad (2\text{-}120)$$

and the simple and familiar relation

$$\frac{\partial p}{\partial z} = \rho g \qquad (2\text{-}121)$$

Under conditions of steady state when following the motion, i.e.,

$D\mathbf{v}/Dt = 0$, and under the assumption that the internal frictional effects are small compared with the Coriolis and pressure gradients and that the water is sufficiently deep that there are no bottom frictional effects, the solutions of (2-120) determine one of the more familiar results of deep sea physical oceanography. Equations (2-120) reduce to

$$2\,\omega_0 v_y \sin \varphi = -\frac{1}{\rho}\frac{\partial p}{\partial x}$$

$$2\,\omega_0 v_x \sin \varphi = \frac{1}{\rho}\frac{\partial p}{\partial y}$$

(2-122)

or simply

$$2\,\omega_0 v \sin \varphi = \frac{1}{\rho}\frac{\partial p}{\partial n}$$

(2-123)

where v is the magnitude of the current, and n is a coordinate direction at right angles to $\mathbf{v}$ with its positive direction to the right of $\mathbf{v}$. We see that under these conditions the force term related to the Coriolis effect is balanced by the force term related to the lateral horizontal pressure gradient. We also note that neither of these terms represents a driving force. We have included no driving force term nor frictional resistance term in (2-120). We have assumed frictionless motion. The Coriolis force is the result of the constant water velocity motion. The lateral pressure gradient term is a necessary resultant from the equations of motion to preserve the assumed steady state conditions. If we consider such a current extending to the sea surface, we will see that the sea surface itself is not a level surface. We show these relations diagrammatically in Fig. 2-14 for the northern hemisphere for a current directed into the page with the Coriolis force to the right and the horizontal pressure gradient force to the left. For the southern hemisphere the force relations are reversed. Currents of this type are known as *geostrophic currents*, and Coriolis acceleration is often referred to in physical oceanography as *geostrophic acceleration*.

Up to this point we have been dealing only with the forces related to a perfect fluid. Now we begin to consider *viscosity*, or *internal friction*. In all the physical oceanographic problems of interest to us in this book we are concerned with internal frictional effects due to the turbulence of water

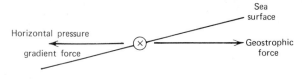

Fig. 2-14

masses, which are referred to as *eddy viscosity*. This is the subject of the next section. However, as an introduction we briefly review as a reference background the classical theory of the viscous effects of a fluid related to its internal properties, which we refer to as *molecular viscosity*. The reader will note in this and the following discussions a close comparison between the phenomena of diffusion and viscosity.

The basis of the classical theory of the dynamics of a viscous fluid is very similar to that of an elastic solid as defined by Hooke's law. The main points of difference are that the shearing stresses in a fluid are proportional to the time rates of change in strain rather than the strain itself, and that the tensional stresses involved are those in excess of the hydrostatic pressure. We may then write for the stress rate of strain relations, following the corresponding relations for an elastic solid, in scalar form,

$$\sigma'_{xx} = \sigma_{xx} + p = 2\mu \frac{\partial v_x}{\partial x}$$

$$\sigma'_{yy} = \sigma_{yy} + p = 2\mu \frac{\partial v_y}{\partial y}$$

$$\sigma'_{zz} = \sigma_{zz} + p = 2\mu \frac{\partial v_z}{\partial z}$$

$$\sigma_{zy} = \sigma_{yz} = \mu \left(\frac{\partial v_z}{\partial y} + \frac{\partial v_y}{\partial z} \right)$$

$$\sigma_{xz} = \sigma_{zx} = \mu \left(\frac{\partial v_x}{\partial z} + \frac{\partial v_z}{\partial x} \right)$$

$$\sigma_{yx} = \sigma_{xy} = \mu \left(\frac{\partial v_y}{\partial x} + \frac{\partial v_x}{\partial y} \right)$$

$$(2\text{-}124)$$

where σ_{mn} are the internal stresses per unit area (the first subscript indicating the normal to the unit area on which the stress is acting and the second subscript indicating the coordinate direction of the stress), and v_x, v_y, and v_z are the components of the fluid velocity. In the first three equations the hydrostatic pressure is numerically negative to the tensional stresses σ_{mm}, and we have also assumed that the fluid is incompressible so that (2-26) and (2-27) apply. The coefficient μ is referred to as the *coefficient of viscosity*, and in this instance we have also assumed that the medium is isotropic, so that μ is a constant independent of coordinate direction; this is generally true for molecular viscosity but not, as in the examples for diffusion, for eddy viscosity.

Let us now consider how these internal stresses per unit area relate to the internal forces per unit volume, or per unit mass, in our equations of motion. Consider a rectangular parallelepiped of the medium, as shown in

Fig. 2-15, of dimensions dx, dy, dz, with the point $P(x, y, z)$ lying at its center. The component of the force on the face *EFGH* acting in the positive x direction is

$$\left(\sigma_{xx} + \frac{\partial \sigma_{xx}}{\partial x} \frac{dx}{2}\right) dy\, dz$$

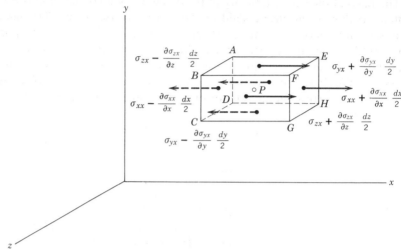

Fig. 2-15

and that on the face *ABCD* acting in the negative x direction is

$$\left(\sigma_{xx} - \frac{\partial \sigma_{xx}}{\partial x} \frac{dx}{2}\right) dy\, dz$$

The net force in the positive x direction due to the forces on these two faces is then simply

$$\frac{\partial \sigma_{xx}}{\partial x} dx\, dy\, dz$$

Similarly the net force in the positive x direction on the two faces *ABFE* and *DCGH* is

$$\frac{\partial \sigma_{yx}}{\partial y} dx\, dy\, dz$$

and that on *BCGF* and *ADHE* is

$$\frac{\partial \sigma_{zx}}{\partial z} dx\, dy\, dz$$

Therefore the total force per unit volume in the positive x direction due to the stresses on all six faces is simply

$$\frac{\partial \sigma_{xx}}{\partial x} + \frac{\partial \sigma_{yx}}{\partial y} + \frac{\partial \sigma_{zx}}{\partial z}$$

and similar expressions hold for the force components in the y and z directions. Our equations of motion, in scalar form, corresponding to (2-9) are then

$$\frac{Dv_x}{Dt} = F_x + \frac{1}{\rho}\left(\frac{\partial \sigma_{xx}}{\partial x} + \frac{\partial \sigma_{yx}}{\partial y} + \frac{\partial \sigma_{zx}}{\partial z}\right)$$

$$\frac{Dv_y}{Dt} = F_y + \frac{1}{\rho}\left(\frac{\partial \sigma_{xy}}{\partial x} + \frac{\partial \sigma_{yy}}{\partial y} + \frac{\partial \sigma_{zy}}{\partial z}\right) \qquad (2\text{-}125)$$

$$\frac{Dv_z}{Dt} = F_z + \frac{1}{\rho}\left(\frac{\partial \sigma_{xz}}{\partial x} + \frac{\partial \sigma_{yz}}{\partial y} + \frac{\partial \sigma_{zz}}{\partial z}\right)$$

where, as before, F_x, F_y, and F_z are the components of the external force per unit mass. Substituting from (2-124) and remembering that the relation (2-27) applies we have

$$\frac{Dv_x}{Dt} = F_x - \frac{1}{\rho}\frac{\partial p}{\partial x} + \frac{\mu}{\rho}\left(\frac{\partial^2 v_x}{\partial x^2} + \frac{\partial^2 v_x}{\partial y^2} + \frac{\partial^2 v_x}{\partial z^2}\right)$$

$$\frac{Dv_y}{Dt} = F_y - \frac{1}{\rho}\frac{\partial p}{\partial y} + \frac{\mu}{\rho}\left(\frac{\partial^2 v_y}{\partial x^2} + \frac{\partial^2 v_y}{\partial y^2} + \frac{\partial^2 v_y}{\partial z^2}\right) \qquad (2\text{-}126)$$

$$\frac{Dv_z}{Dt} = F_z - \frac{1}{\rho}\frac{\partial p}{\partial z} + \frac{\mu}{\rho}\left(\frac{\partial^2 v_z}{\partial x^2} + \frac{\partial^2 v_z}{\partial y^2} + \frac{\partial^2 v_z}{\partial z^2}\right)$$

or, in vector form,

$$\frac{\partial \mathbf{v}}{\partial t} + \mathbf{v} \cdot \nabla \mathbf{v} = \mathbf{F} - \nabla p + \frac{\mu}{\rho} \nabla \cdot \nabla \mathbf{v} \qquad (2\text{-}127)$$

which is a reduced form of the *Navier Stokes equation* for an incompressible medium. The components of the internal frictional force per unit mass are then, respectively, the third terms on the right hand side of (2-126).

2-8. TURBULENCE AND INTERNAL FRICTION

In the previous derivation for the coefficient of molecular viscosity it was assumed that the fluid was in laminar motion, i.e., that there was no turbulence. We saw in that case that the *velocity shear*, or *distortion rate*, $\partial v_m/\partial n$, was a measure of the internal frictional stresses. In actuality the waters of an estuary are continually in turbulent motion. Because of the eddies thus formed there is a much greater interchange of momentum between adjacent layers of the fluid. The concept of viscosity in this case is much the same as that used for the determination of a coefficient of viscosity from the kinetic theory of gases, except that we are concerned with a much larger scale phenomena. In this case the dimensions are those of the elementary portions of the fluid that can be considered continuous, rather than the molecular mean free path, and consequently we are concerned with a much larger coefficient of viscosity.

It is instructive then to restate briefly here the derivation for a coefficient of viscosity from the kinetic theory of gases. Consider a gas or a fluid streaming in the positive x direction, as shown in Fig. 2-16, such that

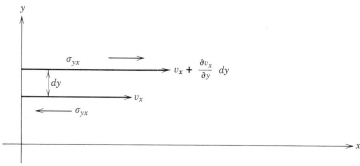

Fig. 2-16

$v_y = v_z = 0$ and in such a manner that different layers have different velocities, where $v_x = v_x(y)$. We anticipate that the shearing stress σ_{yx} is given, from expressions (2-124), by

$$\sigma_{yx} = \mu \frac{\partial v_x}{\partial y} \qquad (2\text{-}128)$$

We explain the viscosity of a gas as being due to the passage of faster moving molecules in the upper layers down into the lower layers which are thereby accelerated, and to the passage of slower moving molecules in the lower layers up into the upper layers which are thereby retarded.

We assume for simplicity that the thermal velocities of all the molecules are the same and are equal to the arithmetic mean velocity Ω. Then the mean free path l is given, in terms of the collision frequency Γ, by $l = \Omega/\Gamma$. For two layers of the gas a distance y apart, the molecules in the lower layer have a translational velocity v_x superposed on a thermal velocity Ω, while those in the upper layer have a translational velocity $v_x + (\partial v_x/\partial y)y$. We assume that v_x is small compared with Ω.

Taking the origin at the center of the unit area perpendicular to the y axis on which we wish to calculate the shearing stress, as shown in Fig. 2-17, consider the molecules whose free paths originate in the volume

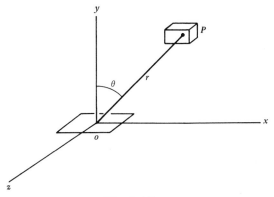

Fig. 2-17

element $dx\,dy\,dz$ at P. If n denotes the number of molecules per unit volume, there are $n\,dx\,dy\,dz$ molecules inside this volume; and these undergo $\Gamma n\,dx\,dy\,dz$ collisions per unit time. Therefore in each unit of time $\Gamma n\,dx\,dy\,dz$ molecules start out on a fresh free path, and we note from elementary statistical mechanics that $e^{-r/l}\,\Gamma n\,dx\,dy\,dz$ travel a distance $r = PO$ before making a collision. As the unit area at O subtends a solid angle $\cos\theta/r^2$ at P, only $\cos\theta/4\pi r^2$ of these molecules are directed so as to pass through the unit area at O. Consequently a number

$$\frac{dN}{dt} = \frac{\Gamma n}{4\pi r^2}\,e^{-r/l}\,\cos\theta\,dx\,dy\,dz \tag{2-129}$$

of molecules whose free paths originate in the volume element at P pass through the unit area at O per unit time.

Now if M is the aggregate mass of these molecules, the force necessary to reduce their translational velocity to that of the gas in the layer at $y = 0$ is, from the familiar relation $F = (d/dt)(Mv)$ between force and the rate of change in momentum,

$$F_x = \frac{dM}{dt}\left[v_x - \left(v_x + \frac{\partial v_x}{\partial y}\,y\right)\right] = -\frac{dM}{dt}\frac{\partial v_x}{\partial y}\,r\cos\theta$$

and the equal reaction of the gas below the XZ plane is

$$d\sigma_{yx} = -F_x = \frac{dM}{dt}\frac{\partial v_x}{\partial y}\,r\cos\theta \tag{2-130}$$

If m is the mass of a single molecule, we have from (2-130) and (2-129), where we have replaced $dx\,dy\,dz$ by the annular volume element

$2\pi r^2 \sin\theta \, d\theta \, dr,$

$$\sigma_{yx} = \tfrac{1}{2}\Gamma nm \frac{\partial v_x}{\partial y} \int_0^\infty e^{-r/l} r \, dr \int_0^\pi \cos^2\theta \sin\theta \, d\theta$$

$$= \tfrac{1}{3}\Gamma nml^2 \frac{\partial v_x}{\partial y} \qquad\qquad\qquad (2\text{-}131)$$

Comparing with (2-128) we then have, for the molecular coefficient of viscosity,

$$\mu = \tfrac{1}{3}\rho\Gamma l^2 \qquad\qquad\qquad (2\text{-}132)$$

as *nm* is the density ρ. The reader should note the similarity of dependence of the molecular coefficient of viscosity determined here to that of the eddy coefficient of diffusion given by (2-71). Again, and as one might expect, the coefficient is proportional to the square of the dimensions and inversely proportional to the time scale.

One can then suppose, within the same sort of limits as discussed in Section 2-5, that the presumed similarity of molecular motion in molecular viscosity and turbulent motion in eddy viscosity leads to a similar description of the internal stresses in terms of the mean velocity shears as given by (2-124). We thus give up any attempt to follow the internal changes in turbulent motion and concern ourselves only with the net effects. In this case, as with turbulent diffusion, the eddy coefficient of viscosity may not be regarded as a scalar or as a physical constant of the fluid. It is a dyadic, or second rank tensor, and its value depends on the type and scale of the motion and often varies considerably from one part of the fluid to another.

The existence of turbulence in a field of motion therefore gives rise to two types of effects, namely, the production of shearing stresses and the process of eddy diffusion. Whereas turbulent shearing stresses react on the mean motion and have an essentially dynamic effect, turbulent diffusing processes affect the distribution of a particular property without reacting directly on the flow.

In the sea, the vertical and horizontal components of turbulence usually differ so much in scale and in intensity that their effects may be considered separately. These differences arise, first, because the horizontal dimensions of the bodies of water concerned are much greater than the vertical dimensions and, second, because of the influence of vertical stability. The presence of a stable vertical gradient of density, as we see later in this section, causes a reduction in the intensity of v_z fluctuations but has no direct effect on the fluctuations in v_x and v_y. Thus the effect of a stable density gradient is to reduce the intensity of the vertical turbulence, hence also the vertical coefficients of eddy viscosity and eddy diffusion.

The general question of turbulent viscosity can also be approached through an alternative method, as originally given by Reynolds (1894), much the same as that followed in the latter part of Section 2-4 for eddy diffusion. We again take the instantaneous velocity $\mathbf{v}$ to be equal to a mean velocity $\bar{\mathbf{v}}$ and a velocity deviation $\mathbf{v}'$ as given by (2-58) with subsidiary relations (2-59) to (2-61). For an incompressible fluid we can then write the equations of motion (2-10) in scalar form as

$$\frac{\partial v_x}{\partial t} = F_x - \frac{1}{\rho}\frac{\partial p}{\partial x} - \frac{1}{\rho}\left[\frac{\partial}{\partial x}(\rho v_x v_x) + \frac{\partial}{\partial y}(\rho v_x v_y) + \frac{\partial}{\partial z}(\rho v_x v_z)\right]$$

$$\frac{\partial v_y}{\partial t} = F_y - \frac{1}{\rho}\frac{\partial p}{\partial y} - \frac{1}{\rho}\left[\frac{\partial}{\partial x}(\rho v_y v_x) + \frac{\partial}{\partial y}(\rho v_y v_y) + \frac{\partial}{\partial z}(\rho v_y v_z)\right] \quad (2\text{-}133)$$

$$\frac{\partial v_z}{\partial t} = F_z - \frac{1}{\rho}\frac{\partial p}{\partial z} - \frac{1}{\rho}\left[\frac{\partial}{\partial x}(\rho v_z v_x) + \frac{\partial}{\partial y}(\rho v_z v_y) + \frac{\partial}{\partial z}(\rho v_z v_z)\right]$$

remembering also that relation (2-27) applies. Using the same relationships as in Section 2-4 for taking mean values, we then have, for the mean value of any such quantity as $\overline{v_x v_y}$,

$$\overline{v_x v_y} = \overline{(\bar{v}_x + v'_x)(\bar{v}_y + v'_y)}$$
$$= \bar{v}_x \bar{v}_y + \bar{v}_y \overline{v'_x} + \bar{v}_x \overline{v'_y} + \overline{v'_x v'_y}$$
$$= \bar{v}_x \bar{v}_y + \overline{v'_x v'_y}$$

Substituting these relations into (2-133), we then have for the mean values of (2-133), using (2-60),

$$\frac{D\bar{v}_x}{Dt} = F_x - \frac{1}{\rho}\left[\frac{\partial \bar{p}}{\partial x} + \frac{\partial}{\partial x}\overline{(\rho v'_x v'_x)} + \frac{\partial}{\partial y}\overline{(\rho v'_x v'_y)} + \frac{\partial}{\partial z}\overline{(\rho v'_x v'_z)}\right]$$

$$\frac{D\bar{v}_y}{Dt} = F_y - \frac{1}{\rho}\left[\frac{\partial \bar{p}}{\partial y} + \frac{\partial}{\partial x}\overline{(\rho v'_y v'_x)} + \frac{\partial}{\partial y}\overline{(\rho v'_y v'_y)} + \frac{\partial}{\partial z}\overline{(\rho v'_y v'_z)}\right] \quad (2\text{-}134)$$

$$\frac{D\bar{v}_z}{Dt} = F_z - \frac{1}{\rho}\left[\frac{\partial \bar{p}}{\partial z} + \frac{\partial}{\partial x}\overline{(\rho v'_z v'_x)} + \frac{\partial}{\partial y}\overline{(\rho v'_z v'_y)} + \frac{\partial}{\partial z}\overline{(\rho v'_z v'_z)}\right]$$

Comparing these relations with (2-125) and (2-124) gives us the relations between the turbulent velocity fluctuations and the eddy coefficient of viscosity. The internal frictional stresses σ_{mn} as given by (2-134) are often referred to as *Reynolds stresses*. For example, if we have a mean current in the positive X direction so that $\bar{v}_y = \bar{v}_z = 0$, and from (2-59) and (2-60) also that $\partial \bar{v}_x / \partial x = 0$, we will have, from (2-124) (2-125), and (2-134),

$$\sigma_{yx} = \rho N_y \frac{\partial \bar{v}_x}{\partial y} = -\overline{\rho v'_x v'_y} = -f_{yx}$$

$$\sigma_{zx} = \rho N_z \frac{\partial \bar{v}_x}{\partial z} = -\overline{\rho v'_x v'_z} = -f_{zx}$$

(2-135)

where we have also made the substitution

$$N = \frac{\mu}{\rho}$$

(2-136)

It is important to note that the *frictional stresses* f_{mn} defined by (2-135) are the negative of the internal stresses σ_{mn} we have been using. The internal or bottom frictional stress or resistance always opposes the motion of the fluid, and it is convenient in a physical sense in equations of motion, in rectangular coordinate or scalar form, to use the stresses f_{mn}, which are positive numerical quantities and which vectorially oppose the motion.

The quantity N is referred to as the *kinematic coefficient of viscosity*, similar to the kinematic coefficient of diffusion of (2-56). We may then write

$$N_y = -\frac{\overline{\rho v'_x v'_y}}{\rho \dfrac{\partial \bar{v}_x}{\partial y}} \qquad N_z = -\frac{\overline{\rho v'_x v'_z}}{\rho \dfrac{\partial \bar{v}_x}{\partial z}}$$

(2-137)

which is similar to relations (2-64) for the kinematic coefficients of eddy diffusion.

We have thus far ignored the energy required to maintain turbulent motion. In particular, under the normal conditions in the ocean of an increase in density with depth, the vertical mixing process leads to an increase in potential energy. This increase in potential energy must therefore be supplied from the turbulent motion itself. We may see, from the following simple example, that there is indeed an increase in potential energy due to the mixing. As shown in Fig. 2-18, suppose that water of uniform density ρ_1 lies to a depth of h_1 over water of greater uniform density ρ_2 and of thickness h_2. Then, when we take the sea surface as the zero surface for gravitational potential energy and the z axis vertically downward, the potential energy per unit horizontal area for the entire column is

$$-\int_0^{h_1} g\rho_1 z \, dz - \int_{h_1}^{h_1+h_2} g\rho_2 z \, dz = -\tfrac{1}{2}g\rho_1 h_1(h_1) - \tfrac{1}{2}g\rho_2 h_2(2h_1 + h_2)$$

If through some process these waters are then completely mixed, their

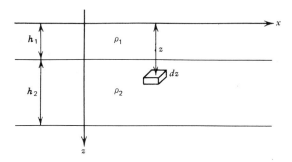

Fig. 2-18

new potential energy per unit horizontal area for the entire column will be

$$-\int_0^{h_1+h_2} g\bar{\rho}z\, dz = -\tfrac{1}{2}g(\rho_1 h_1 + \rho_2 h_2)(h_1 + h_2)$$

where the mean density $\bar{\rho}$ is given by

$$\bar{\rho} = \frac{\rho_1 h_1 + \rho_2 h_2}{h_1 + h_2}$$

The increase in potential energy due to the mixing is then the difference of these two expressions, or

$$\tfrac{1}{2}g(\rho_2 - \rho_1)h_1 h_2$$

We see, as expected, that if there is no vertical variation in density there is no change in potential energy through vertical mixing.

In the case of vertical turbulence we considered in Section 2-4 the salinity to be composed of the sum of a mean salinity and a salinity deviation. As the density increases with increasing salinity, or because of any other causes that might cause a vertical variation in density, we may consider the instantaneous density ρ to be equal to a mean density $\bar{\rho}$ and a density deviation ρ'. Then, following the same reasoning as before, the vertical turbulent mixing gives rise to an increase in potential energy per unit volume per unit time of $-g\overline{\rho'v_z'}$, remembering that the z axis is vertical downward, which from (2-64) we may write as

$$P = -g\overline{\rho'v_z'} = gK_z\frac{\partial\bar{\rho}}{\partial z} \tag{2-138}$$

Now the energy of turbulent motion is continually derived from the mean motion through the action of the Reynolds stresses. As before, in the case of a horizontal flow in the positive x direction such that

$v_y = v_z = 0$ and $v_x = v_x(z)$ only, the stresses on an elemental volume are as shown in Fig. 2-19. We consider the motion to be in a steady state, so

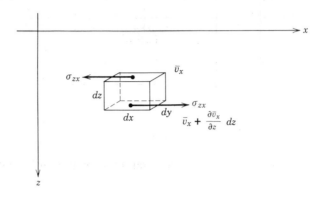

Fig. 2-19

that the horizontal Reynolds stresses on the two faces are the same. (The same result would be obtained if accelerated motion were considered, the net work done by the resultant stress $(\partial\sigma_{zx}/\partial z)\,dz$ on the elemental volume balancing out with a corresponding increase in energy from the mean motion.) The rate at which work is being done on the elemental volume, or energy being supplied to it, by the mean motion is

$$\sigma_{zx}\left(\bar{v}_x + \frac{\partial\bar{v}_x}{\partial z}\,dz\right)dx\,dy - \sigma_{zx}\bar{v}_x\,dx\,dy$$

so that the rate at which energy is being transferred per unit volume is using (2-135),

$$G = \sigma_{zx}\frac{\partial\bar{v}_x}{\partial z} = -\overline{\rho v'_x v'_z}\frac{\partial\bar{v}_x}{\partial z} = \rho N_z\left(\frac{\partial\bar{v}_x}{\partial z}\right)^2 \tag{2-139}$$

Then, if vertical turbulence is to be maintained, clearly the rate at which energy is being supplied from the mean motion, (2-138), must be greater than the rate at which the potential energy of turbulent motion is increased, (2-138). Hence, for the maintenance of vertical turbulence, we may write the criterion

$$gK_z\frac{\partial\rho}{\partial z} < \rho N_z\left(\frac{\partial v_x}{\partial z}\right)^2 \tag{2-140}$$

or

$$Ri < \frac{N_z}{K_z} \tag{2-141}$$

where

$$Ri = \frac{g(\partial\rho/\partial z)}{\rho(\partial v_x/\partial z)^2} \qquad (2\text{-}142)$$

is known as the *Richardson number* and where for convenience we have omitted the overbar signs.

This criterion is of extraordinary importance in estuaries, more in a general sense that in specific numerical application. When $\partial\rho/\partial z$ is very great, there is very little vertical turbulence, hence very little vertical mixing or internal friction across horizontal surfaces. In the limiting case of a surface of discontinuity in density, it is often assumed for the sake of simplicity that there is neither mixing nor friction across the surface. Because of the overriding importance of this stability condition the more common density structures in estuaries are either relatively well mixed waters from surface to bottom or two well mixed bodies of different density separated by a zone of rapid change in density, a halocline or a thermocline. In specific application the simple criterion of (2-141) breaks down. Whereas the values of N_z are always greater than the corresponding measured values of K_z, so that the ratio N_z/K_z usually has values of 2-18, the vertical turbulence is markedly decreased for Ri values as small as 0.2–0.5. Various empirical and semiempirical relations for N_z and K_z as a function of Ri have been obtained and used with varying degrees of success.

For additional information on turbulence, internal friction, and viscosity the reader may refer to Lamb (1932), Proudman (1953), and Bowden (1962).

2-9. BOUNDARY LAYER AND BOTTOM FRICTION

In the previous section we were concerned, in a general way, with eddy viscosity and internal friction. Of considerably greater importance in estuary circulation is the effect of an external friction, namely, bottom friction. This is one of the more fundamental differences in the description of circulation in estuaries or coastal waters and that in deep oceans; and it is very simply related to the depth of water involved, about 10 m in estuaries as contrasted with 5000 m in deep oceans. At the estuary bottom we have a *boundary layer* over which the horizontal velocity changes from a value we designate v_b to zero, since the bottom itself is stationary. This very large velocity shear has associated with it, from relations such as (2-128), a large bottom frictional stress. Further, for estuaries, the energy loss related to this bottom frictional stress, as we shall see later in this section, is usually much larger than the energy loss associated with the internal eddy viscosity stresses integrated over a vertical column from the ocean surface to the bottom boundary layer. For the purposes of this book we are not interested in the boundary layer itself but only in its gross effects on estuary circulation. Suffice it to say that boundary layer theory is an extraordinarily complex aspect of fluid dynamics.

In particular we are interested in obtaining a rather simple relation between the bottom frictional stress f_b and the horizontal velocity v_b at the top of the bottom boundary layer. Consider then a horizontal flow in the positive x direction, the velocity v_x being given by $\bar{v}_x + v'_x$ from (2-58) and the velocity v_z by v'_z. Then, within the boundary layer of thickness δ, there will be a rapid change in velocity from some finite value v_b to zero, as shown in Fig. 2-20. Then, it may be reasoned that a vertical velocity

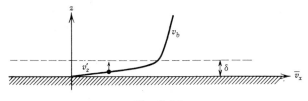

Fig. 2-20

fluctuation v'_z acting over some small distance we may think of as a mixing length results in a momentary vertical flow from a zone of a given velocity into a zone of somewhat higher velocity. This in turn produces a longitudinal velocity fluctuation $-v'_x$ in the upper layer. Further, the magnitude of this velocity fluctuation is related to the product of the vertical mixing length and the mean velocity gradient $\partial \bar{v}_x / \partial z$. Still further, since the bottom itself is fixed and the thickness δ is small, it can be argued from the continuity equation (2-61) that there must be a negative correlation between v'_x and v'_z. We may then write from (2-135), for the bottom frictional stress developed within the boundary layer,

$$f_{zx} = f_b = \overline{\rho v'_x v'_z} \propto \rho \left(\frac{\partial \bar{v}_x}{\partial z} \right)^2$$

Integrating over the layer thickness, for which we may take f_b to be constant, we then have

$$f_b = k\rho v_b^2 \qquad (2\text{-}143)$$

where k is a dimensionless drag coefficient. As the frictional stress is always in a direction opposite that of the flow, relation (2-143) should be corrected to read

$$f_b = k\rho \, |v_b| \, v_b \qquad (2\text{-}144)$$

We may obtain this same quadratic relation between the bottom frictional stress and the near bottom velocity through an alternative consideration taken from engineering hydraulics. Let us presume that we have an inertial flow of uniform velocity in the positive x direction moving over a frictionless bottom to the left of the origin, as shown in Fig. 2-21. Let us then presume that the flow encounters a frictional

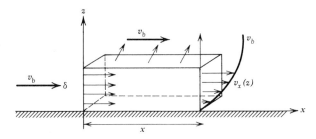

Fig. 2-21

bottom to the right of the origin. Considering then a control volume of unit width, of height δ equal to the boundary layer thickness, and of finite distance x in the positive x direction, the bottom frictional drag force along the length x is given by the rate of change in momentum, or momentum flux, out of the control volume from the usual relation $F = (d/dt)(Mv)$. In this case the gross changes occur in the mean values, and we may ignore the turbulent contributions. There is a momentum influx through the left hand vertical face and a momentum efflux through the right hand vertical face; and since there is a net decrease in the flow out the right hand face as compared to that in the left hand face, there is also a momentum efflux through the upper horizontal face. We then have, for the horizontal drag force F_b,

$$F_b = \rho \int_0^\delta v_b^2\, dz - \rho \int_0^\delta v_x^2\, dz - \rho v_b \int_0^\delta (v_b - v_x)\, dz$$

$$= \rho v_b^2 \int_0^\delta g(z)[1 - g(z)]\, dz$$

where the integral in the third term is the volume efflux through the upper horizontal surface, which when multiplied by ρv_b gives the momentum efflux in the positive x direction through the upper surface, and where we have made the substitution for $v_x(z)$,

$$v_x(z) = g(z)v_b$$

the quantity $g(z)$ varying from zero at $z = 0$ to unity at $z = \delta$. The bottom drag force in terms of the bottom frictional stress per unit area f_b is given simply by

$$F_b = \int_0^x f_b \, dx$$

so that we obtain finally

$$f_b = \rho v_b^2 \frac{d}{dx} \int_0^\delta g(z)[1 - g(z)] \, dz \qquad (2\text{-}145)$$

giving the same quadratic relation as before.

It is of interest at this point to consider a rather simplified example of estuary circulation to illustrate the importance of bottom friction and to ascertain the variation in internal friction and horizontal water velocity as a function of depth. Let us assume that the estuary water is of constant density and that the driving force is a surface water slope of constant inclination in the positive x direction. Under steady state conditions we then have from our equations of motion (2-134), using (2-135) and assuming no lateral contributions,

$$\frac{\partial p}{\partial x} = -\frac{\partial f_{zx}}{\partial z} \qquad (2\text{-}146)$$

We see immediately that, under our assumption of a constant $\partial p/\partial x$, f_{zx} must be a linear function in z varying from a value of zero at the water surface, no wind stress, to a value f_b at the bottom boundary layer, as illustrated in Fig. 2-22. We further see from (2-135) that if the eddy

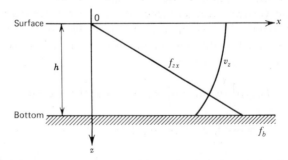

Fig. 2-22

viscosity coefficient N_z is a constant, $v_x(z)$ will be parabolic as a function of the depth. By using (2-106), (2-146) then becomes

$$-g\rho i = \rho N_z \frac{\partial^2 v_x}{\partial z^2}$$

or

$$\frac{d^2 v_x}{dz^2} = -\frac{gi}{N_z} \qquad (2\text{-}147)$$

where i is slope of the water surface. The solution of this equation is

$$v_x = -\alpha z^2 + C_1 z + C_2 \qquad (2\text{-}148)$$

where α is given by

$$\alpha = \frac{gi}{2 N_z} \qquad (2\text{-}149)$$

Our boundary conditions are

$$f_{zx} = -\rho N_z \frac{dv_x}{dz} = 0 \qquad\qquad z = 0$$

$$f_{zx} = -\rho N_z \frac{dv_x}{dz} = f_b = k\rho v_b^2 \qquad z = h \qquad (2\text{-}150)$$

The first boundary condition gives $C_1 = 0$, and the second gives a determined value for C_2, which in turn gives for v_x,

$$v_x = \sqrt{\frac{2\alpha N_z h}{k}} + \alpha(h^2 - z^2) \qquad (2\text{-}151)$$

The velocity distribution then resembles that shown in Fig. 2-22. The radical term in (2-151) is the bottom water velocity v_b, and αh^2 represents the change in velocity throughout the water column. Now for most estuaries the quantity v_b in (2-151) is usually considerably larger than αh_2. This then leads to the approximation often made in engineering hydraulic problems for unstratified flow that the water velocity is constant from surface to bottom and is given, from (2-151) and (2-148), by

$$v_b^2 = \frac{2\alpha N_z h}{k} = \frac{igh}{k} \qquad (2\text{-}152)$$

As mentioned previously and as we shall see later, for stratified flow or one in which there are density gradient forces, such a simple relation does not hold. In this case, although the net circulation velocity is of the order of that given by (2-152), both the upestuary and downestuary components of the stratified flow are much larger than their net.

Considering next the energy losses associated with the flow defined by (2-151), we have from (2-139) that the rate of internal energy loss is

$$I = \int_0^h \rho N_z \left(\frac{dv_x}{dz}\right)^2 dz = 4\rho N_z \alpha^2 \int_0^h z^2 \, dz = \tfrac{4}{3}\rho N_z \alpha^2 h^3 \qquad (2\text{-}153)$$

The rate of energy loss due to the bottom frictional stress is simply that

58 *Hydrodynamics*

stress multiplied by the bottom water velocity, or

$$E = f_b v_b = k\rho v_b^3 = k\rho\left(\frac{2\alpha N_z h}{k}\right)^{3/2} \tag{2-154}$$

The ratio of these two quantities is

$$\frac{E}{I} = \sqrt{\frac{9N_z}{2k\alpha h^3}} \tag{2-155}$$

From (2-151) the ratio of v_b to αh^2 is

$$\frac{v_b}{\alpha h^2} = \sqrt{\frac{2N_z}{k\alpha h^3}} \tag{2-156}$$

which except for the numerical factor is the same as (2-155). We see then that the bottom friction energy loss is the dominant term.

Up to this point we have not mentioned external stress effects that can occur at the upper boundary of an estuary, the water surface. Here there can be winds blowing which create a shearing stress at the water surface. If the wind is blowing in the positive x direction, this shearing stress f_{zx} can be represented in the water column by a corresponding internal frictional term $\rho N_z(\partial v_x/\partial z)$, evaluated at $z = 0$.

For additional information on boundary layer theory the reader may refer to Schlichting (1955), Rouse (1950), Streeter (1958), and Rossby and Montgomery (1935).

2-10. SCALING AND INTEGRATED EQUATIONS

We have in the previous five sections discussed at some length the phenomenon of turbulence and its importance in diffusion and in internal and external friction. We now look at and discuss in a general manner the importance of the various eddy diffusion and eddy viscosity terms in the equations of continuity and motion for an estuary.

We consider the mean motion to be along the x longitudinal axis, with the y axis lateral and the z axis vertically down. Then, from the equation of continuity (2-65) for a conservative substance, which we take for convenience to be the salinity, and from the longitudinal equations of motion (2-134) along with (2-135), we obtain eddy diffusion and eddy viscosity terms of the form

$$K_x\frac{\partial s}{\partial x} \qquad K_y\frac{\partial s}{\partial y} \qquad K_z\frac{\partial s}{\partial z} \tag{2-157}$$

and

$$N_x\frac{\partial v_x}{\partial x} \qquad N_y\frac{\partial v_x}{\partial y} \qquad N_z\frac{\partial v_x}{\partial z} \tag{2-158}$$

Now, if the boundaries of the estuary are essentially parallel to the coordinate axes, the equation of continuity (2-27) indicates that $\partial v_x/\partial x$ will be zero, so that the first expression in (2-158) drops out. Further, if the lateral extent of the estuary is small compared with its longitudinal extent, so that there is essentially no variation in the dependent variables as a function of y, the second expressions in (2-157) and (2-158) will also drop out. We are then left with the first and third expressions of (2-157) and the third expression of (2-158). We now look at these three terms in somewhat more detail.

Although both $K_z(\partial s/\partial z)$ and $N_z(\partial v_x/\partial z)$ represent the effects of vertical mixing or turbulence, there is a fundamental difference in their vertically integrated effect. As illustrated in Fig. 2-23, the eddy diffusion term

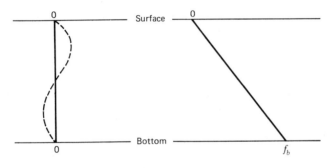

Fig. 2-23

represents the vertical flux of salt due to turbulence, so that this term must necessarily be zero at the upper and lower boundaries. However, the eddy viscosity term represents the shearing stress f_{zx} which, for no wind stress, is zero at the surface but is equal to the external frictional stress f_b at the bottom. We then note from (2-65), (2-134), and (2-135) that a vertical integration of the equations of continuity and motion from the surface to the bottom drop out the vertical diffusion term and leave the vertical viscosity term equal to f_b.

Let us now look at the relative importance of $K_x(\partial s/\partial x)$ and $K_z(\partial s/\partial z)$ in the equation of continuity. If the estuary waters are essentially completely mixed in the vertical direction, $\partial s/\partial z$ will be zero, and we will have only the longitudinal eddy diffusion term.

If the estuary waters are stratified, we must look at the relative magnitudes of the horizontal and vertical eddy diffusion coefficients and the relative magnitudes of the horizontal and vertical salinity variations. For estuaries under the influence of tidal mixing the values of K_x are large, of the order of 10^5–10^6 cm^2 sec^{-1}. The values of K_z are much smaller and more variable and are of the order of 10^0–10^3 cm^2 sec^{-1}. We may then expect that K_x/K_z will be of the order of 10^3–10^5. We must, however, also look at the scale of the salinity variations. Let us change to dimensionless coordinates such that $x_1 = x/l$ and $z_1 = z/h$, where h is the water depth and l is a longitudinal distance such that a plot of s versus x_1 and z_1 shows similar variations. Then the horizontal and vertical eddy diffusion terms in (2-65) become, taking K_x and K_z to be constants,

$$K_x \frac{\partial^2 s}{\partial x^2} = \frac{K_x}{l^2} \frac{\partial^2 s}{\partial x_1^2}$$

$$K_z \frac{\partial^2 s}{\partial z^2} = \frac{K_z}{h^2} \frac{\partial^2 s}{\partial z_1^2}$$

(2-159)

Now, for a characteristic stratified estuary condition with $h = 10$ m, l is of the order of 10 km. We then have $l^2/h^2 = 10^6$, so that the second expression of (2-159) is more important than the first. We then have the

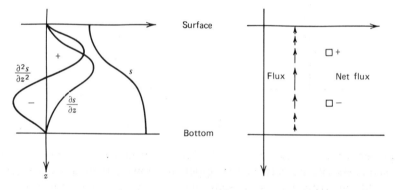

Fig. 2-24

interesting result that, if we consider an incremental scaled volume in an estuary, the vertical diffusion will dominate over the horizontal diffusion, whereas, if we take an integrated volume from surface to bottom, only the horizontal diffusion term will remain.

Some of these aspects of vertical diffusion are illustrated in Fig. 2-24. A characteristic salinity versus depth plot for a stratified estuary is as shown. The first derivative and second derivative graphs are then as sketched. From Fig. 2-24 and (2-48) we see that the salt flux is always upward, varying from zero at the bottom back to zero at the surface. From (2-65) we see that the net flux into an incremental volume is positive in the upper portion of the water column and negative in the lower portion. Let us now include in a similar diagram the advection effects in the upper and

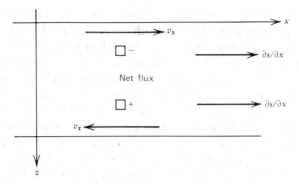

Fig. 2-25

lower parts of the water column. As shown in Fig. 2-25, for a stratified flow there is a downestuary flow in the upper portion and an upestuary flow in the lower portion. For both flows the horizontal salinity gradient increases in a downestuary direction. From Fig. 2-25 and (2-65) we then see that the net flux into an incremental volume is negative in the upper portion of the water column and positive in the lower portion. The horizontal advection effects in both the upper and lower portions of the water column are opposite the vertical diffusion effects. For an incremental volume in a stratified flow under steady state conditions these are the two principal balancing effects.

Let us now consider the next obvious step, that of the integrated effect. As stated previously, the only diffusion contribution is that of the smaller horizontal diffusion term. The advection contribution is that of the net of downestuary over upestuary flow, or the net river flow into the estuary. Again, as mentioned previously, this net advection is much smaller than either the separate downestuary or upestuary flow, corresponding to the smaller horizontal as compared with vertical diffusion term.

In conclusion, let us make a rather obvious but important comment. It should be apparent from this discussion that judicious interpretation of measured salinity values in an estuary, along with a knowledge of the river flow into the estuary and circulation velocities, when considered both over an incremental volume and over an integrated volume, can lead to evaluations of K_x and K_z as a function of longitudinal extent along the estuary and depth.

2-11. HYDRAULICS OF OPEN CHANNEL FLOW

We digress for a moment to the consideration of some aspects of the engineering hydraulics of open channel flow to obtain some concepts which can be useful in estuary circulation. We are interested in rather gross inertial changes having to do with the *control* of the flow, the effects of *transitions* in the flow caused by large changes in channel geometry, and the phenomenon of the *hydraulic jump*. We omit for purposes of illustration the effects of internal and external friction in consideration of the larger inertial changes, and we correspondingly do not include any driving force terms to maintain an unaccelerated flow. Further, the density is taken as a constant; and as a necessary condition to the above, the flow velocity is taken as zero in the vertical and lateral directions.

Our governing equations are those of continuity of mass, energy, and momentum. We see then from the above that we have met the conditions for the derivation of the Bernoulli equation (2-17). We also see from the form of the terms in (2-17) that it is an energy equation. The only external force is that of gravity $\mathbf{g}$ at right angles to the flow which, taking the z axis vertically downward gives for the gravitational potential

$$\mathbf{F} = \mathbf{g} = \mathbf{k}g = -\nabla\Omega = -\mathbf{k}\frac{\partial}{\partial z}(-gz)$$

or

$$\Omega = -gz \tag{2-160}$$

where $\mathbf{k}$ is the unit vector in the z direction. Equation (2-17) may then be written as

$$\frac{p}{\rho} - gz + \tfrac{1}{2}v^2 = D \tag{2-161}$$

where D is a constant. When the origin of the reference coordinate system is at the water surface, the pressure is given by $p = \rho gz$, from which we have the obvious result that the velocity is independent of z.

We are interested in determining what changes in the flow occur when there is a change in channel geometry, either from a change in the bottom depth or in the lateral extent of the channel, as shown in Fig. 2-26. If the coordinate axes is taken as shown, (2-161) will become

$$\tfrac{1}{2}v_1^2 = g(z - \zeta) - gz + \tfrac{1}{2}v_2^2 = D$$

$$\tfrac{1}{2}v_1^2 = -g\xi + \tfrac{1}{2}v_2^2 = D$$

$$\frac{v_1^2}{2g} + h_1 = \frac{v_2^2}{2g} + h_2 + e = E \tag{2-162}$$

where E is a constant. The mass, or flow, continuity relation is simply

$$Q = v_1 b_1 h_1 = v_2 b_2 h_2 = q_1 b_1 = q_2 b_2 \tag{2-163}$$

where Q is the total flow, and q is defined by

$$q = \frac{Q}{b} \tag{2-164}$$

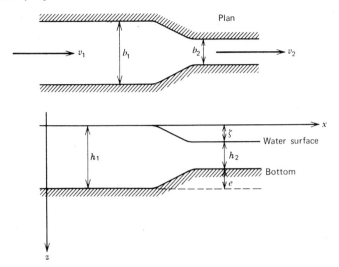

Fig. 2-26

the flow per unit breadth. Substituting (2-163) into (2-162) we obtain

$$\frac{q_1^2}{2gh_1^2}+h_1=\frac{q_2^2}{2gh_2^2}+h_2+e=E$$

or

$$\frac{q^2}{2gh^2}+h+e=E \qquad (2\text{-}165)$$

in generalized form.

Let us see what happens to the flow for a transition in which there is no change in the channel breadth b but only in the channel elevation e. Then, from (2-164) we see that q is a constant; and we have only to consider the variations in h as a function of e. From the form of (2-165) we see that for any given value of e there are two positive roots for h up to a maximum value of e. This maximum value of e and the corresponding values of h and v can be determined simply by setting the derivative de/dh equal to zero or, from (2-165),

$$\frac{de}{dh}=\frac{q^2}{gh^3}-1=0$$

$$h_c=\sqrt[3]{\frac{q_c^2}{g}} \qquad (2\text{-}166)$$

or since, from (2-163), v is given in terms of q by $v=q/h$,

$$v_c=\frac{q_c}{h_c}=\sqrt{gh_c} \qquad (2\text{-}167)$$

The velocity v_c is referred to as the *critical velocity*. A graph of h versus e then resembles that sketched in Fig. 2-27. As v is inversely proportional to h, we see that for values of h greater than h_c, v is less than v_c; and the

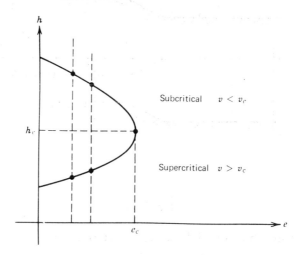

Fig. 2-27

region and velocity values are referred to as being *subcritical*. For values of h less than h_c, v is greater than v_c; and the region and velocity values are referred to as being *supercritical*. If the original flow is subcritical, a decrease in the bottom depth at the transition will lead to a decrease in h and an increase in v. If the original flow is supercritical, there will be an increase in h and a decrease in v. We also see from Fig. 2-27 that the greatest permissible decrease in the bottom depth under the assumed flow conditions is e_c. For a greater magnitude decrease in bottom depth, there has to be a decrease in q or Q before the flow can be physically possible; and this decrease in q will be such that the resultant flow at the transition is critical. In this latter case the transition acts as a control on the flow.

Let us next see what happens to the flow for a transition in which there is no change in channel depth, where $e = 0$, but where there is either a change in channel breadth b or in flow per unit breadth q. Then we see from (2-164) and (2-165) that we are now interested in considering the variations in h as a function of q. Again we see that there are two positive roots for h up to a maximum value of q. This maximum can be determined in the same manner, setting dq/dh equal to zero, or from

(2-165),

$$\frac{q}{gh^2}\frac{dq}{dh} - \frac{q^2}{gh^3} + 1 = 0$$

$$\frac{q_c^2}{gh^3} = 1$$

which gives the same values for h_c and v_c as (2-166) and (2-167). A graph of h versus q then resembles that shown in Fig. 2-28. Again we see that

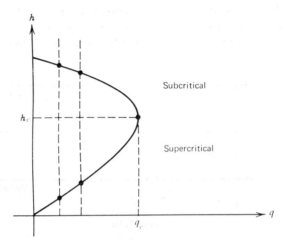

Fig. 2-28

for either an increase in q or a decrease in b there is a decrease in h and an increase in v when the original flow is subcritical, and an increase in h and a decrease in v when the original flow is supercritical up to the critical value q_c. We see that we have the same control condition as before. As the breadth b is decreased for a constant total flow Q, q increases up to a maximum value q_c beyond which any further decrease in breadth must be accommodated by a decrease in the total flow Q maintaining the critical flow conditions. We see further that, for a constant depth and breadth channel and a given value of E, the maximum flow condition is the critical flow condition.

We see from the above discussion that in general, for steady, uniform, open channel flow, there are two flow conditions having the same energy values, as given by the constant E of the Bernoulli equation (2-162). In the supercritical state the kinetic energy term is dominant, and in the subcritical state the potential energy term is dominant. We might then expect that there could occur, under conditions that are not examined

Hydrodynamics

here but in a later section, a change, or jump, from one flow condition to the other. As we see in the discussion that follows, this change, or hydraulic jump as it is generally referred to, can only occur from the supercritical state to the subcritical state, as there is an energy loss in the jump itself. This loss is positive only for a supercritical to a subcritical change, and negative (which represents an increase in total energy and therefore is prohibitive) for a subcritical to a supercritical change. The hydraulic jump in effect converts kinetic energy into potential energy.

The governing equation in this instance is continuity of momentum to maintain an unaccelerated flow. From the relation $\mathbf{F} = (d/dt)(M\mathbf{v})$ and with reference to Fig. 2-29, for a channel of constant depth and lateral

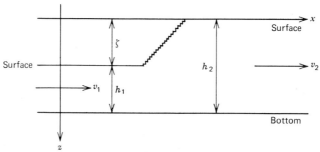

Fig. 2-29

extent the horizontal forces on a face of unit breadth in the supercritical and subcritical states is the integral of the pressure over the depth; and the rate of change in momentum per unit width is the mass flow $\rho q = \rho v_1 h_1$ multiplied by the velocity change $v_2 - v_1$, or

$$\int_0^{h_1} \rho g z \, dz - \int_0^{h_2} \rho g z \, dz = \rho q (v_2 - v_1)$$

$$\tfrac{1}{2}\rho g h_1^2 - \tfrac{1}{2}\rho g h_2^2 = \rho v_1 h_1 (v_2 - v_1) \qquad (2\text{-}168)$$

Substituting for v_2 in terms of v_1, h_1, and h_2 from the equation of continuity (2-163), we may then solve for h_2 in terms of h_1 and v_1, obtaining

$$g(h_1^2 - h_2^2) = \frac{2v_1^2 h_1}{h_2}(h_1 - h_2)$$

$$h_2^2 + h_1 h_2 - 2\frac{v_1^2 h_1}{g} = 0 \qquad (2\text{-}169)$$

$$h_2 = \tfrac{1}{2}\left(-h_1 + \sqrt{h_1^2 + 8\frac{v_1^2 h_1}{g}}\right)$$

where the minus sign is omitted before the radical because a negative value of h_2 has no physical significance.

We could obtain the energy equation directly from the Bernoulli equation (2-162), but it is of slight interest to derive it separately. In taking a unit breadth of the channel the cross sectional area normal to the flow in the supercritical region is h_1, and that in the subcritical region is h_2. Then, the mass passing an incremental cross section dh_1 per unit time is $\rho v_1\, dh_1$, so that the energy passing per unit time is $\rho v_1(\frac{1}{2}v_1{}^2 + \Omega_1)\, dh_1$ and the work done on it per unit time by the pressure p_1 is $p_1 v_1\, dh_1$. As the flow is steady, the sum of these two quantities when integrated over the cross section h_1 must be the same as for the corresponding quantities over h_2, or with reference to Fig. 2-29 and using the continuity relation (2-163) and the potential relation (2-160),

$$\int_{\zeta}^{\zeta+h_1} \rho v_1\left(\frac{p_1}{\rho} + \Omega_1 + \tfrac{1}{2}v_1{}^2\right) dz = \int_0^{h_2} \rho v_2\left(\frac{p_2}{\rho} + \Omega_2 + \tfrac{1}{2}v_2{}^2\right) dz + \epsilon'$$

$$\rho v_1 \int_{\zeta}^{\zeta+h_1} [g(z-\zeta) - gz + \tfrac{1}{2}v_1{}^2]\, dz = \rho v_2 \int_0^{h_2} [gz - gz + \tfrac{1}{2}v_2{}^2]\, dz + \epsilon' \qquad (2\text{-}170)$$

$$(-g\zeta + \tfrac{1}{2}v_1{}^2)\rho_1 v_1 h_1 = (\tfrac{1}{2}v_2{}^2)\rho v_2 h_2 + \epsilon'$$

$$\frac{v_1{}^2}{2g} + h_1 = \frac{v_2{}^2}{2g} + h_2 + \epsilon$$

which is the same as (2-162) with the inclusion of ϵ representing the energy loss associated with the jump and with the exclusion of the bottom elevation term e. Substituting for v_2 from (2-163) and for v_1 from the first of (2-169), we obtain for ϵ, in terms of h_1 and h_2,

$$\epsilon = \frac{v_1{}^2}{2g} - \frac{v_2{}^2}{2g} + h_1 - h_2$$

$$= \frac{h_2}{4h_1}(h_1 + h_2)\left(1 - \frac{h_1{}^2}{h_2}\right) - (h_2 - h_1) \qquad (2\text{-}171)$$

$$= \frac{(h_2 - h_1)^3}{4h_1 h_2}$$

which simply shows that, for a jump from a supercritical state to a subcritical state, $h_2 > h_1$, the energy loss associated with the jump is a positive quantity.

2-12. HYDRODYNAMIC NUMBERS AND COEFFICIENTS

We conclude this chapter with some consideration of various dimensionless numbers and coefficients used in hydrodynamics and hydraulic engineering. Although we have little direct use for these numbers and coefficients in this text, it is considered useful to give at least a brief and generalized description of them at this point.

Because of the large variation in the scale of hydrodynamic phenomena and the variety of stresses that can be important, a set of dimensionless numbers has been developed which are used in a generalized way and in empirical formulas to describe the state of motion. There are various ways to arrive at these dimensionless numbers, from specific examples, from dimensional analysis, and from the form of the equations of motion. We choose the last-mentioned as more instructive in a physical understanding of them. In the equations of motion, considered for a unit volume of the fluid, we have terms of the form on the left hand and right hand side,

$$\rho \frac{\partial v}{\partial t}, \rho v \frac{\partial v}{\partial n} \quad \text{and} \quad \frac{\partial p}{\partial n}, \frac{\partial}{\partial m}\left(\mu \frac{\partial v}{\partial n}\right), \rho \omega v, \rho g \qquad (2\text{-}172)$$

where v is a velocity component, m, n are coordinate directions x, y, z, $\rho(\partial v/\partial t)$ represents the local acceleration, $\rho v(\partial v/\partial n)$ is the inertial force or kinetic reaction, $\partial p/\partial n$ is the pressure gradient force, $\partial/\partial m[\mu(\partial v/\partial n)]$ is the eddy viscosity force, $\rho \omega v$ is the Coriolis force, and ρg is the gravity force. For steady state motion we may then consider integrating the particular, and appropriate, component of v over the flow dimensions, obtaining terms of the form

$$\rho V^2 \quad \text{and} \quad p, \mu \frac{V}{D}, \rho \omega V r, \rho g h \qquad (2\text{-}173)$$

where V is a generalized velocity, h is a characteristic depth, D is a characteristic cross dimension of the flow, and r a characteristic radius of the flow stream lines. The *Froude number* Π_g is defined as the ratio of such a generalized inertial force to the gravity force, or

$$\Pi_g = \frac{\rho V^2}{\rho g h} = \frac{V^2}{g h} \qquad (2\text{-}174)$$

The *Reynolds number* Π_μ is defined as the ratio of such a generalized inertial force to the eddy viscosity force, or

$$\Pi_\mu = \frac{\rho V^2}{\mu(V/D)} = \frac{\rho V D}{\mu} = \frac{V D}{N} \qquad (2\text{-}175)$$

The *Rossby number* Π_ω is defined as the ratio of such a generalized inertial force to the Coriolis force, or

$$\Pi_\omega = \frac{\rho V^2}{\rho \omega V r} = \frac{V}{\omega r} \qquad (2\text{-}176)$$

It is sometimes more appropriate, depending on the type and scale of the turbulence, to substitute for the velocity V in terms of a characteristic oscillation frequency Γ and a length L, similar to (2-71) or (2-132), defining a Reynolds number $\Pi_{\mu'}$ as

$$\Pi_{\mu'} = \frac{\Gamma L^2}{N} \qquad (2\text{-}177)$$

We note further that in some applications the Froude number is defined as the square root of the expression given by (2-174). We also note in passing that a Froude number of unity defines a critical flow condition for the open channel flow characteristics of the previous section. The significance of any one of these numbers is in giving a generalized description of the nature of a flow where the associated force term is dominant, and in hydraulic modeling where for a given dominant force condition it is necessary to have the corresponding Froude, Reynolds, or Rossby number equal in the model to what it is observed to be in the actual state.

The coefficients we wish to outline next are all related to bottom frictional effects. They were developed largely through hydraulic engineering considerations of flow through pipes and conduits. We defined a dimensionless bottom friction coefficient k in (2-143). The *Darcy resistance coefficient f* is defined in terms of k by

$$f = 8k \qquad (2\text{-}178)$$

The *Chezy coefficient C* is defined in terms of f or k by

$$C = \sqrt{\frac{8g}{f}} = \sqrt{\frac{g}{k}} \qquad (2\text{-}179)$$

We have further an empirically derived formula for C,

$$C = \frac{1.49}{n} R^{1/6} \qquad (2\text{-}180)$$

where n is referred to as the *Manning roughness factor* and R is the *hydraulic radius*, defined as the cross sectional area normal to the flow divided by the wetted perimeter of the cross sectional area subject to bottom friction. For flow along a channel with a rectangular cross section of breadth b and depth h, R is given by $bh/(2h+b)$. For most estuaries where b is much greater than h, R can usually be approximated simply by h.

CHAPTER 3

Tidal Phenomena

3-1. INTRODUCTION

One of the more thoroughly studied areas in physical oceanography deals with tides and tidal phenomena. Here we consider only certain selected aspects. In particular we have little interest in the tide generating effects of the moon and sun motions. The observed tidal motions in an estuary are generally caused by the ocean tide at the estuary entrance and not by the tide generating forces of the moon and sun acting on the estuary waters. Further, it is assumed that the reader has a basic knowledge of the description of various physical phenomena in terms of waves; as most of this chapter is concerned with wave motion, reflection, and dissipation, it is treated in abbreviated form.

We discuss one other important aspect, which applies to both this and the following two chapters, at the outset. One can distinguish three prominent phenomena occurring in an estuary—tides, circulation, and mixing; these phenomena are discussed in this and the following two chapters. For the most part we treat them as separate phenomena; we then solve one differential equation rather than a set of simultaneous differential equations. This is usually satisfactory in a description of the measurable quantities in an estuary. Cross effects, when they occur, are usually in the form of nonlinear terms in the defining differential equations; often, fortunately, these terms are also of second order and can be neglected. When they are not, it is often difficult to obtain an analytic solution that can be simply interpreted and applied. When this is the case, numerical solutions can sometimes be obtained with finite difference approximations and computer calculations.

3-2. TIDAL WAVES

The description of *tidal* or *long* waves, as the name implies, is applicable to the wave motion produced by the tide generating forces of the gravitational attraction of the moon and the sun. The basic assumption in the theory of tidal waves is that the wavelengths are so long in comparison with the ocean depth that the water particle motion is mainly horizontal and is essentially the same for all particles in a given vertical plane. The vertical accelerations may then be neglected, and the pressure at any depth taken to be the hydrostatic pressure. We assume that the fluid is incompressible and for the first derivation that the particle motion is sufficiently small that the time derivative D/Dt may be replaced by $\partial/\partial t$.

Let us consider one dimensional wave motion with the x axis in the direction of the wave motion, the z axis vertical downward, and the origin at the sea surface, as shown in Fig. 3-1. Let the displacements of the

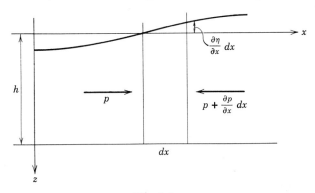

Fig. 3-1

ocean surface away from its horizontal, neutral position be denoted by ξ and η, respectively, in the positive x and negative z directions, and let the depth of the ocean be h. Then the hydrostatic pressure at any depth is directly proportional to the overlying column of water, and the horizontal pressure gradient is given by

$$\frac{\partial p}{\partial x} = \rho g \frac{\partial \eta}{\partial x} \tag{3-1}$$

The horizontal force per unit volume in the positive x direction is the negative of (3-1), so that the equation of motion in the x direction is

$$\rho \frac{\partial^2 \xi}{\partial t^2} = -\rho g \frac{\partial \eta}{\partial x} \tag{3-2}$$

The continuity conditions may be obtained by considering the net volume of water that has entered the column shown in Fig. 3-2 in a time t. The net displacements ξ and η in the x and z directions are the integrals of their respective velocity components, so that we have, for continuity of an incompressible fluid,

$$\eta = -h \frac{\partial \xi}{\partial x} \tag{3-3}$$

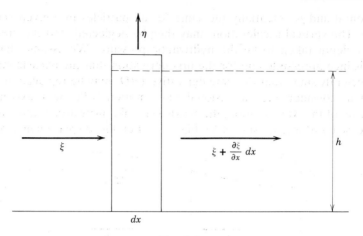

Fig. 3-2

Combining (3-2) and (3-3) we obtain

$$\frac{\partial^2 \xi}{\partial t^2} = gh \frac{\partial^2 \xi}{\partial x^2} \qquad (3\text{-}4)$$

which is simply the one dimensional form of the wave equation. The elimination of ξ from (3-2) and (3-3) gives, as expected, the same wave equation for η. We see then that the wave motion is propagated with a *wave velocity* c given by

$$c = \sqrt{gh} \qquad (3\text{-}5)$$

and with all the other characteristics usually associated with one dimensional wave propagation. If we take a simple harmonic wave for ξ of the form

$$\xi = A \cos \frac{2\pi}{\lambda}(x - ct) \qquad (3\text{-}6)$$

where λ is the *wavelength*, we will have for η, from (3-3),

$$\eta = \frac{2\pi h}{\lambda} A \sin \frac{2\pi}{\lambda}(x - ct) \qquad (3\text{-}7)$$

showing that, when h is much smaller than λ, the displacement, velocity, and acceleration in the vertical direction are small compared with those in the horizontal direction. In general the solution to the wave equation (3-4) for a wave propagating in the positive x direction is of the form

$$\xi = F(x - ct) \qquad (3\text{-}8)$$

From (3-8) and (3-3) we then have for the ratio of the water particle velocity, $\dot{\xi} = v_x$, to the wave velocity c,

$$\frac{v_x}{c} = \frac{\eta}{h} \qquad (3\text{-}9)$$

The ratio v_x to c is the same as the ratio of the wave height η to the water depth h. At a wave crest the horizontal particle velocity is in the direction of wave propagation; at a wave trough the horizontal particle velocity is opposite the direction of wave propagation. It should be obvious but is perhaps of some value to note that the motion and continuity relations of (3-2) and (3-3) could have been obtained directly from the motion and continuity equations of Chapter 2.

We may arrive at the same conclusions with regard to the value of the wave velocity and the ratio of the particle velocity to the wave velocity through an interesting alternative approach. Let us consider a tidal wave propagating in the negative x direction and impose on it a mass movement of the whole ocean with a constant velocity equal and opposite that of the wave velocity, as shown in Fig. 3-3. Then the motion becomes

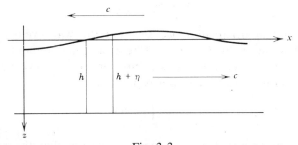

Fig. 3-3

steady. The ocean surface becomes stationary, while the forces acting on the water particles remain the same as before. Equation (2-17) then applies, and we have

$$\frac{p}{\rho} = D - g(h + \eta) - \tfrac{1}{2}v^2 \qquad (3\text{-}10)$$

where D is a constant, and v is the particle velocity, taken as before to be principally in the x direction. Our continuity condition now becomes that the volume transport of water $v(h + \eta)$ through any vertical section of height $h + \eta$ must be a constant. If the wave motion height is gradually reduced to zero, the volume transport through a section of vertical height h will then be ch. The equation of continuity then becomes

$$v(h + \eta) = ch \qquad (3\text{-}11)$$

Substituting (3-11) into (3-10) we obtain

$$\frac{1}{\rho} p(\eta) = D - g(h + \eta) - \tfrac{1}{2} \frac{c^2 h^2}{(h + \eta)^2} \qquad (3\text{-}12)$$

If the ocean surface is to remain stationary, the pressure along this surface must be a constant. This will be achieved if $p'(\eta)$ is zero. From (3-12) we then have

$$\frac{1}{\rho} p'(\eta) = -g + \frac{c^2 h^2}{(h + \eta)^3} = 0$$

or

$$c^2 = gh\left(1 + \frac{\eta}{h}\right)^3 \qquad (3\text{-}13)$$

To a first order approximation for small values of η this is simply

$$c = c_0 = \sqrt{gh} \qquad (3\text{-}14)$$

the same as (3-5). To a second order approximation this is

$$c = c_0\left(1 + \frac{3}{2}\frac{\eta}{h}\right) \qquad (3\text{-}15)$$

In this second approximation we see that the wave velocity is not a constant. A wave of this type cannot be propagated without a change in profile, since the wave velocity is a function of the wave height. From (3-11) we also have to a first approximation,

$$v = c\left(1 - \frac{\eta}{h}\right) \qquad (3\text{-}16)$$

Subtracting the imposed mass velocity c, we see that the ratio of the particle velocity in the undisturbed state to the wave velocity is the same as that given by (3-9) in the direction of propagation.

We now examine briefly in a somewhat more formal manner tidal waves in a body of water whose depth is sufficiently shallow that the approximation of $\eta \ll h$, which we have used up to this point, can no longer be considered valid. The additional terms derived may be considered the *shallow water constituent* of tidal waves. We restrict ourselves to an open sea source condition of a simple harmonic wave of the form

$$\eta = a \cos \omega t \qquad (3\text{-}17)$$

corresponding to the ocean tides, where ω, the *circular frequency*, is given by the usual expression

$$\omega = \frac{2\pi}{T} \qquad (3\text{-}18)$$

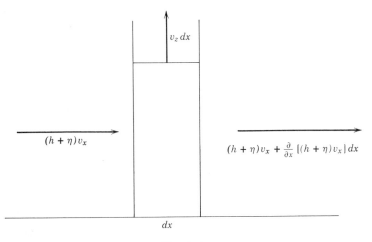

Fig. 3-4

in terms of the *period* T of the semidiurnal or diurnal tide. For this case our continuity relation must be as shown in Fig. 3-4. Equating the rate of mass transport into and out of the vertical column for an incompressible fluid gives

$$\frac{\partial}{\partial x}\left[(h+\eta)v_x\right]=-v_z=-\frac{\partial\eta}{\partial t}$$

or

$$\frac{\partial\eta}{\partial t}+v_x\frac{\partial\eta}{\partial x}=-(h+\eta)\frac{\partial v_x}{\partial x} \tag{3-19}$$

We must correspondingly use D/Dt instead of $\partial/\partial t$ in the equation of motion, so (3-2) now becomes

$$\frac{\partial v_x}{\partial t}+v_x\frac{\partial v_x}{\partial x}=-g\frac{\partial\eta}{\partial x} \tag{3-20}$$

We may solve these two simultaneous nonlinear differential equations by the method of successive approximations. For a first approximation we have, for equations (3-19) and (3-20),

$$\frac{\partial\eta_1}{\partial t}=-h\frac{\partial v_{x1}}{\partial x}$$

and

$$\frac{\partial v_{x1}}{\partial t}=-g\frac{\partial\eta_1}{\partial x} \tag{3-21}$$

whose solutions consistent with (3-17) are

$$\eta_1 = a \cos \omega\left(t - \frac{x}{c}\right)$$

and   (3-22)

$$v_{x1} = \frac{ga}{c} \cos \omega\left(t - \frac{x}{c}\right)$$

where c is given by (3-14) and where we have taken, for convenience in this case, the direction of the x axis to be upestuary. Taking η and v_x to be given by

$$\eta = \eta_1 + \eta_2$$

and   (3-23)

$$v_x = v_{x1} + v_{x2}$$

we then have for the differential equations to be satisfied by η_2 and v_{x2}, substituting (3-23) into (3-19) and (3-20), using (3-21), and neglecting product terms in η_2 and v_{x2},

$$\frac{\partial \eta_2}{\partial t} = -h \frac{\partial v_{x2}}{\partial x} - \eta_1 \frac{\partial v_{x1}}{\partial x} - v_{x1} \frac{\partial \eta_1}{\partial x}$$

and   (3-24)

$$\frac{\partial v_{x2}}{\partial t} = -g \frac{\partial \eta_2}{\partial x} - v_{x1} \frac{\partial v_{x1}}{\partial x}$$

Substituting solutions (3-22) into (3-24), we then have

$$\frac{\partial \eta_2}{\partial t} = -h \frac{\partial v_{x2}}{\partial x} - \frac{ga^2\omega}{c^2} \sin 2\omega\left(t - \frac{x}{c}\right)$$

and   (3-25)

$$\frac{\partial v_{x2}}{\partial t} = -g \frac{\partial \eta_2}{\partial x} - \frac{g^2a^2\omega}{2c^3} \sin 2\omega\left(t - \frac{x}{c}\right)$$

Taking cross derivatives $\partial/\partial t$ of the first and $h(\partial/\partial x)$ of the second equation of (3-25) and subtracting we obtain an equation in η_2 only which can be solved by the usual methods of partial differential equations, giving in conjunction with (3-22) for the final solution,

$$\eta = a \cos \omega\left(t - \frac{x}{c}\right) - \frac{3}{4} \frac{ga^2\omega}{c^3} x \sin 2\omega\left(t - \frac{x}{c}\right)$$ (3-26)

and then, using either equation of (3-25) for v_{x2} and v_x,

$$v_x = \frac{ga}{c} \cos \omega\left(t - \frac{x}{c}\right) - \frac{1}{8} \frac{g^2a^2}{c^3} \cos 2\omega\left(t - \frac{x}{c}\right)$$

$$- \frac{3}{4} \frac{g^2a^2\omega}{c^4} x \sin 2\omega\left(t - \frac{x}{c}\right) \quad (3\text{-}27)$$

Although these results have little direct application in themselves, they are instructive in illustrating the effects that do occur with shallow water tides. First, we note that as the distance x appears as a multiplying factor in the amplitude term of the second part of (3-26), there is a limit to which our approximation that η_2 is small with respect to η_1 holds. Taking the ratio of the amplitudes of η_2 and η_1 and substituting for c from (3-14) and for ω from the familiar relation $\lambda = 2\pi c/\omega$, we note that $2\pi(a/h)(x/\lambda)$ must be small. Hence however small the ratio of the original amplitude to the depth, this fraction ceases to be small when x is a sufficient multiple of the wavelength. If we look at the tidal motion at some distance up the estuary, we will see that the effect of the second term in (3-26) is to shift the position of the maximum and the minimum in such a manner that the fall of the tide takes a longer time than the rise. In a general physical way this is understandable, in that at a wave trough in shallow water the water depth is significantly less than at a wave crest with, from (3-14), a consequent lower wave velocity at the trough than at the crest. This then has the effect that the wave minimum occurs later and the wave maximum earlier than normal, so that the rise time of the tide is shorter than the fall time. In general the inclusion of a second harmonic term such as occurs in (3-26) introduces the concept of *overtides*. These general characteristics of tidal distortion with the inclusion of bottom friction can be of significance in shallow waters; although they are unimportant in most estuaries, they are important in many tidal rivers.

Let us now look at a few rather simple but important properties of wave motion. In many instances we have to consider the reflection of a tidal wave from the upper end of an estuary. In this instance we consider a *standing wave* of the form

$$\eta = a \cos \frac{2\pi x}{\lambda} \cos \frac{2\pi t}{T} \qquad (3\text{-}28)$$

which in itself is a solution of the wave equation (3-4), where the wave velocity is given by the usual and familiar relation

$$c = \sqrt{gh} = \frac{\lambda}{T} \qquad (3\text{-}29)$$

From (3-2) and (3-3), or more simply from (3-21), v_x is then

$$v_x = \frac{ac}{h} \sin \frac{2\pi x}{\lambda} \sin \frac{2\pi t}{T} \qquad (3\text{-}30)$$

We see that in this case the current velocity is $\pi/2$ out of phase with the tide height. The maximum flood and ebb currents occur at half tide, and slack water occurs at high and low tide. This is to be contrasted with the characteristic for a *progressive wave* as given by expression (3-9). The amplitude ratios are the same as that given by (3-30), but in the case of a progressive wave the current velocity and tide height are in phase. The maximum flood and ebb currents occur at high and low tide, respectively, and slack water occurs at half tide.

To take this simple analysis one step further let us consider what the variations in a conservative quantity, such as salinity, are during a tidal cycle. Within a tidal cycle the advection effects of the tidal motion are almost always much greater than the corresponding tidal diffusion effects or the advection effects of runoff or circulation. The equation of continuity (2-65) then reduces to

$$\frac{\partial s}{\partial t} = -v_x \frac{\partial s}{\partial x} \tag{3-31}$$

where v_x is of the form of a simple harmonic tidal variation,

$$v_x = C \sin \frac{2\pi t}{T} \tag{3-32}$$

Taking $\partial s / \partial x$ to be a constant over the variations, we then have for the salinity variation Δs about its mean value, from (3-31) and (3-32),

$$\Delta s = \frac{CT}{2\pi} \frac{\partial s}{\partial x} \cos \frac{2\pi t}{T} \tag{3-33}$$

We then see that the salinity variation at a particular location in the estuary is $\pi/2$ out of phase with the current velocity, which is to be expected. As the current floods, the salinity at a given observation station increases; as the current ebbs, the salinity decreases. These various relations among wave height, current velocity, and salinity variation are shown in Figs. 3-5 and 3-6 for a progressive wave and for a standing

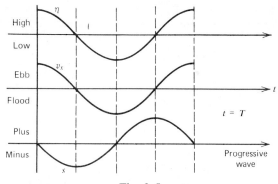

Fig. 3-5

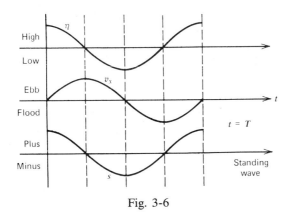

Fig. 3-6

wave, where for convenience we have taken the x axis to be in a downestuary direction so that $\partial s/\partial x$ is positive and an ebb current is positive. Taking the origin of the x axis to be at the upper end of the estuary, corresponding to the reflection point for the standing wave for which $v_x = 0$ for all x and t, we see that the relations of Fig. 3-6 apply specifically to an estuary of less than one quarter wavelength in length. For an estuary of greater length there is a nodal point for η at $x = \lambda/4$, and η is the negative of that shown in Fig. 3-6 for distances of x such that $\lambda/4 < x < \lambda/2$.

We also see from (3-33) that observations of the amplitude of the salinity variations along with a knowledge of $\partial s/\partial x$ allows us to estimate the amplitude of the tidal current variations and in turn, from (3-30) or (3-9) for a rectangular estuary, the amplitude of tidal height variations. We see further that the equation of mass continuity provides us with a method for determining tide heights in a two dimensional body when the tidal currents v_x and v_y are known at selected locations. The solutions can be given in analytic or finite difference form.

To continue let us next consider the energy relations for a progressive wave. For a column of height equal to the water depth h and of unit breadth and unit longitudinal extent, the *kinetic energy* is

$$KE = \tfrac{1}{2}\rho h v_x^2 \qquad (3\text{-}34)$$

where, as before, v_x is a function of x and t only. For the same situation the *potential energy* for a column of elevation η above sea level as compared with a sea level column is

$$PE = \rho g \int_0^\eta z \, dz = \tfrac{1}{2}\rho g \eta^2 \qquad (3\text{-}35)$$

From (3-9) and (3-5) we see that for a progressive wave expressions (3-34) and (3-35) are equal; for a progressive wave the energy at any instant is always half kinetic and half potential. The *total energy* is simply the sum of (3-34) and (3-35), or

$$E = \rho g \eta^2 = \rho h v_x^2 \tag{3-36}$$

If we had a simple harmonic progressive wave, such as that given by (3-7), we could then obtain the average kinetic, potential, or total energy per unit column by intergrating over a period of the oscillation,

$$\begin{aligned}
\bar{E} &= \frac{1}{T} \int_0^T \rho g \eta^2 \, dt \\
&= \frac{\rho g}{T} \int_0^T a^2 \sin^2 2\pi \left(\frac{x}{\lambda} - \frac{t}{T} \right) dt \\
&= \tfrac{1}{2} \rho g a^2
\end{aligned} \tag{3-37}$$

where a is the amplitude of the tidal height variation. Next we note that the time rate of doing work, or *energy flow*, across a transverse section of unit breadth and height h due to the excess pressure resulting from the wave elevation η is this excess pressure expressed as a force over the entire transverse section multiplied by the current velocity v_x, or

$$P = \rho g \eta h v_x = \rho g^{3/2} h^{1/2} \eta^2 = \rho g c \eta^2 = cE \tag{3-38}$$

where we have also used (3-9) and (3-36). We see, as expected, that this rate of doing work is also the rate at which energy is propagated along the wave cE, exclusive of any frictional or other loss terms.

For further information on the material in this section the reader may refer to Lamb (1932) and Proudman (1953).

3-3. TIDAL COOSCILLATION

We now look at some of the elementary relations for the tides in an estuary of rectangular dimensions whose upper end is closed and whose lower end is open to the ocean. In the first instance we neglect the effects of bottom friction.

We must first make a distinction between the *independent tide* and the *cooscillating tide*. The independent tide is the tidal motion caused by the gravitational attraction of the moon or the sun on the waters of the estuary. For the dimensions and depths of most estuaries this is usually a negligible effect. This is understandable from physical experience that the tides in most inland bodies of water of moderate depth not connected with the ocean are unimportant. The cooscillating tide is the tide created in an estuary caused by the ocean tide at the entrance to the estuary acting as a driving force. This is the important estuary tidal contribution with which we are concerned.

Taking the x axis to be in a downestuary-direction with the origin at the upper closed end of the estuary, we have a boundary condition at $x = 0$ that $v_x = 0$. We see that the standing wave solution to the wave equation in the forms (3-28) and (3-30) meets this condition. We also have a boundary condition at the lower end of the estuary, where $x = l$, for which the tidal height variations are given by

$$\eta = a_0 \cos \frac{2\pi t}{T} \qquad x = l \qquad (3\text{-}39)$$

where a_0 is the amplitude of the ocean tide. Substituting (3-28) into (3-39), we obtain for our final solutions,

$$\eta = a_0 \frac{\cos (2\pi x/\lambda)}{\cos (2\pi l/\lambda)} \cos \frac{2\pi t}{T} \qquad (3\text{-}40)$$

and

$$v_x = \frac{a_0 c}{h} \frac{\sin (2\pi x/\lambda)}{\cos (2\pi l/\lambda)} \sin \frac{2\pi t}{T} \qquad (3\text{-}41)$$

We see that *resonance* occurs for the zeros of the denominator of (3-40), or for values of the length of the estuary given by

$$l = (2n - 1)\frac{\lambda}{4} = \frac{2n - 1}{4} T\sqrt{gh} \qquad (3\text{-}42)$$

where we have also used (3-29) to obtain the last term in (3-42). Resonance occurs for estuary lengths equal to $\frac{1}{4}, \frac{3}{4}, \dots$ of the wavelength. These simple results are of course substantially altered by the effects of bottom friction and estuary geometry; but they are interesting and useful criteria to keep in mind. We also see that the tidal current velocity of (3-41), which is important in determining the scale of the tidal diffusion, varies as $\sin (2\pi x/\lambda)$. This means that, for an estuary whose length is less than $\lambda/4$, the amplitude of the current velocity varies from zero at the head of the estuary to a maximum at its mouth. For longer estuaries the tidal current has zeroes at $x = 0$, $\lambda/2$, $\lambda, \dots$, and maxima at $x = \lambda/4$, $3\lambda/4, \dots$.

We now look at the geostrophic effects on a progressive tidal wave in an estuary and on a tidal cooscillation. As mentioned previously, geostrophic effects are usually of secondary importance in estuarine hydrodynamics. Nevertheless, it is important to have some understanding of what one might expect in this regard. We treat the geostrophic effect here in an elementary and what is clearly an incomplete manner. In particular we ignore the important consideration of the boundary and source conditions. For a more complete description the reader may refer to Poincaré (1910), Taylor (1920), and Proudman (1925, 1946).

We consider the geostrophic effect on a progressive tidal wave in an infinitely long, straight canal for which the velocity component in the transverse direction is zero, $v_y = 0$. Then the equations of motion (2-120) and the equation of continuity reduce to the following, using relations (3-1) and (3-3),

$$\frac{\partial v_x}{\partial t} = -g \frac{\partial \eta}{\partial x} \tag{3-43}$$

$$2\omega_0 v_x \sin \phi = g \frac{\partial \eta}{\partial y} \tag{3-44}$$

$$\frac{\partial \eta}{\partial t} = -h \frac{\partial v_x}{\partial x} \tag{3-45}$$

We see that (3-43) and (3-45) define the wave equation relation for x and t we had before in Section 3-2 and also the two important subsidiary relations (3-5) and (3-9). We also see from the form of (3-44) along with relation (3-9) that the y dependent term is in the form of an exponential. We then correctly assume a solution,

$$\eta = ae^{my} \cos \frac{2\pi}{\lambda} (x - ct) \tag{3-46}$$

in simple harmonic form. Substituting (3-46) into (3-44) and using (3-9) and (3-5), we then obtain for m,

$$m = \frac{2\omega_0 \sin \phi}{c} \tag{3-47}$$

The wave velocity is unaffected by the earth's rotation, but the wave height is not the same everywhere in a section normal to the wave motion. The variation in wave amplitude, and consequently from (3-9) in current velocity, is greater to the right of the wave motion than to the left for the northern hemisphere. In a section normal to the direction of propagation the wave height slopes down from right to left, facing the direction of propagation at a wave crest, or high water, and slopes down from left to right at a wave trough, or low water. These results are in accordance with what is expected for the geostrophic effect from the discussion of Section 2-7, with particular reference to Fig. 2-14; in order to maintain a linear current direction the force related to the transverse surface slope must balance the geostrophic force, remembering that the

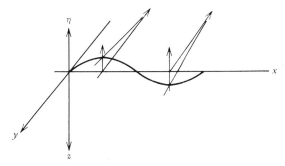

Fig. 3-7

current velocity is in the direction of the wave motion at a wave crest and opposite the direction of wave motion at a wave trough. These relations are illustrated in Fig. 3-7. Such geostrophic wave motions are referred to as *Kelvin waves*. Considering the x axis to be in a downestuary direction, the ebb tide has higher wave height amplitudes and higher current velocity amplitudes on the right hand side of the estuary than on the left hand side; and the opposite relations hold for the flood tide. We may then expect that there will be a tendency toward an asymmetric tidal flow in the transverse direction as shown in Fig. 3-8.

Let us next consider the interference effect, or standing wave effect, of two simple harmonic Kelvin waves of equal amplitude progressing in opposite directions. From (3–46) their combined effect may be represented by

$$\eta = ae^{my}\cos\frac{2\pi}{\lambda}(x-ct) - ae^{-my}\cos\frac{2\pi}{\lambda}(x+ct) \qquad (3\text{-}48)$$

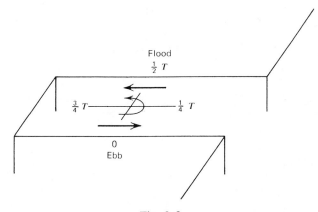

Fig. 3-8

where we have now taken the origin, for convenience, such that the y axis would have been a nodal line for two such interfering waves without the geostrophic term. As we watch this steady state wave motion as a function of time, the crests of the waves, or high water, are defined by the relation $\partial\eta/\partial t = 0$, which from (3-48) gives

$$e^{my} \sin\frac{2\pi}{\lambda}(x - ct) + e^{-my} \sin\frac{2\pi}{\lambda}(x + ct) = 0$$

$$\cosh my \sin\frac{2\pi x}{\lambda}\cos\frac{2\pi t}{T} - \sinh my \cos\frac{2\pi x}{\lambda}\sin\frac{2\pi t}{T} = 0$$

$$\frac{\tanh my}{\tan\frac{2\pi x}{\lambda}} = \cot\frac{2\pi t}{T} \qquad (3\text{-}49)$$

For locations near the origin this reduces to

$$y = \frac{2\pi x}{m\lambda}\cot\frac{2\pi t}{T} \qquad (3\text{-}50)$$

Equation (3-50) defines a straight line centered at the origin which rotates counterclockwise about the origin with the angular frequency $2\pi/T$. Such lines of equal tidal phase are referred to as *cotidal lines*. For such a case the time of high tide progresses counterclockwise in a regular manner around the estuary or bay, as illustrated in Fig. 3-8. Instead of having nodal lines of no motion we are now reduced to a central region, referred to as an *amphidromic region*, of no motion at the origin. The interference effect of two geostrophically controlled simple harmonic waves is to change the observed wave pattern from that of a linear standing wave to a rotary wave.

For further information on the material in this section the reader may refer to Proudman (1953) and Defant (1961).

3-4. TIDAL WAVES WITH FRICTION

We have pointed out that the tidal wave motion in an estuary can usually be treated in the manner of a relatively simple mechanical system with a driving force at one end, and we have discussed some of the aspects of the resultant motion. We now consider the effect of the primary resistance force to that motion, namely, that of bottom friction.

We assume as before that the water density may be taken as a constant and that the motion is one dimensional. From (2-134), along with (2-135) and (3-1), we have for the equation of motion,

$$\frac{\partial v_x}{\partial t} = -g\frac{\partial\eta}{\partial x} - \frac{1}{\rho}\frac{\partial f_{zx}}{\partial z} \qquad (3\text{-}51)$$

If the assumption can then be made that f_{zx} is a continuous and well behaved function of z from the surface to the bottom for the type of motion being considered, an extraordinarily useful simplification of (3-51) can be made by integrating with respect to z from the sea surface to a depth z or to the bottom. In essence this amounts in part to establishing for a control volume of the estuary extending from the estuary surface to its bottom that the frictional energy losses related to the bottom friction are much greater than the energy losses related to internal friction. This is the case for the water depths of most estuaries. Integrating (3-51) from the sea surface to a depth z and dividing by z, we have

$$\frac{\partial \langle v_x \rangle}{\partial t} = -g \frac{\partial \eta}{\partial x} - \frac{f_{zx}}{\rho z} \tag{3-52}$$

where we have also assumed that there is negligible wind stress at the ocean surface, $f_{zx} = 0$ at $x = 0$, and where $\langle v_x \rangle$ is the vertically averaged value of v_x, given by

$$\langle v_x \rangle = \frac{1}{z} \int_0^z v_x \, dz \tag{3-53}$$

as contrasted with the time average value $\bar{v}_x$ of Sections 2-4 and 2-8. Integrating (3-51) from the surface to the bottom, we obtain

$$\frac{\partial \langle v_x \rangle}{\partial t} = -g \frac{\partial \eta}{\partial x} - \frac{f_b}{\rho h}$$

$$= -g \frac{\partial \eta}{\partial x} - \frac{k |v_b| v_b}{h} \tag{3-54}$$

where we have also used (2-144), and where $\langle v_x \rangle$ is the vertically averaged value of v_x from the surface to the bottom. If we look at a time in the tidal cycle for which the mass acceleration $\partial \langle v_x \rangle / \partial t$ is zero, (3-54) reduces further to

$$\frac{\partial \eta}{\partial x} = -\frac{k |v_b| v_b}{gh} \tag{3-55}$$

for the surface slope in terms of the bottom friction. Expression (3-55) can then be used from observations of $\partial \eta / \partial x$ and v_b under the conditions of $\partial \langle v_x \rangle / \partial t = 0$ to obtain an evaluation of k.

We see that the bottom friction term in (3-54) is in the form of a quadratic in v_b. It would simplify the solution if this term could be linearized. Fortunately this can be done for the simple harmonic tidal motion with which we are concerned. If the bottom velocity at any specific location is

$$v_b = C_b \cos \frac{2\pi t}{T} \tag{3-56}$$

the bottom frictional stress is

$$f_b = k\rho C_b^2 \left| \cos \frac{2\pi t}{T} \right| \cos \frac{2\pi t}{T} \tag{3-57}$$

We may then expand (3-57) in a Fourier cosine series for which we obtain for the coefficient of the first cosine term $\cos(2\pi t/T)$,

$$a_1 = \frac{4}{\pi} \int_0^{\pi/2} C_b^2 \cos^2 x \cos x \, dx = \frac{8}{3\pi} C_b^2 \tag{3-58}$$

The coefficients $a_3, a_5, \ldots$, are indeed much smaller than a_1, as can be seen by continuing the expansions, so that we may approximate (3-57) by

$$f_b = \frac{8}{3\pi} k\rho C_b^2 \cos \frac{2\pi t}{T}$$

$$= \frac{8}{3\pi} k\rho C_b v_b \tag{3-59}$$

The additional terms like the shallow water constituent terms of Section 3-2 give rise to overtides. Further, this evaluation is valid only if the runoff or circulation velocities are substantially smaller than the amplitude of the tidal velocity. This is generally true in estuaries but not in the upper reaches of tidal rivers.

Although the vertical variation in v_x for estuarine tides is usually sufficiently slight that we can approximate v_b in (3-54) by $\langle v_x \rangle$, we take a brief look at the vertical variations to obtain a useful relation between N_z and k. If the bottom current is given by (3-56), the tidal current at any shallower depth will be given by

$$v_x = C \cos\left(\frac{2\pi t}{T} - \delta\right) \tag{3-60}$$

where the phase lag is chosen such that $\delta = 0$ at $z = h$. From (2-135) and (3-59) we then have

$$f_b = (f_{zx})_{z=h} = -\rho N_z \left(\frac{\partial v_x}{\partial z}\right)_{z=h}$$

$$\frac{8}{3\pi} k\rho C_b^2 \cos \frac{2\pi t}{T} = -\rho N_z \left[\left(\frac{dC}{dz}\right)_{z=h} \cos \frac{2\pi t}{T} + C_b \left(\frac{d\delta}{dz}\right)_{z=h} \sin \frac{2\pi t}{T} \right]$$

from which we obtain

$$\left(\frac{d\delta}{dz}\right)_{z=h} = 0 \tag{3-61}$$

and

$$N_z = -\frac{8k}{3\pi} \frac{C_b^2}{(dC/dz)_{z=h}} \tag{3-62}$$

From measured values of C_b and $(dC/dz)_{z=h}$ and a knowledge of k from other determinations, we can obtain a value of N_z for the tidal motion. However, evaluations of N_z from such a formula may well be more related to an N_z applicable to the bottom boundary layer to which k applies than to the water column in general.

We can obtain an additional relation for N_z anywhere in the water column in terms of physically observable quantities. Subtracting (3-54) from (3-51), we have

$$\frac{\partial}{\partial t}[v_x - \langle v_x \rangle] = -\frac{1}{\rho}\frac{\partial f_{zx}}{\partial z} + \frac{f_b}{\rho h} \tag{3-63}$$

Integrating (3-63) from the surface to any depth z, we then have

$$f_{zx} = f_b \frac{z}{h} - \rho \int_0^z \frac{\partial}{\partial t}[v_x - \langle v_x \rangle]\, dz \tag{3-64}$$

in terms of the bottom friction and the observed tidal velocity profile as a function of depth. The integral term in (3-64) may be thought of as a correction factor to a linear stress profile. From the calculated values of f_{zx} in (3-64) we then have from the definition of N_z,

$$N_z = -\frac{f_{zx}}{\rho}\left(\frac{\partial v_x}{\partial z}\right)^{-1} \tag{3-65}$$

As we have just stated, the observed vertical variations of the tidal current v_x in an estuary are usually sufficiently slight that we can approximate v_b by $\langle v_x \rangle$. Using the linearized form of f_b from (3-59), (3-54) becomes

$$\frac{\partial \langle v_x \rangle}{\partial t} = -g\frac{\partial \eta}{\partial x} - \frac{8k}{3\pi}\frac{C_b}{h}\langle v_x \rangle \tag{3-66}$$

which along with the same continuity relation

$$\frac{\partial \eta}{\partial t} = -h\frac{\partial \langle v_x \rangle}{\partial x} \tag{3-67}$$

are the governing equations for tidal motion with friction. Combining (3-66) and (3-67), we obtain

$$\frac{\partial^2 \langle v_x \rangle}{\partial t^2} = c_0^2 \frac{\partial^2 \langle v_x \rangle}{\partial x^2} - \beta\frac{\partial \langle v_x \rangle}{\partial t} \tag{3-68}$$

or

$$\frac{\partial^2 \eta}{\partial t^2} = c_0^2 \frac{\partial^2 \eta}{\partial x^2} - \beta\frac{\partial \eta}{\partial t} \tag{3-69}$$

where c_0 is the wave velocity (3-5) without friction, and

$$\beta = \frac{8k}{3\pi}\frac{C_b}{h} \tag{3-70}$$

Equation (3-68) or (3-69) is simply the wave equation with a frictional damping term proportional to the water particle velocity.

We note, however, in (3-70) that the coefficient β is a function of the amplitude C_b of the tidal current. For simple analytic solutions we must then restrict ourselves to regions for which C_b is sufficiently slowly varying that an averaged value of β may be taken. For actual examples, using the equations in finite difference form with a computer or other numerical procedures, the value of β can be altered appropriately as the iteration process is followed. Considering β a constant then, we need a solution in simple harmonic form whose circular frequency is that of the semidiurnal or diurnal tide. We may take a solution of the form

$$\eta = ae^{-\mu x}\cos(\omega t - \kappa x) \tag{3-71}$$

where κ is the *wave number*, defined by

$$\kappa = \frac{2\pi}{\lambda} = \frac{\omega}{c} \tag{3-72}$$

Substituting (3-71) into (3-69) and separating sine and cosine terms, we obtain the conditional relations for μ and κ that

$$\kappa^2 - \mu^2 = \frac{\omega^2}{c_0^2} \tag{3-73}$$

and

$$2\mu\kappa = \beta\frac{\omega}{c_0^2} \tag{3-74}$$

We see from (3-73) that the wave number is increased, and the wave velocity decreased, for wave motion with friction as compared with wave motion without friction. From (3-74) we can obtain further, under the normal condition that the frictional damping is sufficiently small for the wave velocity c to be approximated by c_0,

$$\mu = \frac{\beta}{2c_0} \tag{3-75}$$

From (3-71) and (3-67) we see that, since the variable x enters in both the exponential and cosine terms, the simple relation (3-9) that v_x and η are in phase for a progressive tidal wave no longer holds. There is a phase difference such that the maximum current velocity precedes the maximum tide height. After a slight algebraic reduction, we obtain for $\langle v_x \rangle$,

$$\langle v_x \rangle = \frac{a\omega}{h\sqrt{\mu^2 + \kappa^2}} e^{-\mu x} \cos(\omega t - \kappa x + \alpha) \qquad (3\text{-}76)$$

where

$$\tan \alpha = \frac{\mu}{\kappa} = \frac{\beta}{2\omega} \qquad (3\text{-}77)$$

and where the third term of (3-77) is valid to the same degree of approximation as (3-75).

For the tidal cooscillation of the previous section, we have again the boundary condition at the upper end of the estuary that $\langle v_x \rangle = 0$, or what is the same that the incoming, upestuary progressive tidal wave is reflected at $x = 0$ and progresses downestuary with a tide height at the boundary equal to that of the incoming wave. We may then represent the combined tidal motion by

$$\eta = ae^{-\mu x} \cos(\omega t - \kappa x) + ae^{\mu x} \cos(\omega t + \kappa x) \qquad (3\text{-}78)$$

The value of a is determined by the source condition (3-39) at the estuary mouth, $x = l$.

There is a number of physically measurable quantities related to η or $\langle v_x \rangle$ for a damped, tidal cooscillation which we should like to derive. First, let us derive the time of high water t_H, related to a zero time base of the tidal motion at $x = 0$. The time of high water occurs when $\partial \eta / \partial t = 0$, or from (3-78) under the condition

$$e^{-\mu x} \sin(\omega t - \kappa x) + e^{\mu x} \sin(\omega t + \kappa x) = 0$$

$$(e^{\mu x} + e^{-\mu x}) \sin \omega t \cos \kappa x + (e^{\mu x} - e^{-\mu x}) \cos \omega t \sin \kappa x = 0$$

or for the local time angle of high water, ωt_H, such that

$$\omega t_H = \tan^{-1}(-\tan \kappa x \tanh \mu x) \qquad (3\text{-}79)$$

Figure 3-9 is a graph of ωt_H versus the phase difference κx for various values of μ. We note from (3-40) in the undamped case that the local time angle of high water changes from 0 to π at the first node for η at $x = \lambda/4$, or from (3-72) at $\kappa x = \pi/2$. From Fig. 3-9 we see that ωt_H is altered from this discontinuity relation to a more gradual change as a function of κx as μ increases.

Next, we obtain a relation for the relative height of the tide at high water referred to the high water tide height at $x = 0$. It is more convenient with physically observable quantities to normalize the tide height amplitude to the observed tide height amplitude at $x = 0$ rather than to the

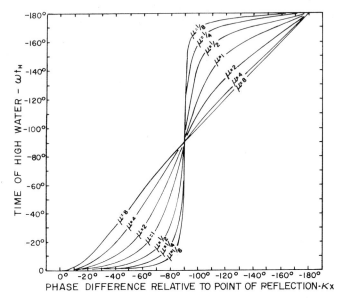

Fig. 3-9 Reproduced with permission from A. C. Redfield, 1950, *Papers in Physical Oceanography and Meteorology* (Massachusetts Institute of Technology and Woods Hole Oceanographic Institution), **11**(4) 1–36 Fig. 1.

driving source amplitude at $x = l$. From (3-78) we may rewrite η as

$$\eta = 2a(\cos \omega t \cos \kappa x \cosh \mu x - \sin \omega t \sin \kappa x \sinh \mu x) \qquad (3\text{-}80)$$

From (3-79) we have for the ωt_H relations at high water,

$$\cos \omega t_H = \frac{1}{\sqrt{1 + \tan^2 \kappa x \tanh^2 \mu x}}$$

and

$$\sin \omega t_H = \frac{-\tan \kappa x \tanh \mu x}{\sqrt{1 + \tan^2 \kappa x \tanh^2 \mu x}}$$

from which we obtain

$$\eta = 2a \frac{\cos \kappa x \cosh \mu x + \tan \kappa x \tanh \mu x \sin \kappa x \sinh \mu x}{\sqrt{1 + \tan^2 \kappa x \tanh^2 \mu x}}$$

$$= 2a\sqrt{\cos^2 \kappa x \cosh^2 \mu x + \sin^2 \kappa x \sinh^2 \mu x}$$

$$= 2a\sqrt{\cos^2 \kappa x - \cosh^2 \mu x - 1}$$

$$= 2a\sqrt{\tfrac{1}{2}(\cosh 2\mu x + \cos 2\kappa x)}$$

Since the tide height at high water at $x = 0$ is, from (3-78), simply $\eta_0 = 2a$, we then have for the relative amplitude of the tide at high water at any distance along the estuary referred to the height at $x = 0$,

$$\frac{\eta}{\eta_0} = \sqrt{\tfrac{1}{2}(\cosh 2\mu x + \cos 2\kappa x)} \qquad (3\text{-}81)$$

Figure 3-10 is a graph of η/η_0 versus κx for various values of μ. We see now that the node for η at $\kappa x = \pi/2$ disappears and instead a region with minimal values of η/η_0 is found.

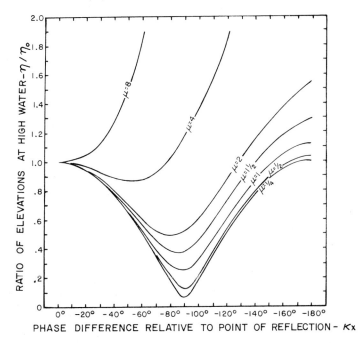

Fig. 3-10   Reproduced with permission from A. C. Redfield, 1950, *Papers in Physical Oceanography and Meterology* (Massachusetts Institute of Technology and Woods Hole Oceanographic Institution), **11**(4) 1–36 Fig. 2.

Finally, we should like to obtain an expression for the time of slack water t_s, again related to the same zero time base at $x = 0$ as applied to t_H. From expression (3-76) we shall then have for the value of $\langle v_x \rangle$ for a damped, tidal cooscillation,

$$\langle v_x \rangle = \frac{a\omega}{h\sqrt{\mu^2 + \kappa^2}} \left[e^{-\mu x} \cos(\omega t - \kappa x + \alpha) - e^{\mu x} \cos(\omega t + \kappa x + \alpha) \right] \qquad (3\text{-}82)$$

At $x = 0$, $\langle v_x \rangle = 0$ for all values of t; at any other distance along the estuary we shall have

$$e^{-\mu x} \cos(\omega t + \alpha - \kappa x) - e^{\mu x} \cos(\omega t + \alpha + \kappa x) = 0$$

or, following the procedures used in obtaining (3-79),

$$\omega t_s = \tan^{-1}\left(\frac{\tanh \mu x}{\tan \kappa x}\right) - \alpha \qquad (3\text{-}83)$$

Figure 3-11 is a graph of the local time angle ωt_s versus κx for various values of μ. We see from expression (3-41) in the undamped case that the

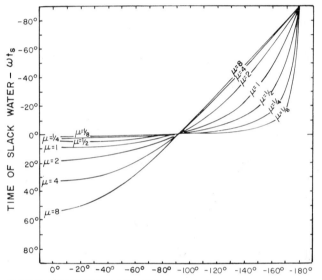

Fig. 3-11 Reproduced with permission from A. C. Redfield, 1950, *Papers in Physical Oceanography and Meteorology* (Massachusetts Institute of Technology and Woods Hole Oceanographic Institution), **11**(4) 1–36 Fig. 3.

local time angle remains unchanged at $\omega t_s = 0$ from $x = 0$ to $x = \lambda/2$. In the damped case we see that there is a gradual change in ωt_s passing through zero from positive to negative at $x = \lambda/4$.

We have mentioned that an evaluation of k can be made through the use of expression (3-55). Such an evaluation usually only produces an averaged value over a broad area. Grace (1931) introduced another method which permits incremental calculations to be made along an estuary where the tidal current is essentially in a longitudinal direction. From the equation of continuity (3-67), expressed in finite difference form and including the appropriate terms for changes in the estuary cross section, and a complete knowledge of the variations and of the value of $\langle v_x \rangle$ at one cross section, the amplitude and phase of $\langle v_x \rangle$ can be determined section by section. The equation of motion (3-66), also expressed in finite difference form, can then be used to obtain the amplitude and phase variations of the frictional term, from which the values of k can in turn be calculated. In carrying out such calculations with a longitudinal current, the geostrophic equation (3-44) must also be used to correct the η values measured on shore to what they are along the median line of the estuary.

For further information on the material covered in this section the reader may refer to Redfield (1950), Dronkers (1964), Ippen and Harleman (1966), and Proudman (1953).

3-5. TIDAL ENERGIES AND ENERGY LOSSES

Taylor (1919) elaborated some rather simple but extremely useful energy and energy loss formulations for coastal tidal bodies. His application of these formulations to the Irish sea demonstrated that the tidal energy flow from the ocean tides into a partially enclosed body of water was dissipated principally in bottom frictional losses. He also showed that the work done by the moon's gravitational attraction on such a body of water was small in comparison with the ocean tidal energy flow into the body or the bottom frictional losses and that in the case of the Irish sea it was a negative quantity, or simply that work was being done by the tides of the Irish sea on the moon.

Let us derive these relations. We have from expression (3-38) that the energy flow across a transverse section of unit breadth and height h is given by the product of the wave height η and the current velocity $\langle v_x \rangle$. If we take the measured wave height at the ocean entrance to be given by

$$\eta = a \cos \frac{2\pi t}{T} \tag{3-84}$$

and the average current velocity then by

$$\langle v_x \rangle = C \cos\left(\frac{2\pi t}{T} - \delta\right) \tag{3-85}$$

where δ is the phase difference between the two, the average value of the product taken over a tidal period is

$$\overline{\eta \langle v_x \rangle} = \tfrac{1}{2} aC \cos \delta \tag{3-86}$$

The average rate at which energy flows into the water body is then

$$P = \tfrac{1}{2}g\rho \int haC \cos \delta \, ds \qquad (3\text{-}87)$$

where s is the transverse distance along the cross section A at the entrance. If, as is usually the case for any set of observations, the cross section A is not normal to $\langle v_x \rangle$, expression (3-87) must be multipled by $\sin \theta$, where θ is the angle between ds and $\langle v_x \rangle$. It is of interest to note that, for a simple progressive wave, $\delta = 0$ and there is a maximum energy flow, and that, for a simple standing wave without any frictional effects, $\delta = \pi/2$ and there is no net energy flow.

We note from expression (2-154) that the rate of energy loss due to bottom friction is given as the product of the bottom frictional stress f_b and the bottom current velocity v_b. From (3-56) and (3-57) we then have for the rate of bottom frictional energy loss per unit area, averaged over a tidal cycle,

$$W = \frac{k\rho C_b{}^3}{T} \int_0^T \left| \cos \frac{2\pi t}{T} \right| \cos^2 \frac{2\pi t}{T} \, dt$$

$$= \frac{4k\rho C_b{}^3}{T} \int_0^{T/4} \cos^3 \frac{2\pi t}{T} \, dt$$

$$= \frac{4k}{3\pi} \rho C_b{}^3 \qquad (3\text{-}88)$$

To find the total rate of bottom frictional energy loss expression (3-88) must then be multiplied by the surface area B of the coastal water body, using a spatially averaged value of $C_b{}^3$.

To find the work done by the moon's gravitational attraction we proceed as follows. First we know that the attraction of the moon may be expressed in terms of a potential function Ω. Now consider the increase, or decrease, in potential energy of a small volume within the waters of the coastal water body. If this volume remains filled with water during an orbit of the moon around the earth, considering the earth stationary, there will be no increase in potential energy as the potential function Ω returns to its initial value. However, if the volume element is located within the space that is sometimes filled with water and sometimes not during the moon orbit, there may be an increase or decrease in the potential energy of the water body. This region of course is the space between high tide water and low tide water. We then have for the average work done during a lunar orbit, or two semidiurnal tidal cycles, per unit surface area,

$$W_M = \frac{1}{2T} \int \rho\eta \, d\Omega \qquad (3\text{-}89)$$

where the integral is taken over a complete lunar orbit and where, as before, T is the period of the semidiurnal tide. We note in passing, for the equilibrium theory of the tides without consideration of frictional effects, that the tidal amplitude due to the moon's attraction is given by $\eta_M = -\Omega/g$ so that expression (3-89) is an exact differential in Ω and its line integral over a closed loop is zero.

From any standard reference, such as Lamb (1932) or Melchoir (1966), we note that the potential function is given by

$$\Omega = \frac{\gamma M r^2}{2R^3}(1 - 3\cos^2\theta)$$

$$= \frac{3}{2}g\frac{M}{E}\left(\frac{r}{R}\right)^3 r\left(\frac{1}{3} - \cos^2\theta\right) \tag{3-90}$$

where M is the mass of the moon, E is the mass of the earth, r is the radius of the earth, R is the radius of the moon's orbit, γ is the universal gravitational constant, g is the earth's gravitational attraction, and θ is the angle between the line joing the center of the earth to the moon and the radius of the earth, which passes through the point on the earth's surface being considered. Again with reference to any standard text, such as Melchoir (1966), we note from the spherical triangle formed on the celestial sphere by the north pole, the moon's zenith, and the zenith at the observing point that

$$\cos\theta = \cos\lambda\,\cos\varphi \tag{3-91}$$

where λ is the latitude of the observing point, and φ is the local hour angle of the moon. In terms of φ we can then write for the tide height amplitude η,

$$\eta = a\,\cos 2(\varphi + \varphi_0) \tag{3-92}$$

where φ_0 is the phase of the tide at the time when the moon crosses the meridian of the observing station. In other words, when φ_0 is positive, high water occurs before the moon crosses the meridian; high water leads the moon. When φ_0 is negative, high water lags the moon. Substituting (3-90) to (3-92) into (3-89), we obtain

$$W_M = -\frac{3a}{4}\frac{\rho g}{T}\frac{M}{E}\left(\frac{r}{R}\right)^3 r\,\cos^2\lambda\int_0^{2\pi}\cos 2(\varphi + \varphi_0)\,d(\cos^2\varphi)$$

$$= -\frac{3\pi a}{4}\frac{\rho g}{T}\frac{M}{E}\left(\frac{r}{R}\right)^3 r\,\cos^2\lambda\,\sin 2\varphi_0 \tag{3-93}$$

To find the total rate of work being done by the moon expression (3-93) must then be multiplied by the surface area B of the coastal water body, using a spatially averaged value of $a\,\sin 2\varphi_0$. When high water leads the moon, $\varphi_0 > 0$, work is done by the tides of the coastal water body on the moon, as indicated by the negative sign in expression (3-93). When high water lags the moon, $\varphi_0 < 0$, work is done by the moon on the coastal water body.

3-6. BORES

One of the more unusual phenomena that can occur in some estuaries and tidal rivers, particularly where the water depths are shallow near the mouth, is a tidal bore. When the tide propagates up a river of shallow depth, distortion results in the simple harmonic wave form, as discussed briefly in Section 3-2, resulting in a shortening of the time interval between low water level and high water level and a corresponding lengthening of the time between the high water level and the subsequent low water level. Under particular circumstances the steepness of the tidal curve between low water and high water may become so large that a finite discontinuity in the water level occurs and a wall of water rushes up the river. This phenomenon is called a *bore*.

Under these conditions the motion is governed by the formulations obtained previously in Section 2-11 for a hydraulic jump. With reference to Fig. 2-29 let us impose a movement of the whole water mass to the left with a velocity $-v_1$. We note that this does not change the governing equation of continuity (2-163) or momentum (2-168). Then the waters to the left of Fig. 2-29 are still; the discontinuity propagates upstream with a velocity v_1; and the water particle velocity of the water mass to the right of Fig. 2-29 is $v_1 - v_2$ in an upstream direction. From (2-169) we then note that the velocity of propagation of the bore discontinuity is

$$v_1 = \sqrt{g h_2 \frac{h_1 + h_2}{2 h_1}} \tag{3-94}$$

Again the energy condition of (2-171) holds, so that we can only have propagation of a bore such that $h_2 > h_1$, or a condition in which the water depth on the front side of the bore is less than that on the back side. We also see that when referring to any hydraulic jump phenomena we should refer to the supercritical or subcritical velocity state relative to the discontinuity so that the velocity relative to the discontinuity is supercritical on the front side and subcritical on the back side.

For further information on the material covered in this section the reader may refer to Stoker (1957) and Dronkers (1964).

CHAPTER 4

Circulation

4-1. CONTROLS, TRANSITIONS, AND JUMPS

In this chapter we are concerned principally with the net circulation effects, over and beyond the oscillatory tidal effects, that occur in an estuary. Linear tidal effects, such as that of the tidal current v_{xt}, necessarily balance out over a tidal cycle. Nonlinear tidal effects, such as bottom friction $v_{xt}|v_{xt}|$ and salt flux $v_{xt}s_t$, do not balance out over a tidal cycle; and as we saw in the last chapter and shall see in the next chapter, these nonlinear tidal effects are not only important but can be the determining factors for some physical phenomena we wish to investigate. In this chapter, however, we look mainly at examples for which the tidal terms are linear and thus balance out of the defining equations. As a gross but nevertheless useful index in estuaries with stratified flow, the amplitudes of the tidal velocities v_{xt} are often about an order of magnitude greater than the circulation velocities $\overline{v}_x$ averaged over a tidal cycle, which in turn are usually an order of magnitude greater than the net circulation velocities $\langle \overline{v}_x \rangle$ or runoff velocities averaged over the vertical water column.

In the first instance, in this section we consider some of the gross inertial effects of the hydraulics of open channel flow, such as those discussed in Section 2-11, but in this case for stratified flow. We consider the densities ρ_1 and ρ_2 and the longitudinal velocities v_1 and v_2 of the stratified flow to be constant; and we omit the effects of internal and bottom friction in considering the larger inertial changes. The developments discussed in this section were originally presented by Stommel and Farmer (1952, 1952a).

We saw in Section 2-11 that, for an unstratified flow, the maximum volume of flow occurs at the critical velocity. For many unstratified or partially stratified estuaries the runoff volumes from river, stream, and groundwater flow and the circulation flows usually do not approach the critical velocity condition. For some strongly stratified estuaries with a net circulation out in the upper layer and a net circulation in in the lower layer, or for a system with a large runoff volume, critical velocities may be approached, in which case the abrupt change in opening at the mouth of

97

the estuary or bay may act as a control on the net circulation volumes. We look at the governing conditions for such a control.

Consider for the moment a bay, lagoon, or estuary with a restricted exit at its mouth opening into an unbounded region, the ocean. We then expect that, if sufficient flow volumes are available as discussed in the preceding paragraph, the flow will adjust to the maximum flow volume attainable under the given *available hydraulic energy*, as expressed by the Bernoulli relation (2-161). Under these conditions, for an unstratified flow, Fig. 2-28 applies. Maximum flow occurs at critical velocity conditions. We note that the critical flow velocity (2-167) is the same as the wave velocity (3-5) for a progressive gravity wave. We might then expect that we can arrive at the critical flow conditions for a stratified flow by determining the conditions for a stationary internal gravity wave or, as we have done before, by simply determining the maximum flow for a given available hydraulic energy. Both give the same result. We follow the latter derivation, being the more direct and physically understandable one.

For an abrupt change in the estuary geometry, breadth, we have the stratified flow conditions shown in Fig. 4-1. Applying the Bernoulli relation (2-161), or, equivalently, the continuity of available energy flow relation of (2-170), to each section of the stratified flow, we obtain

$$\frac{p_1}{\rho_1} - gz + \tfrac{1}{2}v_1{}^2 = \frac{p_1'}{\rho_1} - gz + \tfrac{1}{2}v_1'^2 = E_1$$

$$gz - gz + \tfrac{1}{2}v_1{}^2 = g(z - \zeta) - gz + \tfrac{1}{2}v_1'^2 = E_1$$

$$\tfrac{1}{2}v_1{}^2 = -g\zeta + \tfrac{1}{2}v_1'^2 = E_1 \tag{4-1}$$

and

$$\frac{p_2}{\rho_2} - gz + \tfrac{1}{2}v_2{}^2 = \frac{p_2'}{\rho_2} - gz + \tfrac{1}{2}v_2'^2 = E_2$$

$$\frac{\rho_1}{\rho_2} gh_1 + g(z - h_1) - gz + \tfrac{1}{2}v_2{}^2 = \frac{\rho_1}{\rho_2} gh_1' + g(z - \zeta - h_1') - gz + \tfrac{1}{2}v_2'^2 = E_2$$

$$g\frac{\rho_1 - \rho_2}{\rho_2} h_1 + \tfrac{1}{2}v_2{}^2 = g\frac{\rho_1 - \rho_2}{\rho_2} h_1' - g\zeta + \tfrac{1}{2}v_2'^2 = E_2 \tag{4-2}$$

Eliminating ζ from (4-1) and (4-2), we then obtain

$$\tfrac{1}{2}v_1{}^2 + g\frac{\rho_2 - \rho_1}{\rho_2} h_1 - \tfrac{1}{2}v_2{}^2 = \tfrac{1}{2}v_1'^2 + g\frac{\rho_2 - \rho_1}{\rho_2} h_1' - \tfrac{1}{2}v_2'^2 = E \tag{4-3}$$

where E_1, E_2, and E are all constants.

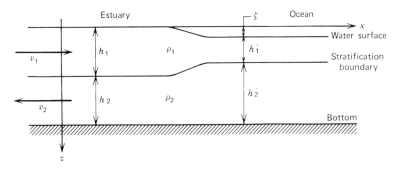

Fig. 4-1

Using the same definitions for q and Q as in Section 2-11, we have

$$v_1 = \frac{q_1}{h_1} = \frac{Q_1}{bh_1} \tag{4-4}$$

and

$$v_2 = \frac{q_2}{h_2} = \frac{Q_2}{bh_2} \tag{4-5}$$

where b is the breadth of the estuary. We assume that ρ_1 is approximately equal to ρ_2, so that we may use a volume continuity relation in place of the mass continuity relation, giving

$$Q_1 = Q_2 + R \tag{4-6}$$

or

$$q_1 = q_2 + r \tag{4-7}$$

where R is the total runoff and r is the runoff per unit breadth. We also have the simple relation

$$h_1 + h_2 = h \tag{4-8}$$

where h is the total depth, a constant. We then have

$$\frac{dq_1}{dh_1} = \frac{dq_2}{dh_2} \tag{4-9}$$

and

$$\frac{dh_2}{dh_1} = -1 \tag{4-10}$$

Substituting (4-4) and (4-5) into (4-3) and taking the derivative of (4-3) with respect to h_1, we obtain, using (4-9) and (4-10),

$$\frac{q_1^2}{2h_1^2} + g\beta h_1 - \frac{q_2^2}{2h_2^2} = E$$

$$\frac{q_1}{h_1^2}\frac{dq_1}{dh_1} - \frac{q_1^2}{h_1^3} + g\beta - \frac{q_2}{h_2^2}\frac{dq_1}{dh_1} - \frac{q_2^2}{h_2^3} = 0 \qquad (4\text{-}11)$$

where β is defined by

$$\beta = \frac{\rho_2 - \rho_1}{\rho_2} \qquad (4\text{-}12)$$

As before, the condition for maximum flow q_1, or q_2, is attained when $dq_1/dh_1 = 0$, which gives

$$\frac{q_1^2}{h_1^3} - g\beta + \frac{q_2^2}{h_2^3} = 0$$

or   (4-13)

$$G_1 + G_2 = \beta$$

where G_1 and G_2 are defined by

$$G_1 = \frac{v_1^2}{gh_1} \qquad (4\text{-}14)$$

and

$$G_2 = \frac{v_2^2}{gh_2} \qquad (4\text{-}15)$$

Under the conditions of an arrested salt wedge or a fjord for which v_2 is small, the depth of the stratification interface can be approximated from (4-13) by

$$h_1^3 = \frac{q_1}{g\beta} \qquad (4\text{-}16)$$

under critical flow conditions. We also see from (4-16) and (4-4) that the critical velocity v_c under these conditions is

$$v_c = v_1 = \sqrt{g\beta h_1} \qquad (4\text{-}17)$$

Under the conditions of a stratified estuary flow for which the net circulation flows q_1 and q_2 are much larger than r, we may take as an approximation that $q_1 = q_2$ from (4-7). From the salt continuity equation, as derived in Section 5-2, we also have under these conditions the subsidiary relation $h_1 = h_2$ approximately. Then, from (4-13) we have for q_1, in terms of the estuary depth,

$$q_1^2 = \frac{g\beta h_1^3}{16} \qquad (4\text{-}18)$$

under critical flow and overmixing conditions.

Let us next consider a simple transition along the estuary in which there is no change in the total available energy but in which there is either a change in the channel breadth b or the total flow Q. We also assume, corresponding to arrested flow conditions, that the velocity in the lower layer is sufficiently small that it may be neglected. Equation (4-3) then reduces to

$$\tfrac{1}{2}v_1^2 + g\beta h_1 = \tfrac{1}{2}v_1'^2 + g\beta h_1' \qquad (4\text{-}19)$$

which we may rewrite as

$$h_1 - h_1' = \frac{v_1'^2 - v_1^2}{2g\beta} \qquad (4\text{-}20)$$

for the change in the thickness of the upper layer in terms of the flow velocities. We also note, following the reasoning given in Section 2-11, that for an increase in Q_1 or a decrease in b there is a decrease in h_1 and an increase in v_1 when the original flow is subcritical, which is the usual case.

For completeness we can also consider the case of an internal hydraulic jump. As before, we assume that $v_2 = 0$. Then, an internal jump from a supercritical to a subcritical condition is as shown in Fig. 4-2. For the

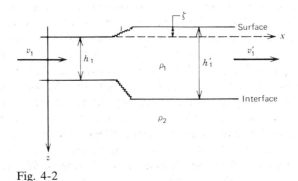

Fig. 4-2

velocity in the lower layer to be zero the pressure in the lower layer must be the same at the same depth on either side of the jump, which gives

$$\rho_1 g \zeta = (\rho_2 - \rho_1) g (h_1' - h_1 - \zeta)$$

or

$$\zeta = \frac{\rho_2 - \rho_1}{\rho_2} (h_1' - h_1)$$

$$h_1' - h_1 - \zeta = \frac{\rho_1}{\rho_2} (h_1' - h_1) \qquad (4\text{-}21)$$

Corresponding to (2-168), we have in this instance

$$\int_0^{h_1} \rho_1 gz \, dz + \int_0^{h_1'-h_1-\zeta} (\rho_1 gh_1 + \rho_2 gz) \, dz - \int_0^{h_1'} \rho_1 gz \, dz = \rho_1 q_1(v_1' - v_1)$$

$$\tfrac{1}{2}\rho_1 gh_1^2 - \tfrac{1}{2}\rho_1 gh_1'^2 + [\rho_1 gh_1 + \tfrac{1}{2}\rho_2 g(h_1 - h_1 - \zeta)](h_1' - h_1 - \zeta) = \rho_1 q_1^2\left(\frac{1}{h_1'} - \frac{1}{h_1}\right)$$

$$\tfrac{1}{2}(h_1' - h_1)(h_1 + h_1') - \left(\frac{h_1 + h_1'}{2}\right)\frac{\rho_1}{\rho_2}(h_1' - h_1) = \frac{q_1^2}{g}\left(\frac{h_1' - h_1}{h_1 h_1'}\right)$$

$$\beta\frac{h_1 + h_1'}{2} = \frac{q_1^2}{gh_1 h_1'} \tag{4-22}$$

using (4-4) and (4-21). From a knowledge of q_1 and h_1 we can therefore compute the layer thickness h_1' downstream of the hydraulic jump.

4-2. FJORD TYPE ENTRAINMENT FLOW

Because of the variety of physical conditions that can and do exist in any given estuary there is a corresponding variety of circulation effects that occur. We start by considering one of the simpler circulation effects that can occur. We take a stratified flow for which the velocity v_2 in the lower layer is sufficiently small that it may be neglected and for which the density ρ_2 of the lower layer is constant. We include upward movement of the deep water into the top layer, or *entrainment*, where it is mixed with a vertical entrainment velocity w, which is much smaller than the longitudinal flow velocity v_1 in the top layer. Thus the density ρ_1 of the upper layer has a value approaching that of fresh water at its head and approaches ρ_2 as the longitudinal distance x approaches infinity. We also assume that the mixing in the upper layer is so strong vertically that ρ_1 and v_1 are independent of the depth coordinate z. Further, as in the previous section, we consider only the gross inertial effects and ignore the frictional effects. The developments discussed in this section were originally presented by Stommel (1951) and Stommel and Farmer (1952b). This type of circulation may be taken as an approximation to what occurs in fjords.

We have two equations defining this type of motion. First, as the velocity in the lower layer is assumed to be zero, the pressure in the lower layer must be a constant as a function of x. Second, as we have assumed no frictional losses, the momentum equation stating that the net force on a control volume of the upper layer is equal to the rate of change in momentum must apply. With reference to Fig. 4-3 in which the z axis is vertical downward and the x axis is in a downestuary direction, we have for the pressure in the lower layer,

$$p_2 = \rho_1 gh_1 + \rho_2 g(z - h_1 - \zeta_1) \tag{4-23}$$

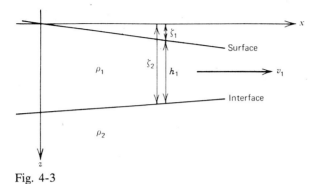

Fig. 4-3

or, from our first condition,

$$\frac{d\zeta_2}{dx} = \frac{1}{\rho_2}\frac{d}{dx}(\rho_1 h_1)$$ (4-24)

As we have assumed that the entrainment velocity w is small in comparison with v_1, we may neglect the added momentum to the upper layer by the addition of the entrained water from the lower layer, so that the flux of momentum per unit breadth is $\rho_1 q_1 v_1 = \rho_1 v_1^2 h_1$, using the definition for q_1 given by (4-4). Then, considering a small control volume for the upper layer of unit width and of length Δx, the net longitudinal force from the forces on the two faces normal to the flow must be equal to the longitudinal momentum flux change through the same volume. Using Leibnitz' rule for differentiation under an integral when the limits of integration are functions of the differentiation variable, we have

$$\frac{d}{dx}(\rho_1 v_1^2 h_1) = -\int_{\zeta_1(x)}^{\zeta_2(x)}\frac{\partial p_1}{\partial x}\,dz$$

$$= -\frac{\partial}{\partial x}\int_0^{h_1}\rho_1 gz\,dz + p_1(\zeta_2)\frac{\partial \zeta_2}{\partial x} - p_1(\zeta_1)\frac{\partial \zeta_1}{\partial x}$$

$$= -\frac{d}{dx}(\tfrac{1}{2}\rho_1 g h_1^2) + \rho_1 g h_1\frac{\partial \zeta_2}{\partial x}$$

$$= -\frac{d}{dx}(\tfrac{1}{2}\rho_1 g h_1^2) + \frac{\rho_1 g h_1}{\rho_2}\frac{d}{dx}(\rho_1 h_1)$$

$$= -\frac{d}{dx}(\tfrac{1}{2}\rho_1 g h_1^2) + \frac{d}{dx}\left(\frac{1}{2}\frac{\rho_1^2}{\rho_2}g h_1^2\right)$$

$$= -\frac{d}{dx}(\tfrac{1}{2}\rho_1 g\beta h_1^2)$$

or

$$\frac{d}{dx}[\rho_1 h_1(v_1^2 + \tfrac{1}{2}g\beta h_1)] = 0$$ (4-25)

where we have also used relation (4-24), and where β is as defined before by (4-12). We thus have a relation similar to the ones obtained in the previous section; the quantity in the brackets of (4-25) must remain constant. As before, it is more convenient in a physical sense to investigate the motion in terms of the variable q_1 rather than v_1. Using (4-4), we obtain

$$\frac{d}{dx}\left[\rho_1\left(\frac{q_1^2}{h_1}+\tfrac{1}{2}g\beta h_1^2\right)\right]=0 \qquad (4\text{-}26)$$

Let us first examine expression (4-26) for conditions along an estuary in which the variations in ρ_1 and β are sufficiently small that these two quantities may be considered a constant. We look at the variations in h_1 as a function of q_1. From (4-26) we then have

$$\frac{q_1^2}{h_1}+\tfrac{1}{2}g\beta h_1^2 = B \qquad (4\text{-}27)$$

where B is a constant, and by taking the derivative with respect to q_1,

$$\frac{2q_1}{h_1}-\frac{q_1^2}{h_1^2}\frac{dh_1}{dq_1}+g\beta h_1\frac{dh_1}{dq_1}=0$$

$$\frac{dh_1}{dq_1}=\frac{2v_1}{v_1^2-g\beta h_1} \qquad (4\text{-}28)$$

We see then that, for conditions of subcritical flow, $v_1<g\beta h_1$, h_1 decreases as q_1 increases, much the same as we had for the discussion of (4-19) in the previous section. We note in passing, however, the difference in the expression for the flow quantity that must remain constant in the two cases, (4-19) and the parentheses in (4-25). In the former case the total energy remains constant, and the variations in q_1 come about through changes in b. In the latter case the total flow Q_1 and total energy of the upper layer change following the flow, and the governing condition is that of the momentum equation. The same analysis as above for (4-19) would have yielded a value of dh_1/dq_1 half that given by (4-28).

Let us now continue the examination of (4-26), including the variations in ρ_1 and β. We have approximately for the mass continuity relation,

$$\rho_1 q_1 + \Delta q\rho_2 = q_1'\rho_1'$$

$$\rho_1 q_1 + (q_1'-q_1)\rho_2 = q_1'\rho_1'$$

$$q_1'\frac{\rho_2-\rho_1'}{\rho_2} = q_1\frac{\rho_2-\rho_1}{\rho_2}$$

or

$$\beta q_1 = \beta_0 q_{10} \qquad (4\text{-}29)$$

where β_0 and q_{10} are the value of these respective quantities at $x = 0$. Rewriting ρ_1 in terms of β from (4-12), we have for (4-26),

$$\rho_2(1-\beta)\left(\frac{q_1^2}{h_1}+\tfrac{1}{2}g\beta h_1^2\right) = D \tag{4-30}$$

Substituting for β from (4-29) and using the relation that β is small with respect to unity in the first parentheses, we have

$$\frac{q_1^2}{h_1}+\frac{g\beta_0 q_{10}}{2}\frac{h_1^2}{q_1} = \frac{D}{\rho_2} \tag{4-31}$$

from which we then obtain

$$\frac{2q_1}{h_1}-\frac{q_1^2}{h_1^2}\frac{dh_1}{dq_1}-\frac{g\beta q_1}{2}\frac{h_1^2}{q_1^2}+g\beta q_1\frac{h_1}{q_1}\frac{dh_1}{dq_1} = 0$$

$$\frac{dh_1}{dq_1}=\frac{2v_1-(g\beta h_1/2v_1)}{v_1^2-g\beta h_1} \tag{4-32}$$

We now have the result that, for conditions of subcritical flow for which $v_1 < \tfrac{1}{2}v_c$, where as before the critical velocity v_c is given by (4-17), dh_1/dq_1 is positive and, for velocity conditions in the interval $\tfrac{1}{2}v_c < v_1 < v_c$, dh_1/dq_1 is negative.

This leads to an interesting result. If the total flow is such that the mouth of the estuary acts as a control, $v_1 = v_c$ at the estuary mouth. The thickness of the upper layer will be a maximum at $v_1 = \tfrac{1}{2}v_c$, and the thickness will decrease in both an upestuary and a downestuary distance from this position.

4-3. ARRESTED SALT WEDGE FLOW

To continue with the type of analysis used in the last section let us consider another example of stratified flow in which the flow velocity in the upper layer is again much larger than that in the lower layer. In this case we attempt to duplicate the conditions that correspond to an arrested salt wedge flow.

For an arrested salt wedge flow in a tidal river or an estuary, the runoff velocities in the upper layer are large. There is a sharp interface between the overlying fresh and the underlying saline waters. The salt wedge itself is bounded at its upestuary end by the river bottom, so that the net flow across a section normal to the flow in the salt wedge must be zero exclusive of the small entrainment contribution to the upper layer. As a contributing factor to the momentum equation for the upper layer, the entrainment contribution to the momentum appears to be small in comparison with the interfacial frictional resistance associated with the high velocity flow in the upper layer as contrasted across the interfacial boundary with the nearly zero velocity flow in the salt wedge. We then assume that the flow in the upper layer is sufficiently turbulent that its flow velocity v_1 is independent of the depth coordinate z. We further assume that there is no entrainment flow across the interfacial boundary, so that the flow volume q_1 in the upper layer is a constant and both densities ρ_1 and ρ_2 are constants. For the lower layer the principal stress is that of the frictional resistance stress at the interfacial boundary from the longitudinal flow in the overlying upper layer; in accordance with the condition that the net longitudinal flow in the salt wedge is zero, the flow velocities in the salt wedge decrease abruptly from the interfacial boundary, and we may take for our boundary condition at the river bottom without significant error that $v_2 = 0$. The developments discussed in this section were originally presented by Stommel and Farmer (1952b).

For the equation of motion, or the momentum equation, in the upper layer, following the procedure used in deriving (4-25), where we must add the interfacial frictional force $f_i \, \Delta x$, on the lower boundary of the control volume in a direction opposite the flow, we have

$$\frac{d}{dx}(\rho_1 v_1{}^2 h_1) = -\int_{\zeta_1(x)}^{\zeta_2(x)} \frac{\partial p_1}{\partial x}\, dz - f_i$$

$$\frac{d}{dx}(\rho_1 v_1{}^2 h_1 + \tfrac{1}{2}\rho_1 g h_1{}^2) - \rho_1 g h_1 \frac{d\zeta_2}{dx} = -f_i$$

$$\frac{d}{dx}\left(\rho_1 \frac{q_1{}^2}{h_1} + \tfrac{1}{2}\rho_1 g h_1{}^2\right) - \rho_1 g h_1 \frac{d\zeta_2}{dx} = -f_i$$

$$-\frac{\rho_1 q_1{}^2}{h_1{}^2}\frac{dh_1}{dx} + \rho_1 g h_1 \frac{dh_1}{dx} - \rho_1 g h_1 \frac{d\zeta_2}{dx} = -f_i$$

$$\left(\frac{\rho_1 q_1{}^2}{h_1{}^2} - \rho_1 g h_1\right)\frac{d\zeta_1}{dx} - \frac{\rho_1 q_1{}^2}{h_1{}^2}\frac{d\zeta_2}{dx} = -k\rho_1 v_1{}^2$$

$$\left(1 - \frac{g h_1{}^3}{q_1{}^2}\right)\frac{d\zeta_1}{dx} - \frac{d\zeta_2}{dx} = -k \tag{4-33}$$

where the various quantities are as defined in Fig. 4-4, where we have used the fact that

$$q_1 = v_1 h_1 \tag{4-34}$$

is a constant, and where we have taken the form of f_i, following (2-143), to be

$$f_i = k\rho v_1{}^2 \tag{4-35}$$

The expression (4-35) for the interfacial frictional resistance is strictly valid only if the velocity v_2 in the salt wedge is zero. This is not the case,

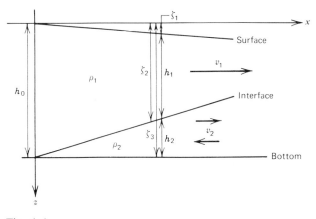

Fig. 4-4

but the velocities v_1 of the river flow are considerably larger than the mean values $|v_2|$ in the salt wedge, so that we may consider (4-35) a good approximation to the actual interfacial frictional resistance. Equation (4-33) may be simplified from the following consideration. At the mouth of the tidal river we may anticipate that the flow conditions are critical, so that the relation (4-16) holds. From this relation we note that q_1^2/gh_1^3 is of the order of β, which is a small quantity as compared with unity. At any distance up the river this quantity is less than it is at the mouth, so that (4-33) can be given approximately by

$$\frac{gh_1^3}{q_1^2}\frac{d\zeta_1}{dx}+\frac{d\zeta_2}{dx}=k \tag{4-36}$$

For the equation of motion in the lower layer we do not now have the simple relation that the horizontal pressure gradient $\partial p_2/\partial x$ is zero as we did in the previous section, but rather that this force is balanced by the internal frictional resistance force. From the basic equations of motion (2-134) and (2-135) under steady state conditions and under the assumption that the velocities in the lower layer are small so that the inertial term $v_2(\partial v_2/\partial x)$ can be neglected, we have

$$\frac{\partial p_2}{\partial x}=\frac{\partial \sigma_{zx}}{\partial z}=\rho_2 N_z\frac{\partial^2 v_2}{\partial z^2} \tag{4-37}$$

where we have also taken the viscosity coefficient N_z to be constant. For the pressure p_2 in the lower layer we have

$$p_2=\rho_1 g(\zeta_2-\zeta_1)+\rho_2 g(z-\zeta_2) \tag{4-38}$$

or

$$\frac{\partial p_2}{\partial x} = -\rho_1 g \frac{\partial \zeta_1}{\partial x} - (\rho_2 - \rho_1)\frac{\partial \zeta_2}{\partial x} \qquad (4\text{-}39)$$

As $\partial p_2/\partial x$ is independent of the depth coordinate z, we may integrate (4-37) over the depth interval from ζ_2 to ζ_3, obtaining

$$-gh_2\left[\rho_1 \frac{d\zeta_1}{dx} + (\rho_2 - \rho_1)\frac{d\zeta_2}{dx}\right] = (\sigma_{zx})_{\zeta_2} - (\sigma_{zx})_{\zeta_3} \qquad (4\text{-}40)$$

We know the value of $(\sigma_{zx})_{\zeta_2}$ from (4-35). To find the value of $(\sigma_{zx})_{\zeta_3}$ let us look at the velocity variation in the lower layer. Since $\partial p_2/\partial x$ is not a function of z, (4-37) tells us that v_2 is a quadratic in z. We then assume v_2 in the form

$$v_2 = a + by + cy^2 \qquad (4\text{-}41)$$

where we have taken a y coordinate, for convenience, in the direction opposite z and with its origin at the bottom. To determine a, b, and c we have the three conditions that v_2 is equal to v_1 at the interfacial boundary, that v_2 is equal to zero at the bottom, and that the integral of v_2 with respect to z over a section of the salt wedge normal to the flow is zero, or simply

$$v_2 = v_1 \qquad y = h_2$$
$$v_2 = 0 \qquad y = 0 \qquad (4\text{-}42)$$
$$\int_0^{h_2} v_2 \, dy = 0$$

From the second condition we have that $a = 0$. From the other two we obtain

$$v_1 = bh_2 + ch_2^2$$
$$0 = \tfrac{1}{2}bh_2^2 + \tfrac{1}{3}ch_2^3$$

from which we obtain for v_2,

$$v_2 = -\frac{2v_1}{h_2}y + \frac{3v_1}{h_2^2}y^2 \qquad (4\text{-}43)$$

A graph of v_2 would resemble that shown in Fig. 4-5, as we might have expected. Further, we note that

$$\frac{\partial v_2}{\partial y} = -\frac{2v_1}{h_2} + \frac{6v_1}{h_2^2}y$$

so that the stress σ_{zx} at $y = 0$ is one half that at $y = h_2$ and of the opposite sign. We have finally for (4-40), remembering that the frictional stress at

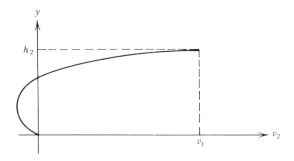

Fig. 4-5

the interface in the lower layer is the negative of (4-35),

$$-gh_2\left[\rho_1\frac{d\zeta_1}{dx}+(\rho_2-\rho_1)\frac{d\zeta_2}{dx}\right]=\tfrac{3}{2}f_i$$

$$-gh_2\rho_2\left[\alpha\frac{d\zeta_1}{dx}+\beta\frac{d\zeta_2}{dx}\right]=\tfrac{3}{2}k\rho_1v_1^2$$

$$-\frac{gh_1^2h_2}{\alpha q_1^2}\left[\alpha\frac{d\zeta_1}{dx}+\beta\frac{d\zeta_2}{dx}\right]=\tfrac{3}{2}k \qquad (4\text{-}44)$$

where we have defined α and β by

$$\alpha=\frac{\rho_1}{\rho_2} \qquad (4\text{-}45)$$

$$\beta=\frac{\rho_2-\rho_1}{\rho_2} \qquad (4\text{-}46)$$

It is of some interest to note that, had we used a bottom boundary condition similar to (2-143), which logically might be considered more correct, we would have obtained the same value for the right hand side of (4-44), provided that v_1 was substantially greater than $2v_{2b}$, where v_{2b} is the bottom velocity. For the salt wedge we may then assume that we have made the correct formulation.

We now have two equations, (4-36) and (4-44), in terms of the first derivatives of the variables ζ_1 and ζ_2. We can reduce this down to one equation in terms of one of the variables, and we do so by eliminating ζ_1, obtaining

$$-h_2\frac{d\zeta_2}{dx}+\frac{\beta}{\alpha}\frac{gh_1^3h_2}{q_1^2}\frac{d\zeta_2}{dx}=-k(h_2+\tfrac{3}{2}h_1)$$

$$h_2\frac{dh_2}{dx}-\frac{\beta}{\alpha}\frac{g(h_0-h_2)^3}{q_1^2}h_2\frac{dh_2}{dx}=-k(\tfrac{3}{2}h_0-\tfrac{1}{2}h_2)$$

$$n_2\frac{dn_2}{dx}-\frac{\beta}{\alpha}\frac{gh_0^3(1-n_2)^3}{q_1^2}n_2\frac{dn_2}{dx}=-\frac{1}{2}\frac{k}{h_0}(3-n_2)$$

$$\frac{(1-n_2)^3-\gamma}{3-n_2}n_2\,dn_2=\frac{1}{2}\frac{k}{h_0}\gamma\,dx \qquad (4\text{-}47)$$

where we have made a change in variable from h_2 to n_2, defined by

$$h_2=n_2h_0 \qquad (4\text{-}48)$$

where we have taken

$$h_1+h_2=h_0 \qquad (4\text{-}49)$$

where we consider h_0 a constant along the length of the tidal river, and where we have defined γ by

$$\gamma=\frac{\alpha q_1^2}{g\beta h_0^3} \qquad (4\text{-}50)$$

The left hand side of (4-47) is a simple algebraic expression which may be integrated directly. With reference to Fig. 4-4 we have chosen the origin of our coordinate system at the upper end, or tip, of the salt wedge. Integrating then from $n_2=0$ to n_2 and from $x=0$ to x, we obtain

$$\frac{1}{2}\frac{k\gamma}{h_0}x=\tfrac{1}{4}n_2^4+\tfrac{3}{2}n_2^2+(8+\gamma)\left(n_2+3\log\frac{3-n_2}{3}\right) \qquad (4\text{-}51)$$

This expression may be simplified through the following consideration. Again invoking the condition that the flow conditions at the river mouth are critical, so that relation (4-16) holds, we see that γ is less than unity and may be considered small as compared to the number 8. Further, we note that n_2 is less than unity, so that we can expand the natural logarithm in the last parentheses in a power series form, obtaining

$$\frac{1}{2}\frac{k\gamma}{h_0}x=\tfrac{1}{4}n_2^4+\tfrac{3}{2}n_2^2+8\left(-\frac{n_2^2}{6}-\frac{n_2^3}{27}-\frac{n_2^4}{108}-\cdots\right)$$

$$=\tfrac{1}{6}n_2^2-\tfrac{8}{27}n_2^3+\tfrac{19}{108}n_2^4-\cdots$$

$$=\tfrac{1}{6}n_2^2(1-\tfrac{16}{9}n_2+\tfrac{19}{18}n_2^2-\cdots) \qquad (4\text{-}52)$$

approximately. Using (4-50) we may rewrite this expression as

$$\frac{3k\alpha q_1^2}{g\beta h_0^4} x = n_2^2(1 - \tfrac{16}{9}n_2 + \tfrac{19}{18}n_2^2 - \cdots)$$ (4-53)

where the reciprocal of the fraction multiplying x on the left hand side of the equation may be considered the scaling factor for the x coordinate distance. Designating the value of h_2 at the river mouth by n_{2m} and the corresponding value of x by x_m, the total length of the salt wedge, we may rewrite (4-53) as

$$\frac{x}{x_m} = \left(\frac{n_2}{n_{2m}}\right)^2 \frac{1 - \tfrac{16}{9}n_2 + \tfrac{19}{18}n_2^2}{1 - \tfrac{16}{9}n_{2m} + \tfrac{19}{18}n_{2m}^2}$$ (4-54)

We see from expression (4-53) that the shape of the salt wedge starts out parabolic in form at the bottom and gradually becomes linear and then curved in the opposite direction as the cubic term becomes important.

We may carry this analysis one step further using the critical condition (4-16) at the mouth of the estuary to obtain the value of n_{2m},

$$(h_0 - h_{2m})^3 = \frac{q_1^2}{g\beta}$$

$$n_{2m} = 1 - \sqrt[3]{\frac{q_1^2}{g\beta h_0^3}} = 1 - \sqrt[3]{\frac{\gamma}{\alpha}}$$ (4-55)

where we have also used relations (4-48) to (4-50). From (4-55) we can then obtain, if we wish, the corresponding value of x_m, using (4-52). For a characteristic value of $n_{2m} = 0.5$, the graph of x/x_m versus n_2/n_{2m}, as determined from (4-54), resembles that shown in Fig. 4-6.

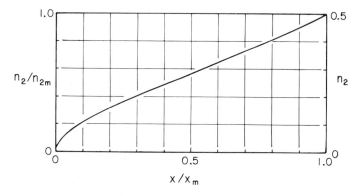

Fig. 4-6

It is of some interest to note that there are two other theoretical solutions to the salt wedge problem, which include additional assumptions over those used here. In one, presented by Schijf and Schonfeld (1953) and Harleman (1961), the assumption is made that the bottom stress is zero. In the other, given by Farmer and Morgan (1953), the assumption is made that the interfacial stress f_i of (4-35) is constant. It is also interesting to note that both approaches give a result with the same scaling factor as (4-53) but with different numerical coefficient values for the n_2 terms. When plotted for any reasonable case that might occur in nature, all three produce much the same profile as Fig. 4-6.

4-4. STRATIFIED FLOW IN STRAITS

We next consider another example of stratified flow, but one in which there is substantial flow in both the upper and lower layers. In this case we attempt to duplicate the conditions that exist for stratified flow in straits.

Here we are interested in describing the vertical variations in the current velocities rather than the horizontal variations in the current velocities or layer thicknesses. We consider both the velocities and surface and interface slopes constant as a function of the longitudinal distance. In the equations of motion the inertial terms therefore drop out, and in the steady state we are left with balancing the horizontal pressure gradient forces of the slopes with the frictional force terms. The continuity relation is that there is no net flow, or simply that the volume transport of the upper current must be equal to that of the lower current. The developments discussed in this section were originally presented by Defant (1930, 1961).

From (2-134) and (2-135) the equation of motion for the upper layer is

$$\frac{\partial p_1}{\partial x} = \rho_1 N_z \frac{\partial^2 v_1}{\partial z^2} \tag{4-56}$$

and similarly for the lower layer. With reference to Fig. 4-7 the pressure in the upper layer referred to a constant atmospheric pressure is

$$p_1 = \rho_1 g(z - \zeta_1) \tag{4-57}$$

or

$$\frac{\partial p_1}{\partial x} = -\rho_1 g \frac{\partial \zeta_1}{\partial x} \tag{4-58}$$

so that

$$\frac{d^2 v_1}{dz^2} = -\frac{g i_1}{N_z} \tag{4-59}$$

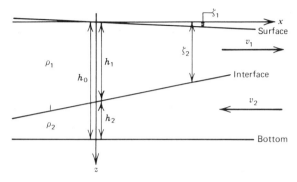

Fig. 4-7

where

$$i_1 = \frac{\partial \zeta_1}{\partial x} \tag{4-60}$$

is the slope of the water surface. The pressure in the lower layer is

$$p_2 = \rho_1 g (\zeta_2 - \zeta_1) + \rho_2 g (z - \zeta_2) \tag{4-61}$$

or

$$\frac{\partial p_2}{\partial x} = -\rho_1 g \frac{\partial \zeta_1}{\partial x} - (\rho_2 - \rho_1) g \frac{\partial \zeta_2}{\partial x} \tag{4-62}$$

so that

$$\frac{d^2 v_2}{dz^2} = -\frac{g(\alpha i_1 + \beta i_2)}{N_z} \tag{4-63}$$

where

$$i_2 = \frac{\partial \zeta_2}{\partial x} \tag{4-64}$$

is the slope of the stratification interface, and α and β are as defined before in (4-45) and (4-46). As the right hand sides of both (4-59) and (4-63) are considered constants, the solutions to both equations are simply quadratics in z.

At the surface, if there is a wind stress T, the surface boundary condition will be

$$T = f_{zx} = -\rho_1 N_z \frac{dv_1}{dz} \qquad z = 0 \tag{4-65}$$

where T is often taken in the form

$$T = k_1 \rho_a |w| w \tag{4-66}$$

similar to (2-144) for the bottom frictional stress, where w is the wind velocity, and ρ_a is the air density. Empirically the constant k_1 is about the same as the constant k for a smooth bottom. If there is no wind stress, the boundary condition (4-65) is simply that $dv_1/dz = 0$. At the interface there is a reversal of current direction so that the boundary condition there is

$$v_1 = v_2 = 0 \qquad z = h_1 \tag{4-67}$$

and also that the interfacial frictional stress is zero. At the bottom we may alternatively take the boundary condition that the velocity is zero,

$$v_2 = 0 \qquad z = h_0 \tag{4-68}$$

or that there is no friction at the bottom,

$$\frac{dv_2}{dz} = 0 \qquad z = h_0 \tag{4-69}$$

or that there is bottom friction in the form we have used previously,

$$f_b = k\rho_2 |v_2| v_2 = -\rho_2 N_z \frac{dv_2}{dz} \qquad z = h_0 \tag{4-70}$$

Equation (4-70) is the preferred form; solutions in terms of (4-68) and (4-69) are of interest in showing the varying effect of bottom friction.

For the upper layer the solution of (4-59) is simply

$$v_1 = -Az^2 + C_1 z + C_2 \tag{4-71}$$

where

$$A = \frac{gi_1}{2N_z} \tag{4-72}$$

The constants C_1 and C_2 are determined from the boundary conditions (4-65) and (4-67), giving for the final solution,

$$v_1 = A(h_1^2 - z^2) + D(h_1 - z) \tag{4-73}$$

where

$$D = \frac{T}{\rho_1 N_z} \tag{4-74}$$

For the lower layer the solution of (4-63) is simply

$$v_2 = -Bz^2 + C_3 z + C_4 \tag{4-75}$$

where

$$B = \frac{g(\alpha i_1 + \beta i_2)}{2N_z} \tag{4-76}$$

Using the boundary conditions (4-67) and (4-68), the final solution is

$$v_2 = B(z - h_1)(h_0 - z) \tag{4-77}$$

The continuity condition that

$$\int_0^{h_1} v_1 \, dz + \int_{h_1}^{h_2} v_2 \, dz = 0 \tag{4-78}$$

can then be used to obtain the determining relation between B and A for no net flow or, equivalently, the determining relation between i_2 and i_1. Carrying out this simple integration gives

$$B = -\frac{4A + 3(D/h_1)}{[(h_0/h_1) - 1]^3} \tag{4-79}$$

When the boundary conditions (4-67) and (4-69) are used, the solution for v_2 is

$$v_2 = B[(h_1^2 - z^2) + 2h_0(z - h_1)] \tag{4-80}$$

and the determining relation for B is

$$B = -\frac{4A + 3(D/h_1)}{4[(h_0/h_1) - 1]^3} \tag{4-81}$$

By using the boundary conditions (4-67) and (4-70), the solution for v_2 may be taken in the form

$$v_2 = B(z - h_1)(H - z) \tag{4-82}$$

and after a fair amount of laborious algebra the determining relation for B,

$$B = -\frac{4A + 3(D/h_1)}{\left(\dfrac{h_0}{h_1} - 1\right)^2 \left(\dfrac{3H - 2h_0}{h_1} - 1\right)} \tag{4-83}$$

and in turn for H,

$$H = \frac{h_1 - h_0\sqrt{4 + (km/N_z)h_1^3}}{1 - \sqrt{4 + (km/N_z)h_1^3}} \tag{4-84}$$

where

$$m = 4A + 3\frac{D}{h_1} \tag{4-85}$$

Several general conclusions can be discerned from these results. We see that B is always the negative of A for moderate values of D; or, equivalently, we see that the slope of the interface is opposite the slope of the surface and of such a magnitude to create a sufficient negative longitudinal pressure gradient in the lower layer to produce the requisite counterflow in that layer. The current velocity profile in both layers is of parabolic form, in accordance with our assumption of a constant N_z

value. In the upper layer the maximum current velocity occurs at the sea surface for the condition of no wind or a wind in the same direction as the surface slope. For a moderate to strong wind against the upper current flow the current velocity maximum is somewhat below the sea surface. For a wind in the direction of the current there is a more rapid decrease in the current velocity as a function of depth than with no wind or a wind against the current flow. In the lower layer the current velocity maximum occurs at some distance above the bottom depending on the magnitude of the bottom friction, being at the bottom for no bottom friction and approaching middepth in the lower layer for very large bottom friction. These relations arc shown graphically for type examples in Fig. 4-8, where the velocities in the lower layer are shown without regard to sign.

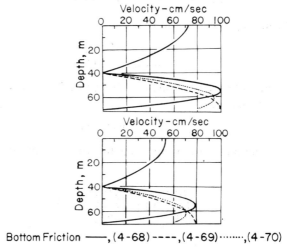

Fig. 4-8   Reproduced with permission from A. Defant, 1961, *Physical Oceanography*, Vol. I, Pergamon Press, New York, Fig. 240.

4-5. HORIZONTAL DENSITY GRADIENT FLOW

Let us continue the form of analysis used in the previous section for another example of estuarine circulation. Here, rather than a stratified flow, we consider a flow for which the density is constant as a function of the depth coordinate and for which the density increases in a gradual manner from the estuary head to the ocean as a function of the longitudinal coordinate, corresponding to the gradual increase in salinity. This may be considered an approximation to the circulation conditions for a well mixed estuary.

Again we are interested in describing the vertical variations in the current velocity rather than its horizontal variation. We consider the surface slope, the horizontal density gradient, and consequently the current velocity to be constant as a function of the longitudinal distance. We also consider, as before, that the viscosity coefficient N_z is a constant. Here again in the equation of motion the inertial terms drop out, and in the steady state we are left with balancing the combined horizontal pressure gradient forces of the surface slope and the horizontal density gradient with the frictional force terms. It should be noted that the force term due to the horizontal density gradient is opposite in direction to the force term due to the surface slope. The continuity relation is that the net circulation flow across any section normal to the flow must be equal to the river runoff flow.

From (2-134) and (2-135) the equation of motion is simply

$$\frac{\partial p}{\partial x} = \rho N_z \frac{d^2 v_x}{dz^2} \qquad (4\text{-}86)$$

The pressure referred to constant atmospheric pressure is then

$$p = \rho g (z - \zeta) \qquad (4\text{-}87)$$

or

$$\frac{\partial p}{\partial x} = -\rho g \frac{\partial \zeta}{\partial x} + g z \frac{\partial \rho}{\partial x}$$

$$= -\rho g i + g \lambda z \qquad (4\text{-}88)$$

where we have ignored the second order term involving $\zeta(\partial\rho/\partial x)$ and have made the substitutions

$$i = \frac{\partial \zeta}{\partial x} \qquad (4\text{-}89)$$

for the surface slope and

$$\lambda = \frac{\partial \rho}{\partial x} \qquad (4\text{-}90)$$

for the horizontal density gradient. Equation (4-86) then becomes

$$\frac{d^2 v_x}{dz^2} = -\frac{g i}{N_z} + \frac{g \lambda}{\rho N_z} z \qquad (4\text{-}91)$$

an ordinary differential equation whose solution is of the form

$$v_x = -\frac{g i}{2 N_z} z^2 + \frac{g \lambda}{6 \rho N_z} z^3 + C_1 z + C_2 \qquad (4\text{-}92)$$

To determine the constants C_1 and C_2 we take as our boundary condition at the surface that there is no wind stress, or

$$\frac{dv_x}{dz} = 0 \qquad z = 0 \tag{4-93}$$

and, in the first instance, at the bottom that the velocity is zero, or

$$v_x = 0 \qquad z = h \tag{4-94}$$

Substituting (4-93) and (4-94) into (4-92) we obtain as our final result

$$v_x = \frac{gi}{2N_z}(h^2 - z^2) - \frac{g\lambda}{6\rho N_z}(h^3 - z^3) \tag{4-95}$$

From our continuity relation we can now find the requisite relation between the surface slope and the horizontal density gradient. This relation gives

$$r = \int_0^h v_x \, dz$$

$$= \frac{gi}{3N_z} h^3 - \frac{g\lambda}{8\rho N_z} h^4 \tag{4-96}$$

where r is the runoff flow per unit breadth. If we now use the observed fact that in many estuaries r is considerably smaller than either of the terms on the right hand side of (4-96), we obtain for i, in terms of λ,

$$i = \frac{3}{8}\frac{\lambda h}{\rho} \tag{4-97}$$

Substituting this relation back into (4-95), we can further reduce (4-95) to

$$v_x = \frac{3}{16}\frac{g\lambda}{\rho N_z}(h^3 - hz^2) - \frac{1}{6}\frac{g\lambda}{\rho N_z}(h^3 - z^3)$$

$$= \frac{1}{48}\frac{g\lambda}{\rho N_z}(h^3 - 9hz^2 + 8z^3)$$

$$= \frac{1}{48}\frac{g\lambda h^3}{\rho N_z}(8n^3 - 9n^2 + 1) \tag{4-98}$$

where we have made the further substitution,

$$n = \frac{z}{h} \tag{4-99}$$

From expression (4-98) we see that the velocity at the surface is given by

$$v_s = \frac{1}{48}\frac{g\lambda h^3}{\rho N_z} \tag{4-100}$$

and from the condition $dv_x/dn = 0$ that the maximum positive and negative current velocities occur at the surface, where $n = 0$, and at $n = \frac{3}{4}$, respectively. Taking characteristic values for the estuary parameters of $Nz = 50 \text{ cm}^2 \text{ sec}^{-1}$, $\lambda = 3 \times 10^{-3}/10 \text{ km}$, and $h = 20 \text{ m}$, we obtain from (4-100) a value for the surface velocity of $v_s = 10 \text{ cm sec}^{-1}$. A graph of (4-98) in terms of v_x/v_s versus n is as shown in Fig. 4-9. One of the more

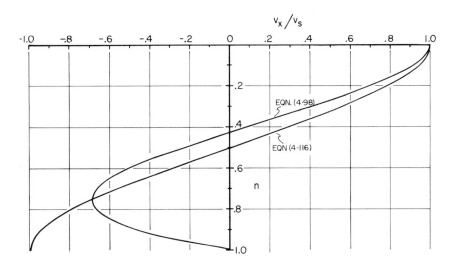

Fig. 4-9

interesting conclusions from this analysis is that, even in the case of a well mixed estuary, with no vertical density stratification, there is a net circulation flow downestuary in the upper portion of the water column and a net circulation flow upestuary in the lower portion of the water column.

If the quantity r is not small with respect to either of the two terms on the right hand side of (4-96), the continuity relation will then give for i in terms of λ, and now also r,

$$\frac{gi}{N_z} = \frac{3g\lambda}{8\rho N_z} h + \frac{3r}{h^3} \qquad (4\text{-}101)$$

Substituting this relation into (4-95), we obtain

$$v_x = \frac{1}{48} \frac{g\lambda h^3}{\rho N_z} (1 - 9n^2 + 8n^3) + \frac{3}{2} v_0 (1 - n^2) \qquad (4\text{-}102)$$

where v_0 is the vertically averaged runoff velocity per unit breadth, defined by

$$v_0 = \frac{r}{h} \tag{4-103}$$

We see that, if the second term of (4-102) is dominant, which can be the case for estuaries associated with rivers with a large runoff flow, the characteristic curve will be a simple parabola starting at $v_x/v_s = 1$ at $n = 0$ and gradually decreasing without any negative velocity section to $v_x/v_z = 0$ at $n = 1$.

We may expand this solution one further step by including a surface wind stress in the form (4-65). Then the solution for v_x is

$$v_x = \frac{gi}{2N_z}(h^2 - z^2) - \frac{g\lambda}{6\rho N_z}(h^3 - z^3) + \frac{T}{\rho N_z}(h - z) \tag{4-104}$$

instead of (4-95). Applying the continuity relation (4-96), we then obtain for the relation for i in terms of λ, r, and T,

$$\frac{gi}{N_z} = \frac{3g\lambda}{8\rho N_z}h + \frac{3r}{h^3} - \frac{3T}{2\rho N_z h} \tag{4-105}$$

instead of (4-101). We then obtain for our final result, substituting (4-105) into (4-104),

$$v_x = \frac{1}{48}\frac{g\lambda h^3}{\rho N_z}(1 - 9n^2 + 8n^3) + \frac{3}{2}v_0(1 - n^2)$$

$$+ \frac{1}{4}\frac{hT}{\rho N_z}(1 - 4n + 3n^2) \tag{4-106}$$

It is now of interest to carry the analysis still one step further to the condition of the more preferred bottom boundary condition, namely,

$$f_b = k\rho |v_x| v_x = -\rho N_z \frac{\partial v_x}{\partial z} \qquad z = h \tag{4-107}$$

Substituting the boundary conditions (4-93) and (4-107) into (4-92), we now obtain for the current velocity,

$$v_x = \frac{gi}{2N_z}(h^2 - z^2) - \frac{g\lambda}{6\rho N_z}(h^3 - z^3) - \sqrt{\frac{gh}{k}\left(\frac{1}{2}\frac{\lambda h}{\rho} - i\right)} \tag{4-108}$$

From the continuity relation we now have for the defining relation between i and λ,

$$r = \frac{gi}{3N_z}h^3 - \frac{g\lambda}{8\rho N_z}h^4 - h\sqrt{\frac{gh}{k}\left(\frac{1}{2}\frac{\lambda h}{\rho} - i\right)} \tag{4-109}$$

instead of (4-96). By using, as before, that r is small, the relation (4-109) reduces to

$$\sqrt{\frac{gh}{k}\left(\frac{1}{2}\frac{\lambda h}{\rho}-i\right)}=\frac{gh^2}{3N_z}\left(i-\frac{3}{8}\frac{\lambda h}{\rho}\right)$$ (4-110)

Although (4-110) is a cumbersome algebraic expression, we can discern from it some of the more important characteristics of the flow. We see first that the value of i must be defined between the limits

$$\frac{3}{8}\frac{\lambda h}{\rho}<i<\frac{1}{2}\frac{\lambda h}{\rho}$$ (4-111)

From the right hand side of (4-110) we see that, for it to be positive, i.e., a net runoff flow downestuary, $i>\frac{3}{8}(\lambda h/\rho)$. From the left hand side of (4-110) we see that, for it to have a real value, i.e., net upestuary bottom flow, $i<\frac{1}{2}(\lambda h/\rho)$. It is not unexpected that i should be greater in this case than as given by (4-97), since the bottom frictional force is reduced from its maximum under the conditions of (4-97) and since the bottom frictional and surface slope forces are in the same direction, balancing the horizontal density gradient force. From (4-108) and (4-110) we obtain for the velocity at the surface,

$$v_s=\frac{gi}{6N_z}h^2-\frac{g\lambda}{24\rho N_z}h^3=\frac{gh^2}{6N_z}\left(i-\frac{1}{4}\frac{\lambda h}{\rho}\right)$$ (4-112)

Using the two limiting values for i, we then have

$$v_{s1}=\frac{1}{48}\frac{g\lambda h^3}{\rho N_z} \qquad i_1=\frac{3}{8}\frac{\lambda h}{\rho}$$

and (4-113)

$$v_{s2}=\frac{1}{24}\frac{g\lambda h^3}{\rho N_z} \qquad i_2=\frac{1}{2}\frac{\lambda h}{\rho}$$

or simply that the surface velocity, with the reduction in the bottom friction, can be as great as twice the velocity given by (4-100). In general then the shape of the curve in Fig. 4-9 is altered to show a greater velocity at the surface, a more rapid decrease in velocity from the surface down, a negative velocity at the bottom, and a lowering of the depth of the maximum negative velocity.

For completeness it is of interest to include also the case for which there is no bottom friction. We then have for the bottom boundary condition,

$$\frac{dv_x}{dz}=0 \qquad z=h$$ (4-114)

instead of (4-94). Following the same procedures as before we obtain successively

$$i = \frac{1}{2}\frac{\lambda h}{\rho} \tag{4-115}$$

instead of (4-97),

$$v_x = \frac{1}{24}\frac{g\lambda h^3}{\rho N_z}(4n^3 - 6n^2 + 1) \tag{4-116}$$

instead of (4-98), and

$$v_s = \frac{1}{24}\frac{g\lambda h^3}{\rho N_z} \tag{4-117}$$

instead of (4-100). A plot of (4-116) has also been included in Fig. 4-9, where it must also be remembered that v_s in this case is twice that of (4-98). It should be noted that (4-116) is symmetric about $n = \frac{1}{2}$.

Let us interject an additional consideration at this point, as it has particular relevance to the previous paragraphs. We have implicitly assumed in the derivations of this section that tidal motions have no effect on the defining quantities in the equation of motion. This is true if the defining terms are linear in the tidal variable, and we consider the equation of motion as relevant to the time-averaged values over a tidal cycle. This is true for the simple reason that tidal variation taken as a sine or cosine term integrates to zero over a tidal cycle. This is not true for nonlinear terms, where the tidal variation in general does not integrate to zero. For the defining equation, as given by (4-86), there is no problem as the terms are linear; and we consider the time-averaged value of the estuary slope i over a tidal cycle and the net circulation velocity v_x, again as averaged over a tidal cycle. There are two exceptions, however. First, the bottom boundary condition (4-107) is not linear in v_x; and the tidal velocity variation should be included. The last term in (4-108) is then incorrect. Second, although the mean values for the inertial term quantity $v_x(\partial v_x/\partial x)$, omitted in (4-86), are usually small, this may not always be the case for the larger tidal variation term. These two items are covered in Section 4-7. The net result of the first nonlinear condition is to alter the effective value of k in (4-107). The net result of the second condition is to add a corrective term to the surface slope force term.

We may look at an additional aspect of horizontal density gradient circulation. If we have physical measurements of v_x as a function of depth so that we can determine $\partial v_x/\partial z$ and, if we have measurements of ρ or, equivalently, measurements of salinity and temperature, as a function of depth and longitudinal distance so that we can determine $\partial \rho/\partial x$, we can then calculate sequentially values for the kinematic viscosity coefficient N_z as a function of depth. In this we also assume that the term $\partial v_x / \partial x$ is sufficiently small that the inertial term in the equation of motion may be neglected. We then have, as before, for the equation of motion,

$$\frac{\partial p}{\partial x} = -\frac{\partial f_{zx}}{\partial z} = \rho N_z \frac{\partial^2 v_x}{\partial z^2} \tag{4-118}$$

In this case the density ρ is given by $\rho = \rho(x, z)$, so that the pressure referred to a constant atmospheric pressure is

$$p = g \int_0^z \rho(x, z) \, dz \tag{4-119}$$

or

$$\frac{\partial p}{\partial x} = -g\rho_s \frac{\partial \zeta}{\partial x} + g \int_0^z \frac{\partial \rho}{\partial x} \, dz \tag{4-120}$$

where ρ_s is the density at the surface and the integral in (4-120) is taken from the surface down. Equation (4-118) then becomes

$$\frac{\partial f_{zx}}{\partial z} = g\rho_s i - g \int_0^z \frac{\partial \rho}{\partial x} \, dz \tag{4-121}$$

We may then integrate this equation with respect to z, using the dummy variable z', obtaining

$$(f_{zx})_{z'} = g\rho_s i z' - g \int_0^{z'} \int_0^z \frac{\partial \rho}{\partial x} \, dz \, dz' \tag{4-122}$$

or

$$\left(\rho N_z \frac{\partial v_x}{\partial z}\right)_{z'} = -g\rho_s i z' + g \int_0^{z'} \int_0^z \frac{\partial \rho}{\partial x} \, dz \, dz' \tag{4-123}$$

where we have also assumed that there is no wind stress at the ocean surface. From the determined values of $\partial v_x / dz$ and $\partial \rho / \partial x$ we can then evaluate (4-123) in finite difference form to obtain values of N_z at various depths $z = z'$. This method was first presented and applied by Jacobsen (1913, 1918).

An expression alternative to (4-123) can be given when we are dealing with a longitudinal tidal current in an estuary. From (2-134), (2-135), and (4-120) we have for the equation of motion,

$$\frac{\partial v_x}{\partial t} = g\rho_s \frac{\partial \zeta}{\partial x} - gP - \frac{1}{\rho} \frac{\partial f_{zx}}{dz} \tag{4-124}$$

where P is defined by

$$P = \frac{1}{\rho} \int_0^z \frac{\partial \rho}{\partial x} \, dz \tag{4-125}$$

Then, integrating from the surface to the bottom and assuming that the surface stress is zero, we obtain

$$\frac{\partial \langle v_x \rangle}{\partial t} = g\rho_s \frac{\partial \zeta}{\partial x} - g\langle P \rangle - \frac{f_b}{\rho h} \qquad (4\text{-}126)$$

Then, on integrating over a tidal period, the left hand sides of (4-124) and (4-126) are zero. Subtracting the two resultant equations and integrating again over the dummy variable z', as previously, we have

$$(f_{zx})_{z'} = \left(-\rho N_z \overline{\frac{\partial v_x}{\partial z}} \right)_{z'} = \bar{f}_b \frac{z'}{h} - g \int_0^{z'} (\bar{P} - \overline{\langle P \rangle})\, dz \qquad (4\text{-}127)$$

where we have also assumed that N_z is a constant over the tidal averaging. We see that expression (4-127) is equivalent to (4-122) and (4-123), as expected. This relation was developed and applied to estuarine circulation problems by Bowden (1960).

Let us now look at still one further point. If in (4-120) $\partial\rho/\partial x$ can be considered constant as a function of depth, we will obtain the same defining relation for $\partial\rho/\partial x$ that we had in (4-88), and consequently the same circulation results. This leads to the useful conclusion that the same results apply to a partially stratified estuary as to a well mixed estuary as long as $\partial\rho/\partial x$ can be considered a constant as a function of both depth and longitudinal distance, regardless of what the vertical density or salinity profile may be. This corresponds to the situation in a partially stratified estuary for which the vertical density or salinity profile has the same shape at various measurement stations along the estuary and for which the incremental increase in density or salinity is the same between equally spaced stations. The isosalinity lines along such an estuary then resemble those shown in Fig. 4-10. Also, we see from relations (4-97), (4-100), and (4-99) that, if there is a gradual increase or decrease in $\partial\rho/\partial x$ from one vicinity along the estuary to another vicinity along the estuary, there will

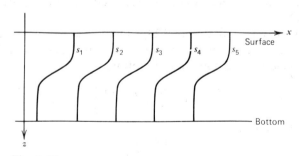

Fig. 4-10

be a corresponding increase or decrease in the surface slope i and the current velocity at the surface v_s, but no change in the shape of the curve defined by Fig. 4-9.

Finally, we may obtain some further general but useful understanding of this type of circulation, which is the general case for estuaries. We note that the longitudinal pressure gradient force due to the surface slope, or *barotropic* force, is constant as a function of depth, whereas the longitudinal pressure gradient force due to the horizontal density gradient, or *baroclinic* force, increases linearly with depth. Therefore, if the net circulation is to be small, the barotropic force will be dominant in the upper portion of the water column and the baroclinic force dominant in the deeper portion, giving a circulation downestuary at the shallower depths and upestuary at the deeper depths.

4-6. TWO DIMENSIONAL DENSITY GRADIENT FLOW

In the previous section we considered longitudinal motion effects as referred to a vertical section normal to the flow. The formulation is quite simple and straightforward. Here we discuss some of the extensions of this in consideration of the interdependency of longitudinal motion and the salt continuity equations and in consideration of vertical circulation effects. We wish to make three specific observations.

First, it is important always to keep clearly in mind the distinction between and the relative importance of the terms involving N_z and N_x and K_z and K_x. As discussed in Section 2-10, the N_z term is usually the dominant one when we are considering the forces related to the equation of motion at a point in the fluid. Similarly, the K_z term is usually the dominant one when we are considering salt fluxes at a point in the fluid. However, when we take the vertically integrated flux for a finite cross section from the surface to the bottom normal to the flow, the K_z term necessarily integrates out; and we are left with only the effect of a vertically averaged K_x term.

Second, we know from the previous section that the magnitude and degree of vertical advection and vertical diffusion in determining vertical density or salinity structure have no effect on the longitudinal velocity distribution in so far as $\partial \rho / \partial x$ remains unaltered. This has the important effect of decoupling the salt continuity equation from the longitudinal equation of motion, so that each may be solved separately rather than being considered two coupled partial differential equations, one of which is nonlinear. This simplification, when appropriate, is of immense help and allows one to obtain simple results which are easily understood in physical terms. This decoupling was observed and applied by Hansen and

Rattray (1965) to what they refer to as the central regime in their general discussion of estuarine circulation and mixing. It was also used by Fisher et al. (1972) in a similar formulation.

It is of significance to look at the formulation by Hansen and Rattray (1965), which is more mathematical in origin of derivation, and compare it with the results of the previous section, which are more physical in origin of derivation. In their central regime theory they assumed that $\partial \rho / \partial x$ was a constant. We anticipate then that their results will be the same as those derived in the previous section. Let us see if this is indeed the case. From the expressions (15) and (17) in Hansen and Rattray (1965) we have in our notation,

$$
\begin{aligned}
v_x = {} & \tfrac{3}{2} v_0 (1 - n^2) + \frac{1}{4} \frac{hT}{\rho N_z} (1 - 4n + 3n^2) \\
& + \tfrac{1}{48} v_0 \nu Ra (1 - 9n^2 + 8n^3)
\end{aligned}
\tag{4-128}
$$

where Ra is defined by

$$
Ra = \frac{g a s_0 h^3}{N_z K_{x0}}
\tag{4-129}
$$

ν is the coefficient of (5-193), a is the coefficient of (5-33), s_0 is $\langle \bar{s} \rangle$ of (5-92) at $x = 0$, and K_{x0} is the tidal diffusion coefficient at $x = 0$. Borrowing then from some of the results and reasoning of Chapter 5, we have from expressions (5-98) and (5-99), where v_x in these expressions is the same as v_0 here and K_x in (5-99) is the longitudinal dispersion coefficient,

$$
v_0 s_0 = K_x \frac{\partial s}{\partial x} = \frac{K_{x0}}{\nu} \frac{\partial s}{\partial x}
\tag{4-130}
$$

and, from (5-33),

$$
\lambda = \frac{\partial \rho}{\partial x} = a \rho \frac{\partial s}{\partial x}
\tag{4-131}
$$

so that

$$
v_0 \nu Ra = v_0 \nu \frac{g a s_0 h^3}{N_z K_{x0}} = \frac{g a h^3}{N_z} \frac{\partial s}{\partial x} = \frac{g \lambda h^3}{\rho N_z}
\tag{4-132}
$$

the same as the coefficient in the first term of (4-106). We then have that the expressions (4-128) and (4-106) are exactly the same.

With regard to the coupling effects of the longitudinal motion and salt continuity equations similarity solutions techniques, as initially introduced and applied by Ratrray and Hansen (1962) and Hansen and Rattray (1965), and as extended by Rattray (1967), Hansen and Rattray (1972),

Winter (1973), and Hansen and Festa (1974), can be used. The treatments are elegant, and the reader is encouraged to consult these references. The problem with similarity solution techniques is that they can soon leave the world of physical reality, or so it seems to us. In order for similarity solutions to be obtained certain conditions have to be imposed, which often do not have any obvious physical reality. Further, rather than taking the coefficients K_x, K_z, and N_z as constants or as variables in terms of some physical process, one or more of them usually have to be specified as a particular function of the longitudinal or depth coordinate in order to be able to obtain a mathematical solution. Still further, the boundaries themselves often are no longer simple rectangular shapes but shapes specified by the solution conditions. Clearly concern has to be given to these considerations. In addition, in practically all physical oceanographic problems we deal with approximated equation forms through the assumption of various defining relations, the elimination of various equational terms, and the assumption of constancy for certain coefficients; it then must always be kept in mind how far the solution procedures can be pushed for any given approximated defining equations. This is not meant to be a criticism of the similarity solutions obtained by Hansen, Rattray, and their colleagues, but only as a word of caution in the use of such techniques; for it is our opinion that their contributions to the literature on the physical oceanography of estuaries are among the most erudite. The nagging question nevertheless remains that perhaps some other equational defining procedure or some other solution technique for such coupled, nonlinear partial differential equations might also contribute to the physical understanding of this coupling relation and its importance in estuarine circulation, mixing, and diffusion.

The essentials of the results obtained by Rattray and Hansen (1962) and Hansen and Rattray (1965), as derived here from more direct physical reasoning and using somewhat more standard solution procedures, were covered in part in Section 4-5 and are covered in the remainder in Section 5-6. The same derivation procedures and comparison apply to what is covered in Section 5-3 and Rattray (1967) and Winter (1973), and to what is covered in Section 5-8 and Hansen and Rattray (1972) and Hansen and Festa (1974).

Third, it is obvious from the results of the previous section that we must have a vertical, upward water movement to complete the circulation loop implied by the downestuary flow in the upper portion of the water column and the upestuary flow in the lower portion of the water column. In the previous section we took as our continuity condition a balance in the flow past any cross section normal to the flow, which is a correct condition; but we have not considered the flow balance for the estuary as

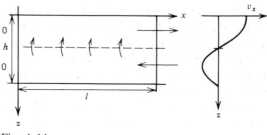

Fig. 4-11

a whole. We can obtain some index of the magnitude of the vertical velocities from the following consideration. With reference to Fig. 4-11 we see that the vertical flow across the internal boundary of zero longitudinal velocity must be such as to balance the net seaward and landward flows in the upper and lower portions of the estuary column at the mouth of the estuary. Taking 10 km as a characteristic length for an estuary and 10 m as a characteristic depth, we then note that the vertical velocities are $10 \text{ m}/10 \text{ km} = 10^{-3}$, or less than that of the horizontal velocities.

To continue, we can obtain some notion as to the longitudinally averaged vertical velocity distribution for circulation in an estuary from the following consideration. With reference to Fig. 4-11 we assume a longitudinal velocity distribution as given by (4-98), so that there is no net circulation out of the estuary at its mouth over a cross section normal to the flow from the surface to the bottom. Taking the longitudinal velocities to be zero at the upper end of the estuary, we next assume that v_x increases linearly along its length to the values observed near its mouth. With no lateral variations we then have, from the equation of continuity,

$$\frac{\partial v_x}{\partial x} + \frac{\partial v_z}{\partial z} = 0$$

or

$$\frac{dv_z}{dz} = -\frac{\partial v_x}{\partial x}$$

$$= -\frac{v_x}{l}$$

$$= -\frac{1}{48} \frac{g\lambda}{\rho N_z l} (h^3 - 9hz^2 + 8z^3) \qquad (4\text{-}133)$$

so that

$$v_z = -\frac{1}{48} \frac{g\lambda h^4}{\rho N_z l} (n - 3n^3 + 2n^4)$$

$$= -v_s \frac{h}{l} (n - 3n^3 + 2n^4) \qquad (4\text{-}134)$$

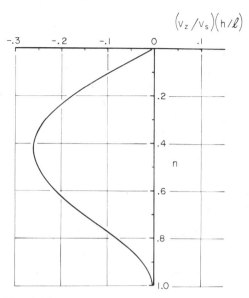

$$\left(v_z / v_s\right)\left(h / \ell\right)$$

Fig. 4-12

where the constant of integration is zero in consideration of the boundary conditions that $v_z = 0$ at $z = 0$ and $z = h$, and v_s is as given in (4-100). The maximum of v_z occurs, as expected, at the depth for which $v_x = 0$. The graph of $(v_z/v_s)(h/l)$ versus n is shown in Fig. 4-12. Although there are gross approximations in this derivation, both the magnitude and the form of the actual vertical variation nevertheless bear a reasonable resemblance to that of equation (4-134).

4-7. TIDAL MOTION AND NEAR BOTTOM EFFECTS

We now examine the nonlinear tidal contributions to a description of horizontal density gradient flow. There are two nonlinear relations, the first being that of the bottom boundary condition, and the second being that of the inertial term in the equation of motion.

The horizontal current velocity, including the tidal term, is given by

$$v_x = \bar{v}_x + C \cos \varphi \qquad (4\text{-}135)$$

where $\bar{v}_x$ is the mean velocity over a tidal cycle, C is the amplitude of the tidal variation taken as a constant as a function of depth as in Chapter 3, and $\varphi = 2\pi t/T$ is the tidal variable (T being as before the period of the semidiurnal or diurnal tide). The time variation of bottom velocity v_b resembles that shown in Fig. 4-13. The bottom boundary condition

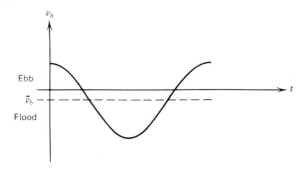

Fig. 4-13

(4-107) is then given by

$$
\begin{aligned}
f_b &= k\rho v_b^2 & 0 < \varphi < \alpha \\
&= -k\rho v_b^2 & \alpha < \varphi < 2\pi - \alpha \\
&= k\rho v_b^2 & 2\pi - \alpha < \varphi < 2\pi
\end{aligned} \tag{4-136}
$$

where α is given by

$$
\cos \alpha = \frac{-\bar{v}_b}{C} \tag{4-137}
$$

The mean value for f_b over a tidal cycle is then given by

$$
\begin{aligned}
\bar{f}_b &= \frac{k\rho}{2\pi}\left(\int_0^\alpha v_b^2\, d\varphi - \int_\alpha^{2\pi-\alpha} v_b^2\, d\varphi + \int_{2\pi-\alpha}^{2\pi} v_b^2\, d\varphi\right) \\
&= \frac{k\rho}{\pi}\left(\int_0^\alpha v_b^2\, d\varphi - \int_\alpha^{\pi} v_b^2\, d\varphi\right) \\
&= \frac{k\rho}{\pi}\left(2\alpha\bar{v}_b^2 + 4\bar{v}_b C \sin\alpha + C^2 \sin\alpha \cos\alpha + \alpha C^2 - \pi\bar{v}_b^2 - \frac{\pi}{2}C^2\right) \quad (4\text{-}138)
\end{aligned}
$$

where we have substituted from (4-135) and evaluated the integrals. By using relation (4-137) for $\cos \alpha$ and the corresponding relation for $\sin \alpha$, expression (4-138) can be rewritten as

$$
\bar{f}_b = \frac{k\rho}{\pi}\left\{3\bar{v}_b\sqrt{C^2 - \bar{v}_b^2} + [C^2 + 2\bar{v}_b^2]\left[\cos^{-1}\left(\frac{-\bar{v}_b}{C}\right) - \frac{\pi}{2}\right]\right\} \tag{4-139}
$$

For $C \gg \bar{v}_b$, which is the usual case in mixed or partially stratified estuaries, relation (4-139) reduces to

$$
\begin{aligned}
\bar{f}_b &= \frac{k\rho}{\pi}\left[3\bar{v}_b C + C^2\left(\frac{\pi}{2} + \frac{\bar{v}_b}{C} - \frac{\pi}{2}\right)\right] \\
&= \frac{k\rho}{\pi}[4\bar{v}_b C] \tag{4-140}
\end{aligned}
$$

We see then that the results of Section 4-5 will be correct if we replace k by $(4/\pi)(kC/|v_b|)$, giving for the radical in (4-108) or, equivalently, for the bottom velocity $\bar{v}_b$,

$$\bar{v}_b = -\frac{\pi}{4}\frac{gh}{kC}\left(\frac{1}{2}\frac{\lambda h}{\rho} - i\right) \qquad (4\text{-}141)$$

The direction of the bottom and near bottom current velocities are of obvious importance in the determination of the sedimentation characteristics for an estuary. The sediment is transported in the direction of the net bottom circulation current, and we anticipate an area of sediment accumulation in the regions for which $\bar{v}_b$ is zero. Now, in general for a mixed or partially stratified estuary we anticipate that, in the upper reaches of the estuary or in its associated tidal river, the runoff contribution v_0 of (4-103) will be dominant, so that the bottom net flow will be downestuary and that in the middle to lower reaches of the estuary the circulation contribution will be dominant so that the net bottom flow will be upestuary. From (4-141) the condition for zero net bottom flow is simply

$$i = \frac{1}{2}\frac{\lambda h}{\rho} \qquad (4\text{-}142)$$

Let us now go back and examine the tidal contribution to the inertial term in the equation of motion. From (2-134) and (4-86) we have an inertial contribution of the form

$$v_x \frac{\partial v_x}{\partial x} \qquad (4\text{-}143)$$

From (4-135) we then have for the mean value of (4-143) over a tidal cycle,

$$\frac{1}{2\pi}\int_0^{2\pi} v_x \frac{\partial v_x}{\partial x}\,d\varphi = \frac{1}{2\pi}\int_0^{2\pi} [\bar{v}_x + C(x)\cos\varphi]\frac{d}{dx}[\bar{v}_x + C(x)\cos\varphi]\,d\varphi$$

$$= \tfrac{1}{2}C\frac{dC}{dx} \qquad (4\text{-}144)$$

where the contribution $\bar{v}_x(d\bar{v}_x/dx)$ may be ignored, as before, as small, and where the cross terms drop out on integration. Considering dC/dx a constant over the longitudinal extent of our investigation, as we did with the surface slope i, we see that the tidal inertial contribution to the equation of motion is of the same form as i, i.e., independent of z. From (2-134) and (4-88) we then see that the results of this section and those of Section 4-5 will be correct if we replace i, where it appears, by

$i+\frac{1}{2}(C/g)(dC/dx)$. The condition for net bottom flow of (4-142) then becomes

$$i+\frac{1}{2}\frac{C}{g}\frac{dC}{dx}=\frac{1}{2}\frac{\lambda h}{\rho} \qquad (4\text{-}145)$$

It is of some interest to note that had we altered the inertial term to a form $(\partial/\partial x)(v_x v_x)$ through the use of the equation of continuity, much as was done in relations (2-62) and (2-63), we might have anticipated a tidal inertial contribution twice that given by expression (4-144); this is incorrect since with either a progressive or standing wave form of the tidal motion there would be a negative contribution of half this amount from the additional introduced inertial term of the form $(\partial/\partial z)(v_x v_z)$.

For further information on the material covered in this section the reader may refer to Abbott (1960, 1960a).

4-8. FRONTS

We are concerned in large part in this and the following two sections with the lateral variations that occur in estuarine circulation. This is undoubtedly one of the least well understood aspects of estuarine circulation, both from an experimental and theoretical point of view. As such variations become better understood, it may be anticipated that they will alter substantially some of the conclusions reached previously from a strictly longitudinal consideration of estuarine circulation and mixing.

One of the more interesting lateral phenomena that occurs in estuaries is a *front*. An oceanic front may be defined as a band along the sea surface across which the density changes abruptly. Presumably then, the frontal surface continues down as a function of depth separating the two water masses, sloping down under the lighter water mass. In the deep ocean the density changes are associated with temperature differences; in estuaries they are generally associated with salinity differences.

In one estuary where fronts have been studied, namely, the Delaware estuary, they have been observed to have the following characteristics. Generally the fronts are parallel to the axis of the estuary and extend for longitudinal distances of tens of kilometers. They are observed on both sides of the estuary and are usually associated with areas of shoaling. In either case the lighter water is toward the central axis of the estuary with the frontal surface in both cases sloping down toward the central axis. They tend to be more prominent on an ebb tide, becoming less distinct or disappearing on a flood tide. Their generation in terms of the lateral juxtaposition of different density water masses has been explained in terms of the higher tidal velocities [as given by (3-5)] and flows for the

deeper, central channel in advecting fresher and lighter water downestuary as compared with the corresponding flows in the adjoining shoal
waters. The fronts are associated with surface salinity changes of a few
parts per thousand over a lateral distance of less than 1 m. The fronts are
also associated with lateral convergence velocities at the surface on both
sides of the front of the order of $10 \, \text{cm sec}^{-1}$. Because of this lateral
convergence the fronts are also often recognizable as foam lines and as
lines of accumulation of floating organic and detrital materials.

Another type of front that can occur in coastal areas is that associated
with a river emptying into an open coastal area, estuary, or sound. In the
case of the Connecticut river a plume of surface, brackish water forms a
density contrast with the more saline water of Long Island sound. The
principal current structure is that of the tides in Long Island sound,
parallel to the longitudinal axis of the sound. At times of high river
discharge a front is formed on the offshore side of the plume, parallel to
the sound axis. The same dynamics applies to this case as to an estuary
front.

Let us take a coordinate system as shown in Fig. 4-14 with the x axis
downestuary, the z axis vertical downward, and the y axis in a lateral

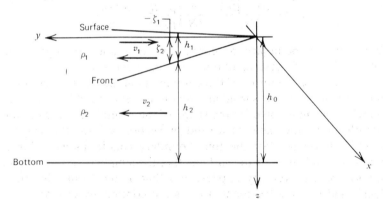

Fig. 4-14

direction, so as to form a right handed coordinate system. We assume that
the densities of the two water masses ρ_1 and ρ_2 are constant, with $\rho_1 < \rho_2$.
We further assume that the lateral velocities and velocity gradients are
sufficiently small that the nonlinear inertial terms may be ignored with
respect to the internal frictional terms. We also assume that the lateral
geostrophic force associated with the longitudinal tidal motion is
constant as a function of depth, so that it may be considered simply an
additive or corrective term to the surface slope force. We are then left in

the steady state with a balance between the lateral surface and frontal slope forces and the internal lateral eddy viscosity resistance force, much the same as we have considered in previous sections. With the other quantities as given in Fig. 4-14 the defining equation and subsidiary relations are the same as those given by (4-56) through (4-60) for the upper layer, where we now use y for x, and the form of solution for the lateral velocity in the upper layer as a function of depth is

$$v_1 = -Az^2 + C_1z + C_2 \tag{4-146}$$

where A is defined by

$$A = \frac{gi_1}{2N_z} = \frac{g}{2N_z}\frac{d\zeta_1}{dy} \tag{4-147}$$

the same as expressions (4-71) and (4-72). For the lower layer relations (4-61) through (4-64) and (4-75) and (4-76) apply, giving

$$v_2 = -Bz^2 + C_3z + C_4 \tag{4-148}$$

where B is defined by

$$B = \frac{g(\alpha i_1 + \beta i_2)}{2N_z} = \frac{g}{2N_z}\left(\frac{\rho_1}{\rho_2}\frac{d\zeta_1}{dy} + \frac{\rho_2 - \rho_1}{\rho_2}\frac{d\zeta_2}{dy}\right) \tag{4-149}$$

We have six boundary or continuity conditions to impose on our solutions. Four of these conditions provide evaluations of the constants C_1, C_2, C_3, and C_4 in terms of the parameters A and B, and the other two conditions provide the subsidiary relations between $d\zeta_1/dy$ and $d\zeta_2/dy$ from which we may solve for the shape of the frontal surface as a function of depth, much the same as we did in Section 4-3 for the shape of an arrested salt wedge. For the four boundary conditions we assume that there is no wind stress at the surface, that the simplified bottom boundary condition of zero velocity applies, and that at the frontal interface the velocity and internal frictional stresses are continuous so that

$$\begin{aligned}
\frac{\partial v_1}{\partial z} &= 0 & z &= 0 \\[4pt]
v_2 &= 0 & z &= h_0 \\[4pt]
v_1 &= v_2 & z &= h_1 \\[4pt]
\frac{\partial v_1}{\partial z} &= \frac{\partial v_2}{\partial z} & z &= h_1
\end{aligned} \tag{4-150}$$

where in the last expression we have taken into consideration that ρ_1 is approximately equal to ρ_2. For the continuity relations we assume that the net lateral flow in the upper layer is zero and that the net lateral flow in the lower layer is equal to a constant,

$$\int_0^{h_1} v_1 \, dz = 0$$

and (4-151)

$$\int_{h_1}^{h_0} v_2 \, dz = q$$

These last two conditions, as stated, are certainly open to criticism in that they are valid only if the net longitudinal flow over the same intervals is invariant as a function of x. In all of this it has also been assumed that the front is stationary; the same conditions apply equally well to a front moving with a constant lateral velocity where the velocities v_1 and v_2 are then referred to the frontal velocity as reference.

Applying the first of conditions (4-150) and the first of conditions (4-151) to (4-146), we obtain

$$v_1 = \frac{A}{3}(h_1^2 - 3z^2) \qquad (4\text{-}152)$$

We see incidentally from this result that the velocity at the interface is twice that at the surface and in an opposite direction. Applying the second and third of conditions (4-150) to (4-148), we obtain

$$v_2 = \frac{B(h_0 - z)(h_0 - h_1)(z - h_1) - \frac{2}{3}Ah_1^2(h_0 - z)}{h_0 - h_1} \qquad (4\text{-}153)$$

The lateral velocity as a function of depth then resembles that sketched in Fig. 4-15, where we have taken a negative value for i_1.

From (4-152) and (4-153) we then have for the fourth of conditions

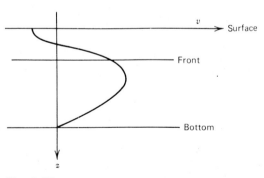

Fig. 4-15

(4-150) and from (4-153) for the second of conditions (4-151) that

$$B(h_0 - h_1)^2 = -\tfrac{2}{3}Ah_1(3h_0 - 2h_1)$$

and   (4-154)

$$q = \tfrac{1}{6}B(h_0 - h_1)^3 - \tfrac{1}{3}Ah_1^2(h_0 - h_1)$$

after a small amount of algebra. Substituting for A and B in terms of $d\zeta_1/dy$ and $d\zeta_2/dy$ from (4-147) and (4-149) we obtain for the first of equations (4-154)

$$\frac{d\zeta_1}{dy} = -3\beta \frac{(h_0 - h_1)^2}{3h_0^2 - h_1^2} \frac{d\zeta_2}{dy}$$ (4-155)

where we have taken α equal to unity and in turn for the second of equations (4-154) in terms of $d\zeta_2/dy$ from (4-155)

$$\frac{6qN_z}{g\beta} = \frac{(h_0 - h_1)^3(3h_0h_1 + h_1^2)}{3h_0^2 - h_1^2} \frac{d\zeta_2}{dy}$$ (4-156)

again after some algebra. Taking the origin of our coordinate system where the frontal surface intersects the ocean surface, considering h_0 to be a constant, using the relation that dh_1/dy is essentially the same as $d\zeta_2/dy$, and making a change of variable from h_1 to n, where n is defined by

$$n = \frac{h_1}{h_0}$$ (4-157)

we may integrate (4-156) with respect to y, obtaining

$$\frac{6qN_z}{g\beta h_0^4} y = \int_0^n \frac{(1-n)^3(3n + n^2)}{3 - n^2} \, dn$$ (4-158)

The right hand side of (4-158) is a simple algebraic expression which may be integrated directly. We obtain

$$\frac{6qN_z}{g\beta h_0^4} y = 8n - \tfrac{3}{2}n^2 + \tfrac{1}{4}n^4 - 6\log\left(1 - \frac{n^2}{3}\right) - \frac{12}{\sqrt{3}}\log\left[\frac{1 + (n/\sqrt{3})}{1 - (n/\sqrt{3})}\right]$$ (4-159)

Following the same procedure used previously for the integrated expression for the shape of an arrested salt wedge, we take the expansion of the natural logarithm in the last two terms of (4-159) obtaining

$$\frac{12qN_z}{g\beta h_0^4} y = n^2(1 - \tfrac{16}{9}n + \tfrac{14}{12}n^2 - \cdots)$$ (4-160)

We note that the right hand side of (4-160) is almost exactly the same as the right hand side of (4-53) for an arrested salt wedge. The shape of the front is simply the invert of Fig. 4-6. The front starts out parabolic in form at the surface with the vertex of the parabola at the surface and then gradually becomes more linear at depth. Again, as in equation (4-53), the reciprocal of the fraction multiplying y on the left hand side of the equation may be considered the scaling factor for the y coordinate distance. We see also that the scaling factor in (4-160) is of a form similar to that in (4-53).

Now for many estuarine frontal situations, particularly those associated with river plumes, the thickness of the upper layer is only a small fraction of the total water depth. The lower layer and bottom boundary effects then become unimportant and can be ignored. Instead of the second of conditions (4-150) and the second of conditions (4-151) we have simply that the lower layer velocity is a constant as a function of z, which implies that both C_3 and B of the solution form (4-148) are zero. The interfacial stress condition is also correspondingly altered to equate the upper layer internal stress to a quadratic boundary layer stress term,

$$\sigma_{zx} = \rho_1 N_z \frac{\partial v_1}{\partial z} = k\rho_2 v_2^2 \qquad z = h_1 \qquad (4\text{-}161)$$

instead of the fourth of conditions (4-150). This relation is valid for the case considered here of a thin surface layer, h_1, in which the first of the conditions (4-151) also applies.

Applying the first of conditions (4-150) and the first of conditions (4-151) to (4-146) we have, as before, (4-152). Applying the third of the conditions (4-150) to (4-148), we obtain

$$v_2 = -\tfrac{2}{3}Ah_1^2 \qquad (4\text{-}162)$$

From the condition (4-161) we then have

$$A = -\frac{9N_z\alpha}{2kh_1^2} \qquad (4\text{-}163)$$

and from the condition that v_2 is a constant that $B = 0$. Then, substituting, as before, for A and B in terms of $d\zeta_1/dy$ and $d\zeta_2/dy$ from (4-147) and (4-149), taking α equal to unity, and using the relation that dh_1/dy is essentially the same as $d\zeta_2/dy$, we obtain

$$\frac{dh_1}{dy} = \frac{9N_z^2}{ky\beta h_1^3} \qquad (4\text{-}164)$$

Integrating as before we have finally that

$$y = \frac{kg\beta}{36N_z^2} h_1^4 \qquad (4\text{-}165)$$

instead of (4-160). We see that in this case the frontal surface is anticipated to follow a fourth power depth dependence with vertex again at the surface but with a more rapid change toward a level surface as a function of depth.

Unquestionably, the theoretical description given here is quite inadequate. Several of the assumptions can be questioned such as the importance of the inertial terms particularly in the vicinity of the intersection of the frontal surface with the ocean surface, the neglect of the Coriolis component, and the invariance in x. At best it is an interesting exercise. Nevertheless it is of some instructive value to see how well this simplified presentation agrees with observations. The only detailed and published series of measurements on the shape of an estuarine frontal surface and on its lateral velocity field are the observations by Garvine (1974) and Garvine and Monk (1974) for the Connecticut river plume extending out into Long Island sound. These measurements are discussed in some detail in Section 9-3. We see from Fig. 9-9 that the observed velocity agrees approximately with that predicted by (4-152) and (4-162) with the velocities convergent into the front from both sides and with the velocities at the surface in the lighter water body being roughly half those of the heavier water body at depth and of opposite sign. We can make an approximate calculation for the shape of the frontal surface from the following considerations. From (4-152), (4-162), and (4-163) we can write the velocity relation as

$$v_2 = -2v_{1s} = \frac{3N_z}{kh_1} \qquad (4\text{-}166)$$

where v_{1s} is the velocity at the surface. Taking characteristic values of $v_{1s} = -25$ cm sec^{-1}, $v_2 = 50$ cm sec^{-1}, and $h_1 = 2$ m at some distance away from the vertex for which this relation might be expected to apply and using a value of $N_z = 8.5$ cm^2 sec^{-1} from other observations in Long Island sound, we calculate a value of $k = 2.6 \times 10^{-3}$. This is a reasonable value for the parameter k and within the range of values obtained in laboratory and field measurements of interfacial flows. Then, using an average value of $\beta = 10 \times 10^{-3}$ from Fig. 9-8, we calculate the shape of the frontal surface from (4-165) with result shown by the heavy line in Fig. 9-8. Considering the approximate nature of the theory, the agreement is quite good.

Garvine (1974a) has taken a somewhat different approach to describing the dynamics of small scale frontal features. He imposes much the same conditions on the problem as used here, with the important exception that he uses a vertically integrated momentum equation including the inertial terms rather than the differential motion equation with the internal frictional term but without the inertial terms. He argues that a frictional contact along the frontal surface, as given by condition (4-150), is necessary for the maintenance of the front rather than a frictionless or shear contact. He then develops certain relations for the front from assumptions regarding the variation of certain of the physical parameters, based in part on the observations of Garvine (1974) and Garvine and Monk (1974) on the front of the Connecticut river plume.

For further information on oceanic and estuary fronts the reader may refer to Cromwell and Reid (1956), Szekielda et al. (1972), Rao and Murty (1973), Garvine (1974, 1974a), and Garvine and Monk (1974).

4-9. TRANSVERSE EFFECTS

For the lateral equation of motion under steady state conditions we have from (2-134) and (2-120), under the assumptions we have used previously that the vertical velocities are small and that the dominant internal frictional stress term is related to f_{zy} as compared with the contributions from the f_{xy} and f_{yy} terms,

$$v_x \frac{\partial v_y}{\partial x} + v_y \frac{\partial v_y}{\partial y} = -\frac{1}{\rho}\frac{\partial p}{\partial y} + 2\omega_0 v_x \sin\varphi - \frac{1}{\rho}\frac{\partial}{\partial z}(\overline{\rho v_y' v_z'}) \qquad (4\text{-}167)$$

where we must now include the Coriolis contribution because of the magnitude of v_x. The terms on the left hand side of (4-167) are the inertial terms. The first term on the right hand side is the pressure gradient term, the second the Coriolis term, and the third the frictional resistance term. Following our previous procedure, we then rewrite (4-167) as

$$v_x \frac{\partial v_y}{\partial x} + v_y \frac{\partial v_y}{\partial y} = -\frac{1}{\rho}\frac{\partial p}{\partial y} + 2\omega_0 v_x \sin\varphi + N_z' \frac{\partial^2 v_y}{\partial z^2} \qquad (4\text{-}168)$$

where N_z' is defined by

$$N_z' = -\frac{\overline{\rho v_y' v_z'}}{\rho \dfrac{\partial v_y}{\partial z}} \qquad (4\text{-}169)$$

and may not necessarily be the same as N_z defined by (2-137) which we have considered previously. In (4-168) we have also assumed that N_z' can be considered a constant.

Now, in (4-167) we see that the Coriolis, or geostrophic, term is linear in v_x, so that the tidal contribution is balanced out as averaged over a tidal cycle. From the form of the frictional resistance term in (4-168) we may also anticipate that this term will be small for essentially longitudinal motion where v_y is approximately equal to zero. The inertial terms are related in this case mainly to curvature of the stream lines. For strictly longitudinal motion these terms can usually be omitted, as we have done before. For curved streamline flow, as determined in large part by the meandering shape of any given estuary, these terms provide a centripetal acceleration contribution which can be significant in comparison with the Coriolis force.

Let us look first at the conditions for longitudinal flow in which the lateral Coriolis force is balanced by a lateral pressure gradient force. Equation (4-167) reduces to

$$2\omega_0 v_x \sin\varphi = \frac{1}{\rho}\frac{\partial p}{\partial y} \tag{4-170}$$

Equation (4-170) is of course the same as the second equation in (2-122). The discussion following (2-123) applies here, and it is important to reemphasize that neither term in (4-170) represents a driving or frictional resistance force. The Coriolis force is a resultant of the longitudinal motion, and the lateral pressure gradient force is needed to maintain the longitudinal motion. We then expect for any estuary a lateral pressure gradient which varies as a function of depth as given by (4-170), where v_x is the determined longitudinal velocity as a function of depth as in (4-98).

Taking the pressure p to be given in the usual form by

$$p = \rho g(z - \zeta) \tag{4-171}$$

so that

$$\frac{\partial p}{\partial y} = -\rho g\frac{\partial \zeta}{\partial y} + gz\frac{\partial \rho}{\partial y} \tag{4-172}$$

we have for the lateral surface slope in terms of the longitudinal velocity at the surface v_s, from (4-170),

$$\frac{\partial \zeta}{\partial y} = -\frac{2\omega_0 \sin\varphi}{g}v_s \tag{4-173}$$

In general we do not have direct measurements of the lateral surface slope nor of the lateral density gradient as a function of depth. However, we often have station measurements of salinity and temperature as a function of depth from which density, or its reciprocal the specific volume α, can be calculated from standard tables. We can then in turn from two adjacent, lateral stations determine $\partial p/\partial y$ as a function of depth, provided we have separate knowledge of the depth of a level of no motion, i.e., $v_x = 0$, from which we then have directly from (4-170) v_x as a function of depth. We thus have a separate means from the simplified lateral equation of motion for determining v_x. This procedure has been one of the more basic methodologies for the investigation of deep sea currents. It

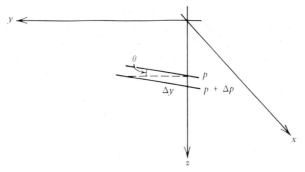

Fig. 4-16

has somewhat more limited application for estuarine investigations because of the generally greater importance of the inertial terms in estuarine circulation. It has been applied successfully to some fjord type
flows.

Let us carry out the derivation related to the above discussion. We note
from (4-170) that the longitudinal velocity at any depth is directly
proportional to the lateral pressure gradient at that depth. We may then
represent graphically the lateral cross section of the constant pressure, or
isobaric, surfaces at any depth as shown in Fig. 4-16. We then have from
relation (2-121) for $\partial p / \partial y$ in terms of the inclination angle of the isobaric
surfaces,

$$\frac{\partial p}{\partial y} = g\rho \tan \theta \qquad (4-174)$$

Consider next two measurement stations a lateral distance apart across a
longitudinal current as shown in Fig. 4-17. We assume that the isobaric
surfaces produce linear traces on a vertical section between these two

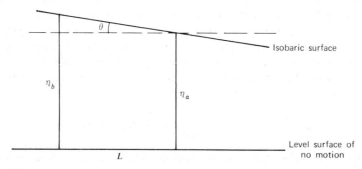

Fig. 4-17

stations and are interested in computing the height η above a level surface of no motion at some depth, where the heights η are measured in a negative z direction with respect to the level of no motion. We cannot state a priori at what depth no motion will occur, so that we are actually computing current velocities relative to the velocity at this depth. From Fig. 4-17 we have

$$g \tan \theta = \frac{g(\eta_b - \eta_a)}{L} \tag{4-175}$$

From (2-121) we have for the incremental change in geopotential,

$$g \, \Delta \eta = -\frac{1}{\rho} \Delta p = -\alpha \, \Delta p \tag{4-176}$$

Integrating from 0 to η, we then have

$$g\eta = \int_p^{P_0} \alpha \, dp \tag{4-177}$$

where p_0 is the pressure at the depth of no motion, and p is the pressure at the depth at which the current is to be determined. Substituting (4-174), (4-175), and (4-177) into (4-170), we obtain finally

$$v_x = \frac{1}{2\omega_0 L \sin \varphi} \int_p^{P_0} (\alpha_b - \alpha_a) \, dp \tag{4-178}$$

or for the velocity at the surface

$$v_s = \frac{1}{2\omega_0 L \sin \varphi} \int_0^{P_0} (\alpha_b - \alpha_a) \, dp \tag{4-179}$$

where we have taken the zero pressure level reference to be an assumed constant atmospheric pressure. Equation (4-178) is referred to as the *Helland-Hansen formula*. The quantity $g\eta$ in (4-177) is referred to as the dynamic height. In the application of (4-178) the integral is determined numerically from values of α obtained empirically from measurements of salinity and temperature at each station as a function of depth. The incremental pressure intervals Δp are taken to a sufficient approximation to be given by $g \, \Delta z$, where Δz is the depth interval. We may add one further useful relation, namely, that for the flow volume per unit breadth q, which is simply

$$q = \int_0^H v_x \, dz = \frac{1}{2\omega_0 L \sin \varphi} \int_0^H \int_p^{P_0} (\alpha_b - \alpha_a) \, dp \, dz \tag{4-180}$$

where $z = H$ is the depth of no motion. Application of these results to estuarine circulation has been principally for fjord type flow where there is a well defined downestuary upper layer flow underlain by a much slower and deeper upestuary flow.

Returning to a consideration of the inertial terms, we know that, if the essentially longitudinal flow has a slight lateral curvature, even though we have chosen the direction of the x coordinate at any given location so that $v_y = 0$, $\partial v_y / \partial x$ is not zero. We must then look at the contribution from the first of the inertial terms in (4-167). As this term is nonlinear, we must also include the tidal component and consider v_x and v_y to be given in the form (4-135). Carrying out the corresponding evaluation similar to that of (4-144), we obtain

$$v_x \frac{\partial v_y}{\partial x} = \bar{v}_x \frac{\partial \bar{v}_y}{\partial x} + \tfrac{1}{2} C_x \frac{\partial C_y}{\partial x} \qquad (4\text{-}181)$$

where we have included the net circulation contribution $\bar{v}_x (\partial \bar{v}_y / \partial x)$ as well as the tidal amplitude contribution. From the geometry of Fig. 4-18 we

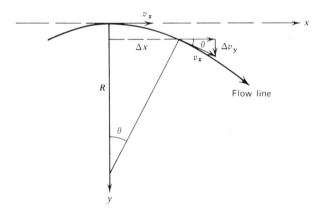

Fig. 4-18

have the relation

$$\Delta v_y = v_x \sin \theta = v_x \frac{\Delta x}{R}$$

or

$$\frac{\partial \bar{v}_y}{\partial x} = \frac{\bar{v}_x}{R}$$

and

$$\frac{\partial C_y}{\partial x} = \frac{C_x}{R} \qquad (4\text{-}182)$$

where R is the radius of curvature of the stream line. Substituting (4-182) into (4-181), we then have for the inertial contribution,

$$v_x \frac{\partial v_y}{\partial x} = \frac{\bar{v}_x^2 + \frac{1}{2} C_x^2}{R^2} \tag{4-183}$$

In taking observed values of R for estuarine flow, it is not uncommon for the inertial term of (4-183) to have values comparable to or greater than the values of the lateral pressure gradient and Coriolis terms.

For further information on the material covered in this section the reader may refer to Proudman (1953), Jakhellyn (1935), Pritchard (1956), and Stewart (1957).

4-10. SOME LATERAL EFFECTS ON CIRCULATION

As we begin to consider lateral as well as longitudinal motion, equations describing the motion become necessarily more complex, and simple, analytic solutions are less easy to obtain. In general we deal with a set of two or more partial differential equations even under steady state conditions. One of the more elegant and instructive solutions for such an estuarine circulation problem is that given by Takano (1954, 1954a, 1955), which has also been summarized by Defant (1961). This problem deals with the flow of river water out into an open coastal area.

We assume that the flow is sufficiently slow that we can neglect the inertial terms in the equations of motion and that there are no nonlinear tidal effects. We also assume that there are no entrainment or diffusion effects across the interface between the river water and the coastal water, and correspondingly that there is no flow in the lower, coastal water. In this case, as there is both lateral and longitudinal spreading, we cannot ignore the horizontal internal frictional resistance terms; but in order to obtain an analytic solution we assume that the horizontal eddy viscosity coefficients are constant and the same in a lateral and longitudinal direction. From (2-120), (2-134), and (2-135) we then have for the equations of motion under steady state conditions,

$$0 = \frac{1}{\rho} \frac{\partial p}{\partial x} - f v_y + N_h \, \nabla^2 v_x + \frac{\partial}{\partial z}\left(N_z \frac{\partial v_x}{\partial z} \right)$$

and   (4-184)

$$0 = \frac{1}{\rho} \frac{\partial p}{\partial y} + f v_x + N_h \, \nabla^2 v_y + \frac{\partial}{\partial z}\left(N_z \frac{\partial v_y}{\partial z} \right)$$

where ∇^2 is the two dimensional Laplacian operator

$$\nabla^2 = \frac{\partial^2}{\partial x^2} + \frac{\partial^2}{\partial y^2} \tag{4-185}$$

and f is the Coriolis parameter, defined by

$$f = 2\omega_0 \sin \varphi \tag{4-186}$$

We also have for the equation of continuity, assuming further that the vertical motions are small and can be neglected,

$$\frac{\partial}{\partial x}(\rho v_x) + \frac{\partial}{\partial y}(\rho v_y) = 0 \tag{4-187}$$

Assuming next that the stress vanishes at the surface $z = \zeta_1$, and at the bottom $z = d$, we can integrate equations (4-184) from the surface to the bottom to eliminate the last term in each equation, obtaining

$$N_h \nabla^2 M_x - f M_y = \frac{\partial P}{\partial x}$$

and

$$\tag{4-188}$$

$$N_h \nabla^2 M_y + f M_x = \frac{\partial P}{\partial y}$$

where M_x, M_y, and P are defined by

$$M_x = \int_{\zeta_1}^{d} \rho v_x \, dz \tag{4-189}$$

$$M_y = \int_{\zeta_1}^{d} \rho v_y \, dz \tag{4-190}$$

and

$$P = \int_{\zeta_1}^{d} p \, dz \tag{4-191}$$

and are, respectively, the longitudinal and lateral mass transports and the vertically integrated pressure, or total force. We also have, from (4-187),

$$\frac{\partial M_x}{\partial x} + \frac{\partial M_y}{\partial y} = 0 \tag{4-192}$$

As is a common practice in solving coupled partial differential equations such as (4-188), we now define a stream function ψ by

$$M_x = \frac{\partial \psi}{\partial y} \qquad M_y = -\frac{\partial \psi}{\partial x} \tag{4-193}$$

which in itself satisfies the continuity relation (4-192). Substituting (4-193) into (4-188), taking cross derivatives to eliminate the pressure gradient and Coriolis terms, and subtracting, we obtain

$$\nabla^4 \psi = 0 \tag{4-194}$$

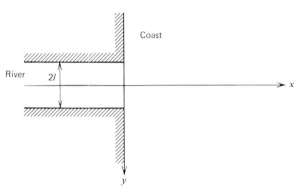

Fig. 4-19

where ∇^4 is the biharmonic operator

$$\nabla^4 = \frac{\partial^4}{\partial x^4} + 2\frac{\partial^4}{\partial x^2 \partial y^2} + \frac{\partial^4}{\partial y^4} \qquad (4\text{-}195)$$

We take a coastal and river geometry as shown in Fig. 4-19. For a constant and uniform river flow at the mouth M_0 our boundary conditions are

$$
\begin{aligned}
M_x &= M_0 & x &= 0 & -l &< y < l \\
M_x &= 0 & x &= 0 & -l &> y, \, y > l \\
M_y &= 0 & x &= 0
\end{aligned}
\qquad (4\text{-}196)
$$

Under these boundary conditions the solution to (4-194) is

$$\psi = \frac{M_0}{\pi}\left[(y+l)\tan^{-1}\frac{y+l}{x} - (y-l)\tan^{-1}\frac{y-l}{x} \right] \qquad (4\text{-}197)$$

This solution is not given here, but the reader may demonstrate to his satisfaction that it is correct by referring to the solution forms of the biharmonic differential equation or by simply substituting (4-197) back into (4-194) and (4-196). Combining the solution (4-197) with (4-188), we have directly the Coriolis contribution to P and, after a fair amount of straightforward but laborious calculus and algebra, the horizontal eddy viscosity contribution to P, giving finally

$$P = \frac{M_0}{\pi}\left\{ f\left[(y+l)\tan^{-1}\frac{y+l}{x} - (y-l)\tan^{-1}\frac{y-l}{x} \right] \right.$$

$$\left. + 2N_h\left[\frac{y+l}{x^2+(y+l)^2} - \frac{y-l}{x^2+(y-l)^2} \right] \right\} \qquad (4\text{-}198)$$

or

$$P = \frac{M_0}{\pi}\left\{ lf\left[(1+\eta)\tan^{-1}\frac{1+\eta}{\xi} - (1-\eta)\tan^{-1}\frac{1-\eta}{\xi} \right] \right.$$

$$\left. + \frac{4N_h}{l}\left[\frac{\xi^2-\eta^2+1}{[\xi^2+(1+\eta)^2][\xi^2+(1-\eta)^2]} \right] \right\} \quad (4\text{-}199)$$

where

$$\xi = \frac{x}{l} \qquad \eta = \frac{y}{l} \qquad (4\text{-}200)$$

Now, P in itself is of little direct use to us; but we can relate it to the depth h of the river flow as it moves out into the coastal water through the following considerations. With reference to Fig. 4-20 we have, as

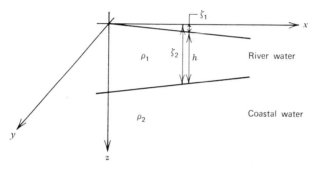

Fig. 4-20

before, that the pressure in the coastal water layer is given by

$$p_2 = \rho_1 g(\zeta_2 - \zeta_1) + \rho_2 g(z - \zeta_2) \qquad (4\text{-}201)$$

As we have assumed no flow in the coastal water, the longitudinal and lateral pressure gradient is zero,

$$\frac{\partial p_2}{\partial x} = \rho_1 g\left(\frac{\partial \zeta_2}{\partial x} - \frac{\partial \zeta_1}{\partial x} \right) - \rho_2 g\frac{\partial \zeta_2}{\partial x} = 0$$

so that

$$\frac{\partial \zeta_2}{\partial x} - \frac{\partial \zeta_1}{\partial x} = \frac{\rho_2}{\rho_1}\frac{\partial \zeta_2}{\partial x} \qquad \frac{\partial \zeta_1}{\partial x} = -\frac{\rho_2-\rho_1}{\rho_1}\frac{\partial \zeta_2}{\partial x} \qquad (4\text{-}202)$$

and similarly for the y variation. We then have for P,

$$P = \int_{\zeta_1}^{d} p\,dz = \int_{\zeta_1}^{\zeta_2} p_1\,dz + \int_{\zeta_2}^{d} p_2\,dz$$

$$= \tfrac{1}{2}\rho_1 g(\zeta_2 - \zeta_1)^2 + \rho_1 g(\zeta_2 - \zeta_1)(d - \zeta_2) + \tfrac{1}{2}\rho_2 g(d - \zeta_2)^2 \qquad (4\text{-}203)$$

where we have used p_1 in the form given by (4-57). For $\partial P / \partial x$ we have in turn from (4-203), using (4-202),

$$\frac{\partial P}{\partial x} = \rho_2 g(\zeta_2 - \zeta_1) \frac{\partial \zeta_2}{\partial x} + \rho_2 g(d - \zeta_2) \frac{\partial \zeta_2}{\partial x} - \rho_1 g(\zeta_2 - \zeta_1) \frac{\partial \zeta_2}{\partial x} - \rho_2 g(d - \zeta_2) \frac{\partial \zeta_2}{\partial x}$$

$$= (\rho_2 - \rho_1) g h \frac{\partial h}{\partial x} = \tfrac{1}{2}(\rho_2 - \rho_1) g \frac{\partial h^2}{\partial x} \tag{4-204}$$

where we have made the usual approximation of $\partial h / \partial x$ for $\partial \zeta_2 / \partial x$. From (4-204) and the similar expression for the y variation, we then have

$$h^2 = \frac{2}{(\rho_2 - \rho_1) g} P \tag{4-205}$$

or simply that the integrated pressure can be taken to represent the thickness of the river flow. It is of interest to note that, had we taken alternative conditions for the underlying coastal water, we would still in general have obtained a functional relation between h and P.

We are now in a position to examine the variation in the thickness of the river flow h away from the river mouth. Let us first examine the case in which the Coriolis parameter is zero. From relation (4-198) or (4-199) we see that, at any distance x away from the source, h vanishes when

$$y^2 - x^2 = l^2 \tag{4-206}$$

In other words the lighter river water fills only the volume between the hyperbolic branches of (4-206) and $x = 0$. For any distance x away from the river mouth h is a maximum along the line $y = 0$ and decreases to zero out to the branches of (4-206). The flow resembles that shown in Fig. 4-21.

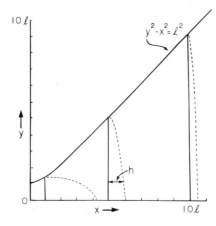

Fig. 4-21 Reproduced with permission from K. Takano, 1954, *Journal of the Oceanographical Society of Japan*, **10**, 92–98, Fig. 2.

Including the Coriolis term, we see again from (4-198) or (4-199) that its effect is to modify the simple symmetric solution represented by (4-206) in such a manner that the river outflow is displaced to the right in the northern hemisphere and further that the flow is thicker on the right hand side than on the left hand side. Using the dimensionless parameter R defined from relation (4-199) as

$$R = \frac{fl^2}{N_h} \tag{4-207}$$

as an index of the relative importance of the horizontal viscosity and Coriolis terms, Fig. 4-22 is a plot of the surface limits of the river flow for

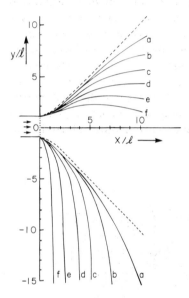

Fig. 4-22   Reproduced with permission from K. Takano, 1955, *Journal of the Oceanographical Society of Japan*, **11**, 147–149, Fig. 2.

various values of R, given by $R = 1, 2, 4, 8, 16$, and $32/500$ for the curves a, b, c, d, e, and f, respectively. In terms of a characteristic value of $f = 10^{-4} \sec^{-1}$ the corresponding values of N_h for various values of the width of the river mouth $2l$ are given in Table 4-1. For values of N_h found in practice and for typical river mouth widths we might expect to observe a noticeable deflection of the river flow to the right in the northern hemisphere. The flow of the river water into the sea at the mouth of the river is shown schematically in Fig. 4-23, and conditions actually found in nature should correspond reasonably well to this.

In conclusion we consider one of the classical examples of lateral effects on currents. We are concerned here with currents produced by the drag

TABLE 4-1

$2l$ in m	a	b	c	d	e	f
200	5.0×10^6	2.5×10^6	1.3×10^6	6.3×10^5	3.1×10^5	1.6×10^5
600	4.5×10^7	2.2×10^7	1.1×10^7	5.6×10^6	2.8×10^6	1.4×10^6
1000	1.3×10^8	6.2×10^7	3.1×10^7	1.6×10^7	7.8×10^6	3.9×10^6
2000	5.0×10^8	2.5×10^8	1.3×10^8	6.3×10^7	3.1×10^7	1.6×10^7

[a] From Defant (1961); N_h given in square centimeters per second.

of the wind passing over the ocean surface under the influence of internal frictional resistance and the Coriolis force. We assume that the sea surface is level and that the water is homogeneous, so that there are no horizontal pressure gradients. The only driving force is the wind stress applied at the ocean surface. We also assume, as we have done before, that the vertical internal frictional stresses are much larger than the horizontal internal frictional stresses. We further assume that the inertial terms can be neglected and that we are dealing with a coastal water body of large longitudinal and lateral extent. Such currents are referred to as

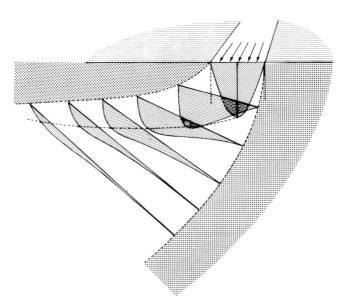

Fig. 4-23 Reproduced with permission from A. Defant, 1961, *Physical Oceanography*, Vol. I, Pergamon Press, New York, Fig. 251A.

drift currents. The original derivation was given by Ekman (1905), concerning which the reader may refer to Defant (1961) and Neumann and Pierson (1966).

For a steady state drift current the equations of motion reduce to, from (4-184) and (4-186),

$$\frac{\partial^2 v_x}{\partial z^2} = \frac{2\omega_0 \sin \varphi}{N_z} v_y$$

and (4-208)

$$\frac{\partial^2 v_y}{\partial z^2} = -\frac{2\omega_0 \sin \varphi}{N_z} v_x$$

If we assume that v_x and v_y are functions of z only and let the complex variable w be defined by

$$w = v_y + iv_x \qquad (4\text{-}209)$$

we obtain, multiplying the first equation in (4-208) by the imaginary i and adding,

$$\frac{d^2 w}{dz^2} = \frac{2i\omega_0 \sin \varphi}{N_z} w \qquad (4\text{-}210)$$

The form of solution of this ordinary differential equation is simply

$$w = e^{\pm(1+i)\alpha z} \qquad (4\text{-}211)$$

where

$$\alpha^2 = \frac{\omega_0 \sin \varphi}{N_z} \qquad (4\text{-}212)$$

We assume that the steady state drift current is maintained by a wind directed in the positive x direction at the sea surface. Such a wind produces a tangential stress T in the ocean at the sea surface given by

$$T = -\rho N_z \frac{\partial v_x}{\partial z}$$

$$z = 0 \qquad (4\text{-}213)$$

$$0 = -\rho N_z \frac{\partial v_y}{\partial z}$$

For our first example we assume that the water depths are sufficiently great that we can take for our lower boundary condition that the velocity components v_x and v_y vanish at great depths. From this lower boundary condition the solution (4-211) reduces to

$$w = Ae^{-(1+i)\alpha z} \qquad (4\text{-}214)$$

In terms of (4-209) the upper boundary condition becomes

$$-iT = \rho N_z \frac{dw}{dz} \qquad (4\text{-}215)$$

Substituting (4-214) into (4-215), we obtain

$$A = \frac{iT}{(1+i)\alpha\rho N_z} = \frac{(1+i)T}{2\alpha\rho N_z} \tag{4-216}$$

or

$$w = \frac{(1+i)T}{2\alpha\rho N_z} e^{-(1+i)\alpha z} \tag{4-217}$$

Separating into real and imaginary parts, we then have for v_x and v_y,

$$\begin{aligned} v_x &= \frac{T}{2\alpha\rho N_z} e^{-\alpha z}(\cos \alpha z - \sin \alpha z) \\ &= \frac{T}{\sqrt{2}\alpha\rho N_z} e^{-\alpha z} \sin\left(\frac{\pi}{4} - \alpha z\right) \\ &= v_0 e^{-\alpha z} \sin\left(\frac{\pi}{4} - \alpha z\right) \end{aligned} \tag{4-218}$$

and

$$\begin{aligned} v_y &= \frac{T}{2\alpha\rho N_z} e^{-\alpha z}(\cos \alpha z + \sin \alpha z) \\ &= \frac{T}{\sqrt{2}\,\alpha\rho N_z} e^{-\alpha z} \cos\left(\frac{\pi}{4} - \alpha z\right) \\ &= v_0 e^{-\alpha z} \cos\left(\frac{\pi}{4} - \alpha z\right) \end{aligned} \tag{4-219}$$

where v_0 is given by

$$v_0 = \frac{T}{\sqrt{2}\alpha\rho N_z} = \frac{T}{(2\rho^2 N_z \omega_0 \sin \varphi)^{1/2}} \tag{4-220}$$

We see that, under these conditions of great water depths, at the sea surface the velocity of a drift current has a magnitude v_0 and is directed at 45° to the right of the wind direction in the northern hemisphere. With increasing depth the angle of deflection increases and the magnitude of the velocity decreases. We may show this diagrammatically as in Fig. 4-24, which when drawn to scale is known as the *Ekman spiral*.

It is sometimes useful to write the results (4-218) and (4-219) in terms of the parameter D rather than α, where D is defined by

$$D = \frac{\pi}{\alpha} = \pi\left(\frac{N_z}{\omega_0 \sin \varphi}\right)^{1/2} \tag{4-221}$$

The results (4-218) and (4-219) become

$$v_x = v_0 e^{-\pi z/D} \sin\left(\frac{\pi}{4} - \frac{\pi z}{D}\right) \tag{4-222}$$

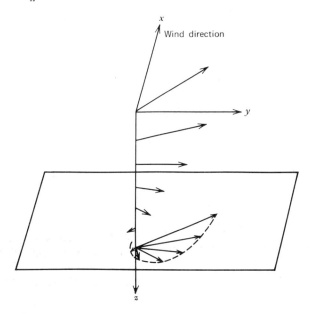

Fig. 4-24

and

$$v_y = v_0 e^{-\pi z/D} \cos\left(\frac{\pi}{4} - \frac{\pi z}{D}\right) \tag{4-223}$$

At the depth $z = D$, the current is in a direction π to the current at the surface and of a magnitude $e^{-\pi} = \frac{1}{23}$ of the surface value. The quantity D then is a convenient index of the effective depth of a drift current and is referred to as the *frictional depth*. For typical values of N_z and at midlatitudes D has values of 50–100 m.

As in most cases we are dealing with water depths less than the value of D, we include a bottom boundary condition other than the one used above. We take the simplified bottom boundary condition

$$v_x = v_y = 0 \qquad z = d \tag{4-224}$$

where d is the bottom depth. Rewriting the solution (4-211) in terms of hyperbolic functions, this solution is, after some algebra,

$$v_x = A \cosh \alpha(d-z) \sin \alpha(d-z) + B \sinh \alpha(d-z) \cos \alpha(d-z) \tag{4-225}$$

and

$$v_y = A \sinh \alpha(d-z) \cos \alpha(d-z) - B \cosh \alpha(d-z) \sin \alpha(d-z) \tag{4-226}$$

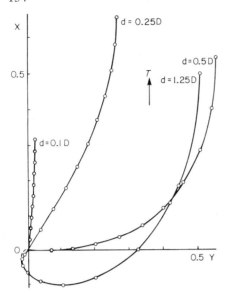

Fig. 4-25 Reproduced with permission from A. Defant, 1961, *Physical Oceanography*, Vol. I, Pergamon Press, New York, Fig. 170.

where

$$A = \frac{T}{\alpha \rho N_z} \frac{\cosh \alpha d \cos \alpha d + \sinh \alpha d \sin \alpha d}{\cosh 2\alpha d + \cos 2\alpha d} \qquad (4\text{-}227)$$

and

$$B = \frac{T}{\alpha \rho N_z} \frac{\cosh \alpha d \cos \alpha d - \sinh \alpha d \sin \alpha d}{\cosh 2\alpha d + \cos 2\alpha d} \qquad (4\text{-}228)$$

The curves in Fig. 4-25 are computed from (4-225) and (4-226) for different ratios of $d/D = \alpha d/\pi$. The 10 small circles on each curve represent the end points of the velocity vectors for the depths 0.0, 0.1d, 0.2d, ..., 0.9d. It is apparent from these curves or from solutions (4-225) and (4-226) that, as d/D becomes substantially less than unity, the bottom has increasingly more influence on the velocity structure of the drift current. In particular, the magnitude of the velocity at any given depth becomes smaller, and the angle of deflection of the surface current away from the wind direction decreases. At small depths of $d \leqslant 0.1D$ the movement shows essentially no effect of the earth's rotation.

CHAPTER 5

Mixing

5-1. SIMPLIFIED MIXING CONCEPTS

One of the more important physical oceanographic processes that occur in an estuary, at least for environmental considerations, is mixing. As mentioned briefly in the previous chapters and discussed in detail in this chapter, mixing takes place through various advection, circulation, and diffusion processes. A thorough knowledge of the natural mixing characteristics of any given estuary is essential to the engineering considerations of the dispersion of a potential thermal, chemical, or biological pollutant.

To start we consider some rather simplified mixing concepts. Here we look at the simple mass, or volume, continuity relations and the salt, freshwater, or conservative concentrate continuity relations. Generally, we consider that the vertical mixing has been sufficient that the waters may be assumed to have the same characteristics over a vertical section normal to the flow. These simplified mixing concepts are not only useful in themselves in giving some indication of the degree of mixing that may occur in any given estuarine situation, but are often sufficient in themselves for application to a particular engineering problem.

To begin let us return to the concept of flushing time discussed in Section 2-4. *Flushing time*, for our purposes, is defined as the time required to replace the existing fresh water in an estuary at a rate equal to the river discharge. If we designate the total river runoff flow, as before, by R, and the total freshwater volume of the estuary by V_f, the flushing time t, will be simply

$$t = \frac{V_f}{R} \tag{5-1}$$

Now the fraction of fresh water f at any given location in an estuary is given in terms of the salinity s at the same location by

$$f = \frac{\sigma - s}{\sigma} \tag{5-2}$$

where σ is the normal ocean salinity of the coastal waters into which the estuary empties. Then the volume of fresh water in an estuary is given by

155

the integral

$$V_f = \int f \, dV = \bar{f} V \qquad (5\text{-}3)$$

where the integral is taken over the entire volume of the estuary. The quantity V in the third expression of (5-3) is the total volume of the estuary, and the average value $\bar{f}$ is as defined by the integral relation of (5-3). From (5-3) and (5-1) we then have simply for the flushing time,

$$t = \frac{\bar{f} V}{R} \qquad (5\text{-}4)$$

in terms of the total volume of the estuary, the river runoff flow, and the average fraction of fresh water in the estuary.

We consider these relations a step further in terms of an *equivalent downestuary transport* Q_d. This equivalent downestuary transport is a fictitious quantity; only under unusual conditions is it a measurable quantity. It is an expression in terms of an advection quantity of the combined advection, circulation, and diffusion effects that take place in an estuary. Consider Fig. 5-1. At the upper end of the estuary where the

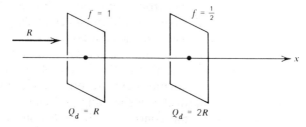

Fig. 5-1

water is entirely fresh and the river runoff flow is R, the equivalent downestuary transport is the same as R. At some further location down the estuary enough of the mixture must escape during each tidal cycle to remove a volume of fresh water equivalent to the total river flow over that tidal cycle. As shown in Fig. 5-1 for a location downestuary where the freshwater fraction is one half, the equivalent downestuary transport must be twice the river runoff flow. We then have the simple relation between Q_d and R,

$$Q_d = \frac{R}{f} \qquad (5\text{-}5)$$

The ratio $Q_d/R = 1/f$ thus is a measure of the combined dispersion effects of advection, circulation, and diffusion in removing a pollutant in an estuary over the simple advection effect.

Another simplified concept that has been applied in the past is the *tidal prism* method. In this method it is assumed that on the flood tide the volume of sea water V_P entering the estuary is entirely of oceanic salinity σ and that it is completely mixed with a corresponding volume of fresh water V_R as measured over the entire tidal cycle. It is further assumed that this entire quantity of mixed water is completely removed from the estuary on the ebb tide and that on the next flood tide the process is repeated with sea water of oceanic salinity entering the estuary. We may envisage this process from Fig. 5-2. The average salinity $\bar{s}$ of the mixed

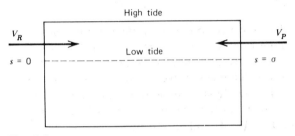

Fig. 5-2

water at high tide is then given by the relation

$$(V_P+V_R)\bar{s} = V_P\sigma$$

or

$$\bar{s} = \frac{V_P}{V_P+V_R}\sigma \qquad (5\text{-}6)$$

From the expression (5-2) the average fraction of fresh water $\bar{f}$ is then given by

$$\bar{f} = 1 - \frac{\bar{s}}{\sigma} = \frac{V_R}{V_P+V_R} \qquad (5\text{-}7)$$

From (5-4) the tidal prism flushing time t_T is then given by

$$t_T = \frac{\bar{f}V}{R} = \frac{V}{V_R/T}\frac{V_R}{V_P+V_R} = \frac{V}{P}T \qquad (5\text{-}8)$$

where T is, as before, the period of the semidiurnal or diurnal tide, and P is the observed intertidal volume, or tidal prism, given by

$$P = V_P+V_R \qquad (5\text{-}9)$$

Now, in general we do not expect complete mixing during each tidal cycle for the entire estuary, and also we anticipate that some of the mixed water will return on each succeeding flood tide. Calculated flushing times t_T are then always less than the observed flushing times t; and we may consider t_T an optimum value.

We may consider these relations alternatively in terms of a flushing rate rather than a flushing time. The flushing rate F is defined as the rate at which the total volume of the estuary is exchanged. From this definition and (5-4) we then have

$$F = \frac{V}{t} = \frac{R}{\bar{f}} \tag{5-10}$$

and, from (5-8),

$$F_T = \frac{V}{t_T} = \frac{P}{T} \tag{5-11}$$

As discussed in the previous paragraph we anticipate, correspondingly, that the flushing rates F_T will be greater than the flushing rates F.

If we perform a dye diffusion experiment in an estuary, or a section of an estuary, in which a mass of dye is released at a constant rate D, we will then have under steady state conditions the same relations that we had for freshwater flow, so that the observed flushing time and flushing rate will be given from (5-4) and (5-10) by

$$t = \frac{\bar{c}\rho V}{D} \tag{5-12}$$

and

$$F = \frac{D}{\bar{c}\rho} \tag{5-13}$$

where $\bar{c}$ is the average dye concentration in the estuary expressed in units of mass of dye per unit mass of sea water (similar to the units used for salinity), and ρ is the density of the estuary waters.

We may make a modification in the tidal prism method by dividing the estuary into segments over which the mixing takes place rather than assuming that there is complete mixing over the length of the entire estuary during each tidal cycle. This development was given by Ketchum (1951). In it he assumes that the segment over which the mixing may be considered complete corresponds in an approximate manner to the excursion of a water particle during the tide. Here again of course there is the assumption, which is not necessarily valid, that the small scale turbulence is sufficiently great that there will be complete mixing over the tidal excursion.

The innermost segment of the estuary is defined as that for which the entire intertidal volume P_0 is supplied by the river flow over a tidal cycle,

$$P_0 = V_R \tag{5-14}$$

The low tide volume of this innermost, or zero, segment is designated V_0. For the succeeding segments downestuary the assumption is then made that the high tide volume of a landward segment is equal to the low tide volume of the adjacent seaward segment, as though the entire volume of the seaward segment was pushed in the manner of a piston action to fill the high tide volume of the landward segment on the flood tide. We then have

$$V_1 = V_0 + P_0$$
$$V_2 = V_1 + P_1 = V_0 + P_0 + P_1$$
$$V_3 = V_2 + P_2 = V_0 + P_0 + P_1 + P_2 \qquad (5\text{-}15)$$
$$\cdots\cdots\cdots\cdots\cdots\cdots\cdots\cdots$$
$$V_n = V_0 + \sum_{n=0}^{n-1} P_n$$

or

$$V_n = V_0 + V_R + \sum_{n=1}^{n-1} P_n \qquad (5\text{-}16)$$

The limits of each volume segment, as defined above, are then an estimate of the average excursion of a particle of water on the flood tide. If it is next assumed that the water within each segment is completely mixed at high tide, the proportion of water removed on the ebb tide will be given by the ratio between the local intertidal volume and the high tide volume of the segment. Thus an *exchange ratio* r_n can be defined for each segment n,

$$r_n = \frac{P_n}{P_n + V_n} \qquad (5\text{-}17)$$

From (5-8) we see that the tidal flushing time for the given segment expressed in tidal cycles is simply the reciprocal of the exchange ratio,

$$t_n = \frac{T}{r_n} \qquad (5\text{-}18)$$

As we have assumed that mixing is complete within each volume segment, the flushing time of (5-18) should correspond with that of (5-1) from which we then have for the accumulated fresh water in each volume segment V_{fn},

$$V_{fn} = t_n R = \frac{T}{r_n} \frac{V_R}{T} = \frac{V_R}{r_n} \qquad (5\text{-}19)$$

From (5-19) and (5-3) we can then calculate the freshwater fraction f_n of each segment, and from (5-2) the salinity s_n of each segment. The total flushing time for the entire estuary with this modified tidal prism method is the sum of the flushing times for the separate segments.

We may arrive at the expression (5-19) through another, more physical line of reasoning, which is instructive in itself. If the river flow is constant, each segment will receive a volume V_R of river water per tidal cycle. The amount of river water removed on the ebb tide is $r_n V_R$, and the amount remaining is $(1-r_n)V_R$. As this process has already been going on for many tidal cycles, there will be contributions from the river flow at those times both to the water removed and to that remaining. This can be summarized as shown in Table 5-1. The total volume of river water

TABLE 5-1

Age in tidal cycles	River water removed	River water remaining
1	$r_n V_R$	$(1-r_n)V_R$
2	$r_n(1-r_n)V_R$	$(1-r_n)^2 V_R$
3	$r_n(1-r_n)^2 V_R$	$(1-r_n)^3 V_R$
. . . .		
m	$r_n(1-r_n)^{m-1} V_R$	$(1-r_n)^m V_R$

accumulated in the segment is the sum of the last column plus 1 volume of river flow at high tide which has not yet been removed,

$$V_{fn} = V_R[1+(1-r_n)+(1-r_n)^2+\cdots+(1-r_n)^m+\cdots]$$
$$= \frac{V_R}{r_n} \qquad (5\text{-}20)$$

as the bracketed expression in (5-20) is an infinite geometric progression whose sum is $1/r_n$. Similarly the amount of river water removed on each tidal cycle is the sum of the second column in Table 5-1, which is simply V_R, as expected.

Ketchum (1955) has further extended these simplified mixing concepts to estimate the distribution of a conservative pollutant in an estuary where the outfall for the pollutant may be at any given location along the estuary. If we had a simple river with advection only, the cross sectional average concentration of the pollutant at the outfall would be given from the continuity condition sketched in Fig. 5-3 as

$$W = c_0 R$$

or

$$c_0 = \frac{W}{R} \qquad (5\text{-}21)$$

where W is rate of supply of pollutant in mass of pollutant per unit time,

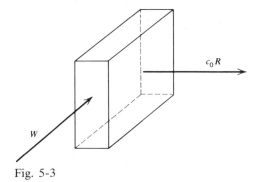

Fig. 5-3

and c_0 is the cross sectional average pollutant concentration at the outfall expressed in units of mass of pollutant per unit volume of sea water. Now, in the case of an estuary the simple river advection must be replaced by the equivalent downestuary transport of (5-5) at the outfall,

$$c_0 = \frac{W}{Q_{d0}} = \frac{W}{R} f_0 \tag{5-22}$$

where f_0 is the freshwater fraction at the outfall. Downstream of the outfall the pollutant must pass a cross section at the same rate it is discharged from the source,

$$c_x = \frac{W}{R} f_x = c_0 \frac{f_x}{f_0} \tag{5-23}$$

Upestuary of the outfall there is no net exchange past a cross section. The quantity carried upestuary by tidal diffusion and circulation is balanced by the quantity carried downestuary by the net river flow. This is the same criterion that applies to the salt distribution upestuary from the outfall, so that the upestuary distribution of a conservative pollutant is directly proportional to the upestuary salinity distribution,

$$c_x = c_0 \frac{S_x}{S_0} \tag{5-24}$$

From these two very simple and direct relations, (5-23) and (5-24), along with the source condition (5-22), we have a powerful method for estimating the distribution of a potential conservative pollutant in an estuary from a knowledge of the salinity distribution in an estuary.

These relationships are illustrated in Fig. 5-4. The solid curve is the freshwater fraction or salinity distribution for an estuary. Consider four possible outfall locations as shown with the same outfall rate $W/R = 1$ for each. Downestuary from each location the pollutant distribution follows

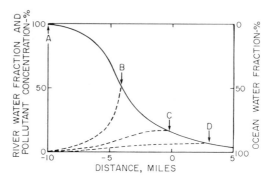

Fig. 5-4   Reproduced with permission from B. H. Ketchum, 1955, *Sewage and Industrial Wastes*, **27**, 1288–1296, Fig. 1.

the freshwater fraction curve. At the outfall the cross sectional average concentration is less for stations further down the estuary. Upestuary from each location the pollutant distribution follows the salinity curve or, equivalently, the reciprocal of the freshwater fraction curve; this is indicated by the dashed curves in Fig. 5-4. As the outfall is moved progressively further down the estuary, the concentration of pollutants upestuary from the outfall decreases while the concentration downestuary remains unchanged.

For further information on the material covered in this section the reader may refer to Ketchum (1950, 1951, 1951a, 1953, 1955), Stommel and Farmer (1952b), Pritchard (1957), and Dyer (1973).

5-2. OVERMIXING

We examined in some detail in Section 4-1 the critical flow conditions for a strongly stratified estuarine flow with a net circulation out in the upper layer and a net circulation in in the lower layer. As discussed in Sections 4-1 and 2-11 we anticipate that such control conditions, if they occur, will be associated with the abrupt change in opening at the mouth of the estuary. We recall that the control itself is associated with a maximum flow condition under the limitation of available hydraulic energy.

There is another effect associated with such a control at an estuary mouth, which we should like to examine here. As the control provides a restriction on the net volume flow out, it also provides through the volume continuity relation a restriction on the net volume flow of sea water in. This in effect limits the amount of salt water available for mixing inside the estuary or bay.

We may envisage that, as the mixing proceeds inside the estuary through whatever processes are dominant, more salt water is added to the net circulation out the estuary and the volume flow out, and the salinity increases up to the critical condition. Beyond this any increased mixing has no further effect on either the discharge flow or on the exiting salinity. We refer to this as a state of *overmixing*. The developments to be discussed here were originally presented by Stommel and Farmer (1953).

We may arrive at a solution for the relations associated with such an overmixing condition from the critical flow equation (4-13),

$$\frac{q_1^2}{h_1^3} - g\beta + \frac{q_2^2}{h_2^3} = 0 \tag{5-25}$$

the volume continuity relation (4-7),

$$q_1 = q_2 + r \tag{5-26}$$

and the salt continuity relation,

$$q_1 s_1 = q_2 s_2 \tag{5-27}$$

where s_1 and s_2 are, respectively, the salinities of the circulation flow out in the upper layer and the circulation flow in in the lower layer. We may easily solve (5-26) and (5-27) for q_1 and q_2 in terms of s_1, s_2, and r, obtaining

$$q_1 = \frac{s_2}{s_2 - s_1} r \tag{5-28}$$

and

$$q_2 = \frac{s_1}{s_2 - s_1} r \tag{5-29}$$

Substituting (5-28) and (5-29) into (5-25), we obtain

$$\frac{s_2^2}{(s_2 - s_1)^2 n^3} + \frac{s_1^2}{(s_2 - s_1)^2 (1-n)^3} = \frac{g\beta h^3}{r^2} \tag{5-30}$$

where n is the fractional water depth defined so that

$$h_1 = nh \tag{5-31}$$

and, from (4-8),

$$h_2 = (1-n)h \tag{5-32}$$

Now, for our purposes we may take the density to be proportional to the salinity, so that we can write for the density ρ of water of salinity s,

$$\rho = \rho_0(1 + as) \tag{5-33}$$

which then gives us for an approximate value of β of (4-12),

$$\beta = \frac{\rho_2 - \rho_1}{\rho_2} = a(s_2 - s_1) \tag{5-34}$$

Defining v by

$$v = \frac{s_1}{s_2} \tag{5-35}$$

and substituting (5-34) and (5-35) into (5-30), we then have

$$\varphi(n) = v^2 n^3 + (1-n)^3 - (1-v)^3 n^3 (1-n)^3 F^{-1} = 0 \tag{5-36}$$

where

$$F = \frac{r^2}{gas_2 h^3} \tag{5-37}$$

A graph of $\varphi(n)$ versus n for a given value of v resembles that shown in Fig. 5-5. The possible solutions to (5-36) then of course correspond to

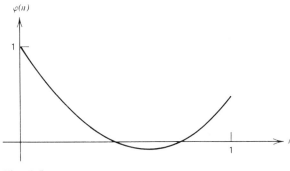

Fig. 5-5

the intersections of this curve with the n axis. As $v \to 1$, however, the real roots cease to exist. The cutoff point, i.e., the point of overmixing, is where there is a double root. The value of n at which there is a double root occurs when

$$\frac{\partial \varphi}{\partial n} = v^2 n^2 - (1-n)^2 - (1-v)^3 n^2 (1-n)^2 (1-2n) F^{-1} = 0 \tag{5-38}$$

in addition to (5-36). Multiplying equation (5-38) by n and subtracting from (5-36) we obtain the relation

$$(1-v)^3 n^4 = F \tag{5-39}$$

or

$$v = \frac{s_1}{s_2} = 1 - \sqrt[3]{\frac{F}{n^4}} \tag{5-40}$$

or

$$\left(\frac{s_2 - s_1}{s_2}\right)^3 = \frac{F}{n^4} \tag{5-41}$$

Substituting (5-39) into (5-36) or (5-38) to eliminate ν, we obtain a relation for n,

$$Fn^2 = (2n - 1)^3 \tag{5-42}$$

For weak river flows as $F \to 0$, we see from (5-42) that $n \to \frac{1}{2}$. Substituting this value of n back into (5-39) and (5-28) or directly into (5-25) gives relation (4-18) for the circulation flow. For most examples it is sufficient to use a value of $n = \frac{1}{2}$ in the applications of relations (5-39), (5-40), or (5-41). When a correction to $n = \frac{1}{2}$ is needed or in general to investigate the effect of $F > 0$, we can introduce the relation

$$n = \tfrac{1}{2} + \epsilon \tag{5-43}$$

into (5-42), where ϵ is a small quantity as long as F is small. Substituting and retaining only the large terms, we obtain

$$\frac{F}{4} = 8\epsilon^3$$

or

$$\epsilon = \sqrt[3]{\frac{F}{32}} \tag{5-44}$$

We can apply relation (5-40) or (5-41), which contains only physically observable quantities or constants, with a value of $n = \frac{1}{2}$ or the correction given by (5-44) if needed to see if the conditions at the mouth of a given estuary or bay represent overmixing.

5-3. ENTRAINMENT MIXING

We examine here in a somewhat simplified manner the entrainment mixing effects associated with two types of flow. In the first we consider the fjord type entrainment flow of Section 4-2. We take a stratified flow for which the velocity v_2 in the lower layer is sufficiently small that it may be neglected and for which the density ρ_2 and salinity s_2 of the lower layer are constant. We include upward movement of the deep water into the top layer, or entrainment, with a vertical entrainment velocity w much smaller than the longitudinal flow velocity v_1 in the upper layer. We assume, as before, that mixing in the upper layer is so strong vertically that v_1, ρ_1, and s_1 may be considered independent of the depth coordinate z. For the second we consider a strongly stratified estuarine flow. Here we consider that there is a substantial circulation flow in both the upper and lower layers. Again we assume that mixing is sufficiently strong in both layers across the interface that the velocity, density, and salinity in each layer may be considered constant as a function of depth. Because of the simplifications in each case not only in the model itself but also in the formulation of the solutions, there is probably little direct applicability of the results of this section; however, they are of interest in showing the form of solution and type of variations. The developments to be discussed here were originally presented by Stommel (1951) and Stommel and Farmer (1952b).

Our emphasis in this chapter is on the effects of entrainment, or interfacial, mixing on the various physically observable quantities in an estuary. It does not include in itself a discussion of the interesting theoretical and experimental investigations that have been made of the phenomena of interfacial instability and the resultant mass transfer across the interface, which we refer to collectively as entrainment mixing, but only the gross results of such mass transfer. For such discussions the reader may refer to Keulegan (1949), Lofquist (1960), Macagno and Rouse (1962), and Vreugdenhill (1969), among others.

For the first example, with reference to Fig. 5-6, we have for the volume continuity relation, which we take as an approximation to the

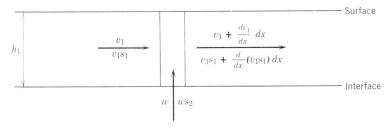

Fig. 5-6

mass continuity relation,

$$h_1\left(v_1 + \frac{dv_1}{dx}\,dx\right) = h_1 v_1 + w\,dx$$

or

$$\frac{dv_1}{dx} = \frac{w}{h_1} \tag{5-45}$$

and for the salt continuity relation

$$h_1\left[v_1 s_1 + \frac{d}{dx}(v_1 s_1)\,dx\right] = h_1 v_1 s_1 + w s_2\,dx$$

or

$$\frac{d}{dx}(v_1 s_1) = \frac{w s_2}{h_1} \tag{5-46}$$

Substituting (5-45) into (5-46), we then have

$$\frac{ds_1}{dx} = \frac{w(s_2 - s_1)}{v_1 h_1} \tag{5-47}$$

Now, from Keulegan (1949) we have the empirical relation from physical model experiments that the entrainment velocity w is given in terms of the longitudinal velocity of the upper layer by

$$w = C(v_1 - 1.15 v') \tag{5-48}$$

where C is a constant of value $C = 3.5 \times 10^{-4}$. For the fjord conditions usually encountered v_1 is sufficiently greater than v' that we may approximate (5-48) by

$$w = C v_1 \tag{5-49}$$

Substituting (5-49) into (5-47) and substituting f for s_1, where f is defined by the relation

$$f = \frac{s_2 - s_1}{s_2} \tag{5-50}$$

so that in our case, for $s_2 = \sigma$, the constant sea water salinity f represents the freshwater fraction in the upper layer, we then have

$$\frac{df}{dx} = -\frac{C}{h_1} f \tag{5-51}$$

whose solution is simply, assuming h_1 to be constant

$$f = f_0 e^{-(C/h_1)x} \tag{5-52}$$

where, if the origin, $x = 0$, is taken at the upper, freshwater end of the estuary, $f_0 = 1$. We then see that, in this simplified case, the freshwater fraction in the upper layer should vary in an exponential manner from unity at the upper end of the estuary toward a value of zero at its mouth.

We can extend this analysis to a control condition at the mouth of the estuary, which may be applicable to a fjord whose breadth remains fairly restricted along its length and which opens abruptly to the ocean at its mouth. As the flow in the lower layer q_2 is small, condition (4-16) applies. Then, from (4-16), (5-28), and (5-34) we have

$$r^2 = g a s_2 h_1{}^3 \left(\frac{s_2 - s_{1l}}{s_2} \right)^3 \tag{5-53}$$

where s_{1l} is the salinity of the upper layer at the mouth of the fjord, $x = l$, and $s_2 = \sigma$ is a constant. The similarity of (5-53) and (5-41) for the overmixing case should be noted. From (5-52) and (5-50) we rewrite (5-53) as

$$r^2 = gas_2 h_1{}^3 e^{-3(C/h_1)x} \tag{5-54}$$

for h_1 in terms of r. We see that as the river flow r increases the thickness of upper layer h_1 increases. For small river flows h_1 is small, and the freshwater fraction f decreases rapidly from the head of the estuary. As river flow increases, the thickness h_1 increases and the exponential decrease in f becomes more gradual. We may also rewrite (5-53) in terms of r and s_{1l}. We have from (5-52) and (5-50),

$$-\frac{C}{h_1} l = \log\left(1 - \frac{s_{1l}}{s_2}\right)$$

$$\frac{1}{h_1{}^3} = \frac{1}{Cl}\left[\left(\frac{s_{1l}}{s_2}\right)^3 + \cdots\right]$$

using the expansion for the natural logarithm and remembering that s_{1l}/s_2 is less than unity so that (5-53) becomes

$$r^2 = gas_2 Cl\left(\frac{s_2 - s_{1l}}{s_{1l}}\right)^3 \tag{5-55}$$

For a control condition at the mouth of the fjord and an entrainment mixing as defined by relation (5-52) we then anticipate that the cube of the salinity difference ratio on the right hand side of (5-55) varies as the square of the river flow.

For the second example, with reference to Fig. 5-7, we have for the volume continuity relation, which we take as before as an approximation

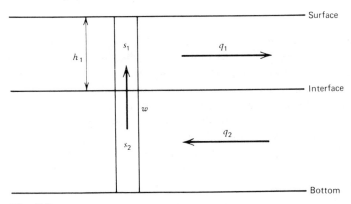

Fig. 5-7

to the mass continuity relation,

$$\frac{dq_1}{dx} = -\frac{dq_2}{dx} = w \tag{5-56}$$

where q_1 and q_2 are the discharges per unit width, and where for convenience we have taken, as before, the entrainment velocity w to be positive in an upward direction. Similarly, for the salt continuity relation we have

$$\frac{d}{dx}(q_1 s_1) = -\frac{d}{dx}(q_2 s_2) = w s_2 + \kappa(s_2 - s_1) \tag{5-57}$$

where we have also included an expression $\kappa(s_2 - s_1)$ to represent the turbulent, or diffusive, salt exchange across the interface. To obtain a simple solution we next make the rather gross assumption that both the entrainment velocity w and the coefficient of turbulent exchange κ are constant along the entire length of the estuary. We can then obtain a solution for s_1 and s_2 in terms of q_1 as the independent variable. Instead we assume that the thickness of the upper layer h_1 is also constant along the entire length of the estuary and obtain a solution directly in terms of x as the independent variable.

With these assumptions in mind and taking our origin, $x = 0$, at the upper freshwater end of the estuary, we can integrate equations (5-56) directly, obtaining

$$q_1 = r + wx \tag{5-58}$$

and

$$q_2 = -wx \tag{5-59}$$

where r is the river discharge per unit width. Substituting (5-56) into (5-57), we have

$$q_1 \frac{ds_1}{dx} = w(s_2 - s_1) + \kappa(s_2 - s_1) \tag{5-60}$$

and

$$q_2 \frac{ds_2}{dx} = -\kappa(s_2 - s_1) \tag{5-61}$$

Multiplying equation (5-60) by q_2 and (5-61) by q_1 and subtracting, we obtain

$$\frac{d}{dx}(s_2 - s_1) = -\left(\frac{w+\kappa}{q_1} + \frac{\kappa}{q_2}\right)(s_2 - s_1) \tag{5-62}$$

or

$$\frac{d(s_2 - s_1)}{s_2 - s_1} = -\left(\frac{1+\gamma}{x'+x} - \frac{\gamma}{x}\right)dx \tag{5-63}$$

on substituting from (5-58) and (5-59), where γ and x' are defined by

$$\gamma = \frac{\kappa}{w} \tag{5-64}$$

and

$$x' = \frac{r}{w} \tag{5-65}$$

Equation (5-63) can then be integrated directly, giving

$$s_2 - s_1 = A \frac{x^\gamma}{(x+x')^{\gamma+1}} \tag{5-66}$$

where A is a constant of integration.

Substituting solution (5-66) back into (5-60), we have

$$\frac{ds_1}{dx} = A \frac{\gamma+1}{x+x'} \frac{x^\gamma}{(x+x')^{\gamma+1}} \tag{5-67}$$

which can also be integrated directly, giving

$$s_1 = \frac{A}{x'} \frac{x^{\gamma+1}}{(x+x')^{\gamma+1}} + B \tag{5-68}$$

where B is also a constant of integration. Our first boundary condition is that the upper layer is entirely fresh water at the upper end of the estuary,

$$s_1 = 0 \qquad x = 0 \tag{5-69}$$

This boundary condition then reduces (5-68) to

$$s_1 = \frac{A}{x'} \frac{x^{\gamma+1}}{(x+x')^{\gamma+1}} \tag{5-70}$$

From (5-66) and (5-70) we then have for s_2,

$$s_2 = \frac{A}{x'} \frac{x^{\gamma+1}}{(x+x')^{\gamma+1}} + A \frac{x^\gamma}{(x+x')^{\gamma+1}}$$

$$= \frac{A}{x'} \frac{x^\gamma}{(x+x')^\gamma} \tag{5-71}$$

Our second boundary condition is that the lower layer is entirely sea water at the lower end of the estuary,

$$s_2 = \sigma \qquad x = l \tag{5-72}$$

where σ is the salinity of the sea water and l is the total length of the estuary. Substituting (5-71) into (5-72), we have

$$A = \sigma \frac{x'(l+x')^\gamma}{l^\gamma} \tag{5-73}$$

Our final solutions for s_1 and s_2 are then, respectively,

$$s_1 = \sigma\left(1 + \frac{x'}{l}\right)^{\gamma}\left(1 + \frac{x'}{x}\right)^{-(\gamma+1)} \qquad (5\text{-}74)$$

and

$$s_2 = \sigma\left(1 + \frac{x'}{l}\right)^{\gamma}\left(1 + \frac{x'}{x}\right)^{-\gamma} \qquad (5\text{-}75)$$

At the mouth of the estuary, $x = l$, the value of s_1 is, from (5-74),

$$s_1 = \sigma\frac{l}{l + x'} \qquad (5\text{-}76)$$

In order to illustrate the properties of these solutions a numerical example is given in Fig. 5-8. It is assumed that the discharge of the upper

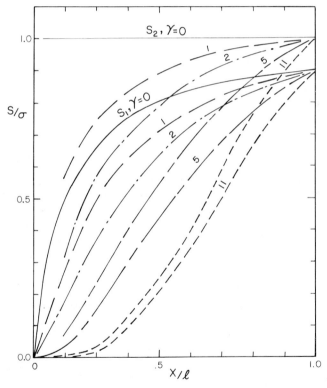

Fig. 5-8 Reproduced with permission from H. Stommel and H. G. Farmer, 1952, Reference 52–88, Woods Hole Oceanographic Institution, Woods Hole, Mass., Fig. 4.6.1.

layer at the mouth is 10 times the discharge of the fresh, river water at the head of the estuary, so that from (5-58), (5-65), and (5-76), $s_1 = 0.9\sigma$ at the estuary mouth.

In conclusion we consider as a third example the case in which there is no interface with uniform conditions above and below, but rather a zone near the surface in which there is a rapid change in longitudinal velocity from large positive, or downestuary, values followed by a region of smaller upestuary values which gradually decrease to zero at great depths. These conditions may be considered characteristic of the circulation flows in British Columbia and Alaska inlets, as contrasted with the conditions of the first example which may be considered an approximation to the summer flows in Norwegian fjords. In this derivation we follow Cameron (1951a), as extended by Rattray (1967) and Winter (1973) with some simplifications and adaptations.

In our equations of motion and continuity we include, as before, the inertial terms and the vertical velocity contributions. We also include here the eddy viscosity and diffusion terms. From (2-134), (2-135), (4-89), (4-120), (5-33), (2-27), and (2-65) we have

$$v_x \frac{\partial v_x}{\partial x} + v_z \frac{\partial v_x}{\partial z} = gi - ga \int_0^z \frac{\partial s}{\partial x} \, dz + N_z \frac{\partial^2 v_x}{\partial z^2} \tag{5-77}$$

$$\frac{\partial v_x}{\partial x} + \frac{\partial v_z}{\partial z} = 0 \tag{5-78}$$

$$v_x \frac{\partial s}{\partial x} + v_z \frac{\partial s}{\partial z} = K_z \frac{\partial^2 s}{\partial z^2} \tag{5-79}$$

where N_z and K_z are constant, and ρ, ρ_s, and ρ_0 are essentially the same. It is convenient to solve these equations in terms of the velocity potential ψ defined by

$$v_x = -\frac{\partial \psi}{\partial z}$$

and $\tag{5-80}$

$$v_z = \frac{\partial \psi}{\partial x}$$

which in themselves satisfy the continuity equation (5-78). We then have for (5-77) and (5-79),

$$\frac{\partial \psi}{\partial z} \frac{\partial^2 \psi}{\partial x \, \partial z} - \frac{\partial \psi}{\partial x} \frac{\partial^2 \psi}{\partial z^2} + N_z \frac{\partial^3 \psi}{\partial z^3} = gi - ga \int_0^z \frac{\partial s}{\partial x} \, dz \tag{5-81}$$

and

$$-\frac{\partial \psi}{\partial z} \frac{\partial s}{\partial x} + \frac{\partial \psi}{\partial x} \frac{\partial s}{\partial z} = K_z \frac{\partial^2 s}{\partial z^2} \tag{5-82}$$

We have that v_x vanishes at great depths, so that the integral of the last term in (5-81) taken to infinity must balance the surface slope term. Taking the derivative of (5-81) with respect to z, we obtain

$$\frac{\partial \psi}{\partial z}\frac{\partial^3 \psi}{\partial x \partial z^2} - \frac{\partial \psi}{\partial x}\frac{\partial^3 \psi}{\partial z^3} + N_z \frac{\partial^4 \psi}{\partial z^4} = -ga\frac{\partial s}{\partial x} \tag{5-83}$$

We next use the argument applied in Section 5-8 and consider a series solution expressed in terms of the small quantity a, so that to a first order solution (5-83) and (5-82) may be considered separately. We are concerned only with the motion equation (5-83) from here on, and we make an approximation for the longitudinal velocity profile and see what salinity distribution that approximation implies.

Having considerable liberty in the choice of the velocity function, we select the following form for ψ,

$$\psi = -\frac{v_s}{b}\frac{x}{l}e^{-bz}(bz + b^2 z^2) \tag{5-84}$$

We see that this implies that v_x varies linearly with x over the extent of the inlet being investigated, where the origin of x is taken at the head of the inlet, l is a characteristic length for the inlet, and v_s is a characteristic surface velocity associated with l. We also have from the exponential term that v_x decreases at great depth as required. The quantity $1/b$ is a depth scaling factor for the zone of high downestuary velocities near the surface. For an imposed condition of small net circulation or river runoff flow, in comparison with downestuary or upestuary flows at any depth, we see from (5-80) and the form of (5-84) that any polynomial in bz for the parentheses of (5-84) satisfies the condition that v_x integrated from the surface to infinity be zero. Finally, for a zero stress condition at the surface the constant term in the corresponding parentheses for $\partial v_x/\partial z$ must vanish; and the simplest polynomial for the parentheses of (5-84) satisfying this condition is the one given above. The resultant assumed velocity distribution as a function of depth is as shown in Fig. 5-9.

Next, in order to obtain a solution of (5-83) separable in x and z, we take s to be of the form

$$s = \sigma - \sigma\frac{l^2 - x^2}{l^2}\theta(z) \tag{5-85}$$

so that

$$\frac{\partial s}{\partial x} = \sigma\frac{2x}{l^2}\theta(z) \tag{5-86}$$

Substituting (5-84) and (5-86) into (5-83), we obtain after a certain

$(v_x/v_s)\,(x/\ell)$

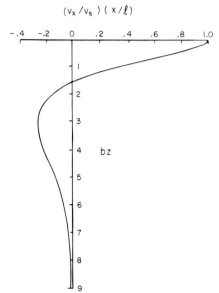

Fig. 5-9

amount of reduction,

$$\theta = \frac{bv_s}{2ga\sigma}\left[v_s e^{-2bz}(4b^2z^2) + N_z b^2 l e^{-bz}(8-7bz+b^2z^2)\right] \qquad (5\text{-}87)$$

The first term in the brackets of (5-87) may be considered the acceleration, or inertial, contribution to the salinity deviation θ, and the second term in the brackets the frictional, or eddy viscosity, contribution. The second term has a maximum at the surface and decreases rapidly with depth. The first term is zero at the surface, reaches a maximum at $z = 1/b$, and then decreases with depth. For a condition of low river runoff a small value of the depth scaling factor $1/b$ is implied, and we anticipate that the second term will be dominant, giving a salinity distribution similar to that of the type 2 inlet distribution described by Pickard (1961) and shown in Fig. 10-6. For larger river runoff conditions and larger values of the scaling factor $1/b$ we anticipate that the first term will also contribute, approaching a distribution similar to that of Pickard's type 1 inlet distribution, also shown in Fig. 10-6.

5-4. LONGITUDINAL DISPERSION ACROSS A VERTICAL SECTION

We are interested here in examining in a general way the dispersion of an index quantity, such as salinity, across a vertical section of an estuary. We use the term *dispersion* to include both diffusion effects from tidal mixing and velocity shear or circulation effects. We examine these two contributing effects separately and in detail in the following two sections.

From (4-135) we may represent the longitudinal velocity by

$$v_x = \bar{v}_x + v_{xt} = \bar{v}_x + C \cos \omega t \tag{5-88}$$

where $\bar{v}_x$ is the mean velocity over a tidal cycle at a particular depth, C is the amplitude of the tidal velocity variation taken as a constant, and $\omega = 2\pi/T$, T being, as before, the period of the diurnal or semidiurnal tide. Similarly, the salinity may be represented by

$$s = \bar{s} + s_t = \bar{s} - B \sin (\omega t - \delta) \tag{5-89}$$

where $\bar{s}$ is the mean salinity over a tidal cycle at a particular depth, B is the amplitude of the tidal salinity variation, and δ is the phase angle between the velocity and salinity variations over that for the simple progressive or standing wave of (3-32) and (3-33). We next separate $\bar{v}_x$ and $\bar{s}$ into depth mean values $\langle \bar{v}_x \rangle$ and $\langle \bar{s} \rangle$, and deviations from this mean v_{x1} and s_1, so that

$$\bar{v}_x = \langle \bar{v}_x \rangle + v_{x1} \tag{5-90}$$

and

$$\bar{s} = \langle \bar{s} \rangle + s_1 \tag{5-91}$$

where

$$\langle \bar{v}_x \rangle = \frac{1}{h} \int_0^h \bar{v}_x \, dz$$

and $\tag{5-92}$

$$\langle \bar{s} \rangle = \frac{1}{h} \int_0^h \bar{s} \, dz$$

Including finally the small scale turbulent fluctuations v_x' and s', we have from (5-88), (5-89), (5-90), and (5-91) for the representations of v_x and s,

$$v_x = \langle \bar{v}_x \rangle + v_{xt} + v_{x1} + v_x' \tag{5-93}$$

and

$$s = \langle \bar{s} \rangle + s_t + s_1 + s' \tag{5-94}$$

Now the salt flux per unit breadth S across a vertical section is the integral over the depth from the surface to the bottom of $v_x s$, and the average value of the salt flux over a tidal cycle $\bar{S}$ is then

$$\bar{S} = \frac{1}{T} \int_0^T \int_0^h v_x s \, dz \, dt \tag{5-95}$$

Looking at the individual cross product terms of (5-93) and (5-94) in this double integration, we see from (5-88) and (5-89) that all the terms involving either v_{xt} or s_t are zero except their product $v_{xt}s_t$. We also see from (5-90) and (5-91) that all the terms involving v_{x1} or s_1 are zero except their product $v_{x1}s_1$. Further, we may assume that there is no correlation between v_x' and s' and any of the other terms, so that we are left with only their product $v_x's'$. We then have finally for (5-95),

$$\bar{S} = h\langle\bar{v}_x\rangle\langle\bar{s}\rangle + h\overline{v_{xt}s_t} + h\langle\overline{v_{x1}s_1}\rangle + h\langle\overline{v_x's'}\rangle$$
$$= S_1 + S_2 + S_3 + S_4 \qquad (5\text{-}96)$$

The first term S_1 represents the contribution of $\bar{S}$ due to *advection* by the mean flow corresponding to the river discharge, since

$$\langle\bar{v}_x\rangle = \frac{R}{A} \qquad (5\text{-}97)$$

where R is the total river runoff flow, and A is the cross sectional area normal to the flow. The second term S_2 represents the *tidal diffusion* contribution. If, for example, the tidal velocity and salinity variations were $\pi/2$ out of phase so that $\delta = 0$, this contribution would be zero. The third term S_3 represents the *velocity shear* or *circulation* contribution related to the variation in the circulation velocity as a function of depth and the tidal averaged salinity as a function of depth. For example, in a stratified estuary v_{x1} is positive and s_1 is negative in the upper portion of the water column, and v_{x1} negative and s_1 positive in the lower portion of the water column, producing a finite value for S_3. The fourth term S_4 represents the small scale turbulent contribution and is usually negligible.

Under steady state conditions we have $\bar{S} = 0$. Continuing then this general analysis one step further, we see that for a well mixed estuary, which implies a large value for K_z, s_1 and consequently S_3 are small. In such a case we anticipate that the S_2 term will be the principal dispersion term to balance the S_1 advection term. We also see, in a general way, that S_3 is inversely proportional to K_z. For a stratified estuary we anticipate that there will be a contribution from both S_2 and S_3 to balance S_1.

It should also be noted that if we had more correctly considered C and B in (5-88) and (5-89) to be functions of the depth coordinate and had included a phase angle as a function of depth in each tidal component, we would still obtain the four terms represented by (5-96). The contributions of the cross products of the depth dependent amplitude and phase factors would be included in the double integral representing S_2.

We could alternatively look at longitudinal dispersion across a vertical section based on the volume and salt continuity relations of Section 2-4.

Considering the vertically averaged values of longitudinal velocity and salinity over a tidal cycle and over a cross section normal to the flow, we would have for the volume and salt continuity relations,

$$Av_x = R \tag{5-98}$$

and

$$v_x s - K_x \frac{ds}{dx} = 0 \tag{5-99}$$

where, for convenience, v_x and s represent $\langle \bar{v}_x \rangle$ and $\langle \bar{s} \rangle$, and K_x is a longitudinal diffusion coefficient as defined by this averaging process. We see from (5-96) that under steady state conditions the contributions from S_2 and S_3, as well as S_4, are included in the diffusion term of (5-99). We see then that we should consider the K_x coefficient of (5-99) an *effective longitudinal dispersion* coefficient rather than a simple eddy diffusion coefficient. For a well mixed estuary K_x is indeed simply the tidal diffusion coefficient; for a stratified estuary it includes both the tidal diffusion and circulation effects.

From (5-98) and (5-99) we have

$$K_x = \frac{Rs}{A(ds/dx)} = \frac{R(f-1)}{A(df/dx)} \tag{5-100}$$

where in the third expression f is the freshwater fraction as defined by (5-2). Thus K_x may be calculated for any position in an estuary if the river flow, cross sectional area, and salinity or freshwater fraction distribution are known.

Let us consider next a conservative pollutant outfall along an estuary whose rate of supply to the estuary is W, a constant, as shown in Fig. 5-3. Then the pollutant continuity relations, corresponding to the conditions for (5-99), are

$$W = Rc - K_x A \frac{dc}{dx} \tag{5-101}$$

in a downestuary direction, and

$$0 = Rc - K_x A \frac{dc}{dx} \tag{5-102}$$

in an upestuary direction. Substituting (5-98) and (5-2) into (5-99), we rewrite (5-99) as

$$R = Rf - K_x A \frac{df}{dx} \tag{5-103}$$

in terms of f, and as

$$0 = Rs - K_x A \frac{ds}{dx} \qquad (5\text{-}104)$$

in terms of s. We see that (5-101) and (5-103) are similar and that (5-102) and (5-104) are similar, the functional relations being the same on the right hand side of the equation in all four cases and this function being equal to a constant in the first case and equal to zero in the second case, so that we have for c in terms of f in a downestuary direction,

$$c = c_0 \frac{f}{f_0} \qquad (5\text{-}105)$$

and for c in terms of s in an upestuary direction,

$$c = c_0 \frac{s}{s_0} \qquad (5\text{-}106)$$

which are the same as relations (5-23) and (5-24), which were obtained previously. We also see that, if we are dealing with a conservative pollutant, we need not go through the calculations of K_x in (5-100) and then to the distribution c from (5-101) and (5-102) but can go directly to the distribution of c from the observed values of salinity, using relations (5-105) and (5-106).

For further information on the material covered in this section the reader may refer to Bowden (1963), Stommel (1953), and Dyer (1973).

5-5. HORIZONTAL, ONE DIMENSIONAL TIDAL MIXING

To begin we consider horizontal, one dimensional tidal mixing in the estuary as a whole. In essence we are making a mathematical formulation of the modified tidal prism concept of Section 5-1. As in Section 5-3, because of the numerous simplifications we make in this beginning derivation, there is probably little direct applicability of the results; however, they are nevertheless of interest in showing the form of solution and the type of variations that may be encountered.

Consider an estuary with rectangular sides or uniform cross sectional area, as shown in Fig. 5-10. We assume that the conditions are stationary and that mixing is complete for any section normal to the x axis. Then, the salinity is a function of x only, $s = s(x)$. We take as the boundary conditions at the upper end of the estuary or river mouth that the salinity

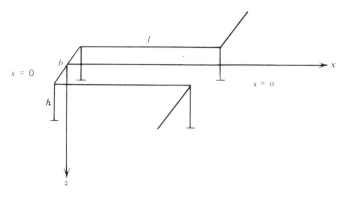

Fig. 5-10

is zero and that the tidal diffusion of salt into the river is zero,

$$s = 0 \qquad x = 0$$

$$K_x \frac{ds}{dx} = 0 \qquad x = 0 \tag{5-107}$$

and at the seaward end of the estuary that the salinity is equal to the ocean salinity σ,

$$s = \sigma \qquad x = l \tag{5-108}$$

For the net circulation velocity averaged over the depth and a tidal cycle we have the relation given by (5-97) and, for convenience, we again use v_x and s to represent $\langle \bar{v}_x \rangle$ and $\langle \bar{s} \rangle$.

If the length of the estuary is small compared with the wavelength of the tidal cooscillation, we may consider the tide to be simultaneous and uniform over the entire estuary. This is equivalent to saying in (3-40) and (3-41) that the spatial cosine terms may be represented by unity and the sine terms by their argument, giving for the tidal amplitude and velocity,

$$\eta = a_0 \cos \omega t \tag{5-109}$$

and

$$v_{xt} = \frac{\omega a_0 x}{h} \sin \omega t \tag{5-110}$$

where we have also used (3-18) and (3-29). Integrating (5-110), we then have for the tidal displacement ξ,

$$\xi = -\frac{a_0 x}{h} \cos \omega t \tag{5-111}$$

where the constant of integration is zero as $\xi = 0$ at $x = 0$.

For steady state the vertically integrated salt continuity equation (2-65) reduces to

$$v_x \frac{\partial s}{\partial x} = \frac{\partial}{\partial x}\left(K_x \frac{\partial s}{\partial x} \right) \qquad (5\text{-}112)$$

Since s is a function of x only, this may be integrated to the following,

$$v_x s = K_x \frac{ds}{dx} \qquad (5\text{-}113)$$

where from the boundary conditions of (5-107) the constant of integration is zero. This equation, as expected, is the same as (5-99).

We are now interested in arriving at some approximate expression for $K_x = K_x(x)$, so that we may solve (5-113) for s. We can assume that the horizontal tidal mixing is such that in an isolated experiment there is complete mixing over a tidal period, $T = 2\pi/\omega$, for the maximum tidal displacement $2\xi_0$. For an original linear salt gradient, as shown in Fig. 5-11, we then have from (2-49) and (2-56) that the salt flux, or salt

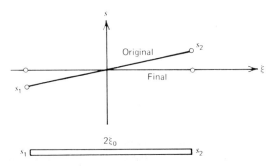

Fig. 5-11

transport over a tidal cycle, is given by the salinity difference between the original and final states divided by the tidal excursion,

$$S = -K_x \frac{ds}{dx}$$

$$\frac{2}{T} \int_0^{\xi_0} \frac{s_2 - s_1}{2\xi_0} x \, dx = -K_x \frac{s_2 - s_1}{2\xi_0}$$

$$\frac{\xi_0^2}{T} = K_x$$

or, from (5-111),

$$K_x = \frac{\omega a_0^2 x^2}{2\pi h^2} \qquad (5\text{-}114)$$

Alternatively, and probably for a more satisfactory approach, we can refer back to expression (2-71) indicating that the diffusion coefficient is proportional to the product of a characteristic velocity and a characteristic length. In the case of a modified tidal prism concept we chose as a characteristic velocity the amplitude of the tidal velocity, and as a characteristic length the total tidal excursion, giving

$$K_x = 2B\xi_0 v_{xt_0}$$
$$= \frac{2B\omega a_0{}^2 x^2}{h^2} \tag{5-115}$$

where B is constant of proportionality, and where we have used (5-110) and (5-111). We see that (5-114) and (5-115) are the same if $B = 1/4\pi$.

Substituting (5-115) into (5-113) and integrating, we obtain

$$v_x s = \frac{2B\omega a_0{}^2 x^2}{h^2} \frac{ds}{dx}$$

$$\frac{ds}{s} = \frac{v_x h^2}{2B\omega a_0{}^2} \frac{dx}{x^2}$$

$$\log s = -\frac{v_x h^2}{2B\omega a_0{}^2} \frac{1}{x} + C$$

Substituting in the boundary condition (5-108), we obtain

$$C = \log \sigma + \frac{v_x h^2}{2B\omega a_0{}^2 l}$$

and finally

$$s = \sigma e^{F[1-(1/\lambda)]} \tag{5-116}$$

where we have made the substitution of the dimensionless parameter

$$\lambda = \frac{x}{l} \tag{5-117}$$

and where

$$F = \frac{v_x h^2}{2B\omega a_0{}^2 l} \tag{5-118}$$

Equation (5-116) defines a set of curves which are asymptotic to $s = 0$ for large values of F and asymptotic to $s = \sigma$ for small values of F, as shown in Fig. 5-12. We also see from Fig. 5-12 that the curves are most sensitive to F in the region $0.1 < F < 10$. In addition, we see from either (5-114) or (5-115) that under the assumptions used here we can anticipate that the longitudinal tidal diffusion coefficient K_x will increase quadratically with distance down the estuary.

It is of interest to carry out the same derivation for the case in which

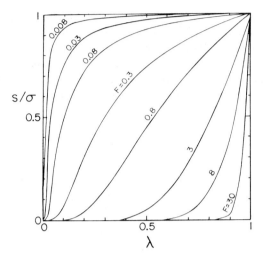

Fig. 5-12   Reproduced with permission from A. B. Arons and H. Stommel, 1951, *Transactions of the American Geophysical Union*, **38**, 1986–2001, Fig. 1.

the diffusion coefficient K_x does not follow a quadratic law but, say, is a constant. This might be considered to correspond to the conditions for which longitudinal mixing is caused by the wind rather than the tide. In this instance as K_x is not zero at $x = 0$, the conditions of (5-107) do not apply at $x = 0$ but approach them some distance up the estuary river. We may still then make the reduction from (5-112) to (5-113). The solution is

$$\log s = \frac{v_x}{K_x} x + C$$

where the constant C is given from the boundary conditions of (5-108) as

$$C = \log \sigma - \frac{v_x}{K_x} l$$

so that

$$s = \sigma e^{(v_x l/K_x)(\lambda - 1)}, \tag{5-119}$$

The salinity variation is then a simple exponential increase from a minimum value at $x = 0$ to σ at $x = l$.

We may continue this type of analysis in another example with the following conditions. We assume, as in the original example, that the longitudinal diffusion is governed by a tidal mixing process with a quadratic diffusion coefficient as given by (5-115). We next assume, however, that the freshwater contribution is governed by evaporation or

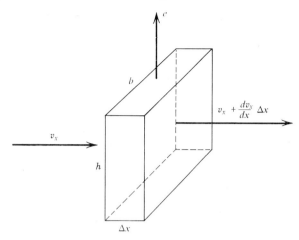

Fig. 5-13

precipitation rather than by river flow. Looking at the freshwater evaporation effect alone on the salinity, we see from Fig. 5-13 that there is a contribution to the rate of increase in salinity in the salt conservation equation (2-65) of se/h, where e is the evaporation rate, and $1/h$ is the ratio of the incremental surface evaporation area to the incremental volume. From the volume continuity relation we also have from Fig. 5-13 a current set up to compensate for the evaporation given by

$$v_x bh = eb\,\Delta x + \left(v_x + \frac{dv_x}{dx}\,\Delta x\right)bh$$

$$\frac{dv_x}{dx} = -\frac{e}{h}$$

or, integrating from $x = 0$ at the head of the estuary, by

$$v_x = -\frac{e}{h}x \tag{5-120}$$

We then have for (2-65), using (5-115) and (5-120),

$$\frac{2B\omega a_0^2}{h^2}\frac{d}{dx}\left(x^2\frac{ds}{dx}\right) + \frac{e}{h}x\frac{ds}{dx} + \frac{e}{h}s = 0$$

$$x^2\frac{d^2s}{dx^2} + (2+b)x\frac{ds}{dx} + bs = 0 \tag{5-121}$$

where b is given by

$$b = \frac{eh}{2B\omega a_0^2} \tag{5-122}$$

Equation (5-121) is a standard form of a linear differential equation whose solution is

$$s = C_1 x^{-1} + C_2 x^{-b} \tag{5-123}$$

The boundary conditions applicable here are somewhat different than the ones we have used previously. As before, we take for one boundary condition that the salinity at the mouth of the estuary is ocean salinity,

$$s = \sigma \qquad x = l \tag{5-124}$$

Since the evaporation or precipitation process itself does not add or subtract any salt from the estuary, we must have as the second boundary condition that the net salt flux at the mouth of the estuary is zero,

$$v_x s = K_x \frac{ds}{dx} \qquad x = l \tag{5-125}$$

Substituting (5-123), (5-115), and (5-120) into (5-124) and (5-125), we have

$$\sigma = C_1 l^{-1} + C_2 l^{-b} \tag{5-126}$$

and

$$b\sigma = C_1 l^{-1} + b C_2 l^{-b} \tag{5-127}$$

Solving equation (5-126) and (5-127) for $C_1 l^{-1}$ and $C_2 l^{-b}$, we then obtain

$$C_1 l^{-1} = 0$$

and $\hspace{8cm}$ (5-128)

$$C_2 l^{-b} = \sigma$$

which gives

$$s = \sigma \lambda^{-b} \tag{5-129}$$

Figures 5-14 and 5-15 are graphs of (5-129) for b positive, evaporation, and for b negative, precipitation.

We may make one further extension of these simplified tidal mixing concepts. If instead of assuming that the length of the estuary is short compared with a tidal wavelength we use the more general tidal cooscillation expressions of (3-40) and (3-41), we will have for K_x of (5-115),

$$K_x = \frac{2 B \omega a_0^{2} \sin^2 kx}{k^2 h^2 \cos^2 kl} \tag{5-130}$$

Substituting this value for K_x into (5-113) and integrating, we then have

$$\frac{ds}{s} = \frac{v_x k^2 h^2 \cos^2 kl}{2 B \omega a_0^{2}} \frac{dx}{\sin^2 kx}$$

$$\log s = -\frac{v_x h^2 \cos^2 kl}{2 B \omega a_0^{2} l} \frac{kl}{\tan kx} + C$$

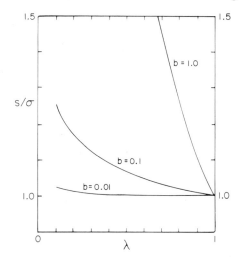

Fig. 5-14 Reproduced with permission from H. Stommel and H. G. Farmer, 1952, Reference 52–88, Woods Hole Oceanographic Institution, Woods Hole, Mass., Fig. 4.4.2.

Using the boundary condition (5-108) to evaluate C then gives for the final solution,

$$s = \sigma e^{Fkl(\cot kl - \cot kx)} \tag{5-131}$$

where F is given by

$$F = \frac{v_x h^2 \cos^2 kl}{2B\omega a_0^2 l} \tag{5-132}$$

We see from solution (5-131) or from (5-130) that s increases from $x = 0$ in much the same manner as shown in Fig. 5-12, then levels out at a distance of $x = \lambda/4$, and increases again toward $s = \sigma$ at $x = l$, leveling out again at distances of $3\lambda/4$, etc., should they exist.

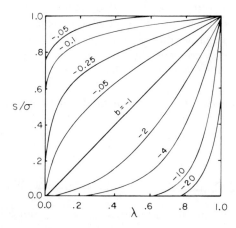

Fig. 5-15 Reproduced with permission from H. Stommel and H. G. Farmer, 1952, Reference 52–88, Woods Hole Oceanographic Institution, Woods Hole, Mass., Fig. 4.4.3.

We now return to the derivation of a more general expression for the tidal diffusion coefficient in terms of physically observable quantities. From the expression (5-95), using (5-88) and (5-89), we have for the mean salt flux per unit breadth due to tidal diffusion,

$$S_2 = \frac{h}{T} \int_0^T v_{xt} s_t \, dt$$

$$= -\frac{hBC}{T} \int_0^T \cos \omega t \sin (\omega t - \delta) \, dt$$

$$= -\tfrac{1}{2} hBC \sin \delta \tag{5-133}$$

Using expression (3-33) for the amplitude of the tidal salinity variation in terms of the longitudinal salinity gradient and the amplitude of the tidal velocity variation, (5-133) may be rewritten as

$$S_2 = -\frac{\sin \delta}{4\pi} hC^2 T \frac{ds}{dx} \tag{5-134}$$

From the definition (2-49) for the salt flux in terms of a diffusion coefficient we then have finally

$$-hK_x \frac{ds}{dx} = -\frac{\sin \delta}{4\pi} hC^2 T \frac{ds}{dx}$$

or

$$K_x = \frac{\sin \delta}{4\pi} C^2 T \tag{5-135}$$

We note that (5-135) has the usual form (2-71) of a coefficient multiplied by a characteristic velocity squared multiplied by a characteristic time.

The $\sin \delta$ term here plays the same role as the $\cos \delta$ term in (3-87) for the tidal energy flow. If v_{xt} and s_t are π out of phase so that $\sin \delta = 1$, then when v_{xt} is negative s_t is positive, and when v_{xt} is positive s_t is negative. The salt flux is always in the same negative direction and is a maximum. Again corresponding to the energy flow concept, the $\sin \delta$ term represents the fraction of water leaving on an ebb tide that is replaced by new water entering on the flood tide.

We may take an alternative approach to horizontal tidal diffusion by considering some of the concepts of Section 2-5. If we take K_x to be a constant, from (2-55) the one dimensional, time dependent Fickian diffusion equation is

$$\frac{\partial c}{\partial t} = K_x \frac{\partial^2 c}{\partial x^2} \tag{5-136}$$

for an index quantity c. Then, for an instantaneous source of finite amount at $x = 0$ and $t = 0$ we have from (2-68) that the solution to (5-136) is

$$c = \frac{1}{\sqrt{4\pi K_x t}} e^{-x^2/4K_x t} \qquad (5\text{-}137)$$

or a familiar Gaussian distribution in x, which spreads out in the x direction as t increases. We also have from (2-69) that the mean square value $\overline{x^2}$ for the concentrate distribution is given by

$$\overline{x^2} = 2K_x t \qquad (5\text{-}138)$$

or that $\overline{x^2}$ increases directly proportional to t.

Applying these concepts to a tidal diffusion we might assume, as a first order approximation, that after one tidal cycle the concentrate is uniformly distributed over the length of the tidal excursion ξ_0 on either side of the origin as shown in Fig. 5-16. Then, using the normalizing relation

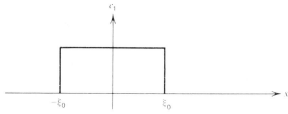

Fig. 5-16

implied in solution (5-137) that we have a source of unit amount,

$$\int_{-\infty}^{\infty} c \, dx = 1 \qquad (5\text{-}139)$$

we have from Fig. 5-16,

$$\int_{-\infty}^{\infty} c_1 \, dx = 1$$

or

$$c_1 = \frac{1}{2\xi_0} \qquad (5\text{-}140)$$

The value of $\overline{x_1^2}$ is then

$$\overline{x_1^2} = 2 \int_0^{\xi_0} \frac{x^2}{2\xi_0} \, dx = \frac{1}{3} \xi_0^2 \qquad (5\text{-}141)$$

In extending this to two tidal cycles, each segment of c_1 concentrate between $-\xi_0$ and ξ_0 is uniformly distributed over a tidal length about the

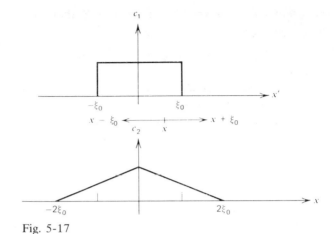

Fig. 5-17

reference point. From Fig. 5-17 we then have for the concentrate c_2 at any positive distance x, and a similar symmetric expression for negative x,

$$c_2 = \frac{1}{2\xi_0} \int_{x-\xi_0}^{\xi_0} c_1 \, dx'$$

$$= \frac{1}{4\xi_0{}^2} \int_{x-\xi_0}^{\xi_0} dx'$$

$$= \frac{1}{2\xi_0} \left(1 - \frac{x}{2\xi_0}\right) \tag{5-142}$$

where the $1/2\xi_0$ factor in front of the first integral is a normalizing factor to satisfy (5-139). We also have for $\overline{x_2{}^2}$,

$$\overline{x_2{}^2} = 2\int_0^{2\xi_0} \frac{x^2}{2\xi_0}\left(1 - \frac{x}{2\xi_0}\right) dx = \frac{2}{3}\xi_0{}^2 \tag{5-143}$$

The same reasoning can be applied to three tidal cycles, obtaining a distribution as sketched in Fig. 5-18. With values for c_3 from 0 to ξ_0 of

$$c_3 = \frac{1}{2\xi_0}\left(\frac{3}{4} - \frac{x^2}{4\xi_0{}^2}\right) \tag{5-144}$$

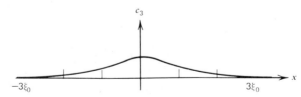

Fig. 5-18

and from ξ_0 to $3\xi_0$ of

$$c_3 = \frac{1}{2\xi_0}\left(\frac{9}{8} - \frac{3x}{4\xi_0} + \frac{x^2}{8\xi_0^2}\right) \tag{5-145}$$

and a value for $\overline{x_3^2}$ of

$$\overline{x_3^2} = \xi_0^2 \tag{5-146}$$

We see then from (5-141), (5-143), and (5-146) that $\overline{x^2}$ increases linearly from one tidal period to the next. Comparing with (5-138), we have a corresponding value for K_x of

$$K_x = \frac{1}{6}\frac{\xi_0^2}{T} \tag{5-147}$$

Once again we have the same form for K_x, namely, that K_x is directly proportional to the square of the tidal excursion and inversely proportional to the tidal period.

We can generalize expression (5-147) through the following consideration. After a long period of time we may have a longitudinal distribution of the index quantity c as indicated in Fig. 5-19. Then, after one tidal cycle we can represent the increase in c for a small incremental unit volume about $x' = 0$ to be

$$\Delta c = \frac{\partial c}{\partial t} T \tag{5-148}$$

Following the same procedure as before, and using a Taylor series expansion for c about $x' = 0$, we have for the new value of the concentrate at $x' = 0$ after one tidal cycle,

$$c = \frac{1}{2\xi_0}\int_{-\xi_0}^{\xi_0}\left(c_0 + \frac{\partial c}{\partial x}x' + \frac{\partial' c}{\partial x^2}\frac{x'^2}{2} + \cdots\right)dx'$$

$$= c_0 + \frac{\partial^2 c}{\partial x^2}\frac{\xi_0^2}{6} + \cdots$$

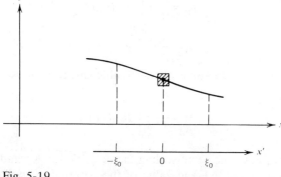

Fig. 5-19

or

$$\Delta c = c - c_0 = \frac{\partial^2 c}{\partial x^2} \frac{\xi_0^2}{6} \tag{5-149}$$

Then, as expression (5-148) must be equal to (5-149), we have

$$\frac{\partial c}{\partial t} = \frac{1}{6} \frac{\xi_0^2}{T} \frac{\partial^2 c}{\partial x^2} \tag{5-150}$$

or, comparing with (5-136),

$$K_x = \frac{1}{6} \frac{\xi_0^2}{T} \tag{5-151}$$

Following further the analysis of Section 2-5, we have from (2-78) and (2-69) for a continuous, turbulent diffusion process,

$$\overline{x^2} = 2K_x T = 2\overline{v^2} \int_0^T \int_0^t R_\tau \, d\tau \, dt \tag{5-152}$$

where we have taken the time scale of the process to be that of the time scale of a tidal period. If we represent the tidal motion ξ by

$$\xi = \xi_0 \cos \frac{2\pi t}{T} \tag{5-153}$$

the tidal velocity v_{xt} will be represented by

$$v_{xt} = -\frac{2\pi\xi_0}{T} \sin \frac{2\pi t}{T} \tag{5-154}$$

so that

$$\overline{v^2} = \overline{v_{xt}^2} = \frac{1}{T} \int_0^T \frac{4\pi^2 \xi_0^2}{T^2} \sin^2 \frac{2\pi t}{T} \, dt = \frac{2\pi^2 \xi_0^2}{T^2} \tag{5-155}$$

and, from (5-152),

$$K_x = \frac{\overline{v^2}}{T} \int_0^T \int_0^t R_\tau \, d\tau \, dt$$
$$= 2\pi^2 \bar{R} \frac{\xi_0^2}{T} \tag{5-156}$$

where $\bar{R}$ is the average value of the double integral for R_τ defined as

$$\bar{R} = \frac{1}{T^2} \int_0^T \int_0^t R_\tau \, d\tau \, dt \tag{5-157}$$

Once again we have the same scaling factor ξ_0^2/T for K_x, and in this case it is multiplied by the average value of the correlation coefficient over a tidal cycle. In this case we treat the tidal motion as a turbulent phenomenon. The eddy size is that of the tidal excursion scale. If there were no small scale turbulence but only a sinusoidal velocity variation, $\bar{R}$ would be zero. The small scale turbulence is necessary to create mixing during the tidal cycle. Formula (5-156) along with (5-157) also provides us with an additional method for determining K_x, providing sufficiently detailed velocity measurements have been made.

For further information on the material covered in this section the reader may refer to Arons and Stommel (1951), Stommel and Farmer (1952b), Bowden (1963), Preddy (1954), and Palmer and Izatt (1971).

5-6. VERTICAL VELOCITY SHEAR, CIRCULATION, AND MIXING EFFECTS

As one begins to look at the vertical mixing effects, there are several items that have to be considered. First, there is the effect of the vertical deviation in the longitudinal velocity v_{x1} in determining the vertical salinity deviation s_1. Second, there is the combined effect of v_{x1} and s_1 in determining the longitudinal circulation salt flux S_3 and in turn the effective longitudinal dispersion coefficient. In addition, there is the complication that it is often inappropriate to consider the vertical diffusion coefficient K_z a constant as a function of depth.

For the last-mentioned effect we may anticipate that, as the bottom and, depending on the type of motion, the free surface are approached, the scale of the vertical eddy diffusion will decrease and consequently K_z as well. Further, depending on the effect of v_{x1} in creating a strong salinity gradient, or halocline, in the middepth region, a potential energy barrier as expressed by relation (2-140) may be created to prevent or diminish vertical mixing with a consequent reduction in K_z in this region as well. In looking at the vertical salinity variation we anticipate in nature that there will be two preferred stable states, either well mixed from the surface to the bottom, or a strong halocline with well mixed water above and below the halocline. This is indeed the case; in general, we have either a well mixed or a stratified estuary. The complication in expressing this relation mathematically comes in how we express K_z as a function of the Richardson number Ri of (2-142).

From the salt continuity equation (2-65) we have, assuming that there are no lateral variations and that the vertical velocity is small,

$$\frac{\partial s}{\partial t} + v_x \frac{\partial s}{\partial x} = \frac{\partial}{\partial z}\left(K_z \frac{\partial s}{\partial z}\right) + \frac{\partial}{\partial x}\left(K_x \frac{\partial s}{\partial x}\right) \tag{5-158}$$

where v_x and s are given from (5-93) and (5-94) by

$$v_x = \langle \bar{v}_x \rangle + v_{xt} + v_{x1} \qquad (5\text{-}159)$$

and

$$s = \langle \bar{s} \rangle + s_t + s_1 \qquad (5\text{-}160)$$

and where we have omitted the small scale turbulence fluctuations v_x' and s'. Then, integrating over depth from the surface to the bottom, we have

$$\frac{\partial s_t}{\partial t} + [\langle \bar{v}_x \rangle + v_{xt}] \frac{\partial s}{\partial x} = \left\langle \frac{\partial}{\partial x}\left(K_x \frac{\partial s}{\partial x} \right) \right\rangle \qquad (5\text{-}161)$$

since the vertical flux of salt must be zero at both the surface and the bottom and where we have also made the important assumption that $\partial s/\partial x$ is independent of z and t. The second term on the left hand side of (5-161) and the term on the right hand side of (5-161) are small in comparison with the other two terms, reducing it to

$$\frac{\partial s_t}{\partial t} = -v_{xt} \frac{\partial s}{\partial x} \qquad (5\text{-}162)$$

which is the same as the relation we had previously in (3-31). Integrating (5-161) over a tidal cycle, we then obtain, using relations (5-88) and (5-89),

$$\langle \bar{v}_x \rangle \frac{\partial s}{\partial x} = \overline{\left\langle \frac{\partial}{\partial x}\left(K_x \frac{\partial s}{\partial x} \right) \right\rangle} \qquad (5\text{-}163)$$

which is the same as the relation we had previously in (5-99). Then, subtracting (5-161) from (5-158), we have further,

$$v_{x1} \frac{\partial s}{\partial x} = \frac{\partial}{\partial z}\left(K_z \frac{\partial s}{\partial z} \right) = \frac{\partial}{\partial z}\left(K_z \frac{\partial s_1}{\partial z} \right) \qquad (5\text{-}164)$$

where we have taken, as before, K_x to be a constant as a function of z. Equation (5-164) provides us with a relation for determining K_z as a function of z from observed values of the salinity deviation as a function of depth s_1, the longitudinal velocity deviation as a function of depth v_{x1}, and the longitudinal salinity gradient $\partial s/\partial x$.

Often in a tidal estuary $\partial s/\partial x$ is not independent of z, and s_1 varies over a tidal period. We then replace (5-161) by

$$\frac{\partial s_t}{\partial t} + \left\langle v_x \frac{\partial s}{\partial x} \right\rangle = \left\langle \frac{\partial}{\partial x}\left(K_x \frac{\partial s}{\partial x} \right) \right\rangle \qquad (5\text{-}165)$$

Referring back to the discussion of Section 2-10, we have as the usual case in an estuary that at any given point the vertical diffusive salt flux is substantially greater than the longitudinal diffusive salt flux. We can then usually ignore the second term on the right hand side of (5-158) and the right hand side of (5-165) in consideration of the first term on the right hand side of (5-158), obtaining

$$\frac{\partial s_1}{\partial t} + v_x \frac{\partial s}{\partial x} - \left\langle v_x \frac{\partial s}{\partial x} \right\rangle = \frac{\partial}{\partial z}\left(K_z \frac{\partial s}{\partial z}\right)$$

or

$$\int_0^T \left(v_x \frac{\partial s}{\partial x} - \left\langle v_x \frac{\partial s}{\partial x} \right\rangle\right) dt = \int_0^T \frac{\partial}{\partial z}\left(K_z \frac{\partial s}{\partial z}\right) dt$$

or

$$\frac{\partial}{\partial z}\left(K_z \frac{\partial s_1}{\partial z}\right) = \frac{1}{T}\int_0^T \left(v_x \frac{\partial s}{\partial x} - \left\langle v_x \frac{\partial s}{\partial x} \right\rangle\right) dt \qquad (5\text{-}166)$$

where K_z in (5-166) is understood to be the effective value of K_z representing the mean mixing conditions during a tidal period. Under these conditions (5-166) can be used in place of (5-164) to determine K_z as a function of z.

Let us now see what we can ascertain as to the vertical distribution of salinity. We here take both K_z and $\partial s/\partial x$ to be constant as a function of depth. Equation (5-164) then reduces to

$$\frac{d^2 s}{dz^2} = \frac{v_{x1}}{K_z} \frac{\partial s}{\partial x} \qquad (5\text{-}167)$$

We use the form of the longitudinal circulation velocity as given by (4-98):

$$v_{x1} = \frac{1}{48} \frac{g\lambda h^3}{\rho N_z}(8n^3 - 9n^2 + 1) = v_s(8n^3 - 9n^2 + 1) \qquad (5\text{-}168)$$

where n is defined by

$$n = \frac{z}{h} \qquad (5\text{-}169)$$

The two boundary conditions are that the vertical salt flux vanishes at the surface and the bottom,

$$K_z \frac{\partial s}{\partial z} = 0 \qquad z = 0$$

and $\qquad (5\text{-}170)$

$$K_z \frac{\partial s}{\partial z} = 0 \qquad z = h$$

Substituting (5-168) into (5-167), we then obtain

$$\frac{ds}{dz} = \frac{v_s h}{K_z} \frac{\partial s}{\partial x}(2n^4 - 3n^3 + n) \qquad (5\text{-}171)$$

where the constant of integration vanishes to satisfy both boundary conditions, and

$$s = \frac{v_s h^2}{K_z} \frac{\partial s}{\partial x} \left(\tfrac{2}{5}n^5 - \tfrac{3}{4}n^4 + \tfrac{1}{2}n^2 + C_2\right) \tag{5-172}$$

Taking the depth average $\langle s \rangle$ of (5-172), we then have for the salinity deviation s_1,

$$s_1 = \frac{v_s h^2}{K_z} \frac{\partial s}{\partial x} \left(\tfrac{2}{5}n^5 - \tfrac{3}{4}n^4 + \tfrac{1}{2}n^2 - \tfrac{1}{12}\right) \tag{5-173}$$

We see from the form of (5-167) that we could alternatively have taken any of the other expressions for v_{x1} from Section 4-5, performing the required double integration with respect to z, to obtain the vertical salinity distribution s_1.

A graph of the parentheses of (5-173), representing the salinity deviation as a function of depth, is given in Fig. 5-20. To give some index of

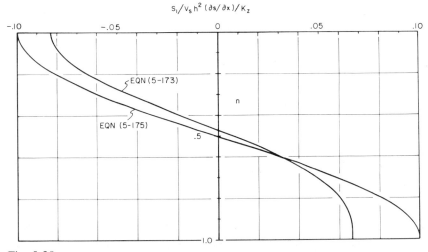

Fig. 5-20

the order of magnitude of quantities involved, we obtain from (5-173) a variation in salinity of 8‰ from the upper portion of the water column to the lower portion, using the same parameter values as given following (4-100) and a value of $K_z = 20 \text{ cm}^2 \text{ sec}^{-1}$.

Using the value for the longitudinal dispersion coefficient as given by expression (5-99), we can rewrite (5-173) as

$$s_1 = \frac{v_s h^2}{K_z} \frac{\langle \bar{v}_x \rangle \langle \bar{s} \rangle}{K_x} \left(\tfrac{2}{5}n^5 - \tfrac{3}{4}n^4 + \tfrac{1}{2}n^2 - \tfrac{1}{12}\right) \tag{5-174}$$

We see then that the vertical salinity deviation is inversely proportional to the vertical mixing coefficient K_z and the longitudinal dispersion coefficient K_x, whether caused by tidal diffusion or by other effects. We then anticipate that in an estuary with strong tidal diffusion the vertical salinity deviation will be slight.

For completeness it is of interest to include the case in which there is no bottom friction and the longitudinal velocity deviation is given by (4-116) rather than by (5-168). Following the same derivation procedures, we obtain

$$s_1 = \frac{v_s h^2}{K_z} \frac{\partial s}{\partial x} \left(\tfrac{1}{5} n^5 - \tfrac{1}{2} n^4 + \tfrac{1}{2} n^2 - \tfrac{1}{10} \right) \tag{5-175}$$

instead of (5-173). A plot of (5-175) is included in Fig. 5-20. It is to be noted that it is symmetric about $n = \tfrac{1}{2}$.

In addition, we include the case corresponding to the preferred bottom friction condition of relation (4-107). From (4-108) and (4-110) the velocity deviation is given by

$$v_{x1} = \frac{1}{6} \frac{gih^2}{N_z} (1 - 3n^2) - \frac{1}{24} \frac{g\lambda h^3}{\rho N_z} (1 - 4n^3) \tag{5-176}$$

from which the surface and bottom velocities are given by

$$v_s = \frac{gh^2}{6N_z} \left(i - \frac{1}{4} \frac{\lambda h}{\rho} \right) \tag{5-177}$$

and

$$v_b = -\frac{gh^2}{3N_z} \left(i - \frac{3}{8} \frac{\lambda h}{\rho} \right) \tag{5-178}$$

Using (5-177) and (5-178), we can then rewrite (5-176) in terms of v_s and v_b as

$$v_{x1} = v_s (1 - 9n^2 + 8n^3) + v_b (-3n^2 + 4n^3) \tag{5-179}$$

Following the same derivation procedures as before, we then obtain

$$s_1 = \frac{h^2}{K_z} \frac{\partial s}{\partial x} \left[v_s \left(\tfrac{2}{5} n^5 - \tfrac{3}{4} n^4 + \tfrac{1}{2} n^2 - \tfrac{1}{12} \right) + v_b \left(\tfrac{1}{5} n^5 - \tfrac{1}{4} n^4 + \tfrac{1}{60} \right) \right] \tag{5-180}$$

Let us now go back and examine what the salinity deviation would be if the river flow advection contribution to v_{x1} in (4-106) were dominant over the net circulation contribution. We have from (4-106) that $\bar{v}_x$ is given by

$$\bar{v}_x = \tfrac{3}{2} \langle \bar{v}_x \rangle (1 - n^2) \tag{5-181}$$

where v_0 in (4-106) is the same as $\langle \bar{v}_x \rangle$ used here. We then have for the corresponding velocity deviation v_{x1},

$$v_{x1} = \tfrac{1}{2} \langle \bar{v}_x \rangle (1 - 3n^2) \tag{5-182}$$

This condition may be considered to apply to the upper reaches and river portions of some estuaries, as compared with (5-168) which applies to the middle and lower reaches of an estuary. Again following the same procedures as before, the salinity deviation is given by

$$s_1 = \frac{\langle \bar{v}_x \rangle h^2}{K_z} \frac{\partial s}{\partial x} \left(-\tfrac{1}{8}n^4 + \tfrac{1}{4}n^2 - \tfrac{7}{120} \right) \tag{5-183}$$

A graph of the parentheses of (5-183) is given in Fig. 5-21. It is not too dissimilar to that of Fig. 5-20, showing a somewhat more linear variation

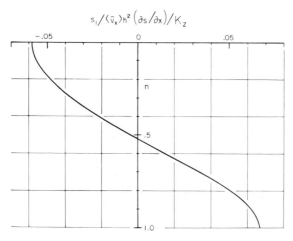

Fig. 5-21

in the salinity deviation as a function of depth. This is not unexpected, as the double integration of a velocity function tends to produce a smoothed result for the determined salinity deviation. Of more importance is the fact that v_s in (5-173) is replaced by $\langle \bar{v}_x \rangle$ in (5-183). Thus, as $\langle \bar{v}_x \rangle$ is usually of the order of $\tfrac{1}{10}$ that of v_s, we expect that an estuary that shows partially stratified conditions in its middle to lower reaches to which (5-173) applies will show essentially well mixed conditions in its upper reaches to which (5-183) applies, which indeed is usually the case.

As the longitudinal velocity terms in the two formulations and the resultant salinity deviations are additive, we have for the case in which both the net circulation and river advection contributions are important,

$$\begin{aligned} v_{x1} &= v_s'(8n^3 - 9n^2 + 1) + \tfrac{1}{2}\langle \bar{v}_x \rangle(-3n^2 + 1) \\ &= v_s(8n^3 - 9n^2 + 1) - \langle \bar{v}_x \rangle(12n^3 - 12n^2 + 1) \end{aligned} \tag{5-184}$$

and, from (5-173) and (5-183),

$$s_1 = \frac{h^2}{K_z} \frac{\partial s}{\partial x} \left[v'_s(\tfrac{2}{5}n^5 - \tfrac{3}{4}n^4 + \tfrac{1}{2}n^2 - \tfrac{1}{12}) + \tfrac{3}{2}\langle \bar{v}_x \rangle (-\tfrac{1}{12}n^4 + \tfrac{1}{6}n^2 - \tfrac{7}{180}) \right]$$

$$= \frac{h^2}{K_z} \frac{\partial s}{\partial x} \left[v_s(\tfrac{2}{5}n^5 - \tfrac{3}{4}n^4 + \tfrac{1}{2}n^2 - \tfrac{1}{12}) - \tfrac{3}{2}\langle \bar{v}_x \rangle (\tfrac{2}{5}n^5 - \tfrac{2}{3}n^4 + \tfrac{1}{3}n^2 - \tfrac{2}{45}) \right]$$

(5-185)

where v'_s is v_s of (5-168), and the net circulation velocity at the surface v_s is now given from (4-106) as

$$v_s = v'_s + \tfrac{3}{2}\langle \bar{v}_x \rangle$$

(5-186)

It is of interest to digress slightly to discuss the formulation by Rattray and Hansen (1962) and Hansen and Rattray (1965) for the vertical salinity distribution as a corollary to the net circulation velocity discussion of Section 4-6. From the discussion in that section we anticipate that their central regime solution results will be the same as those given here. In their procedures they imposed a similarity condition on the longitudinal diffusion coefficient, which is similar to condition (5-163). The term resulting from this is the same as expression (5-183). For the other terms in their results, for comparison with (5-173), we have from expressions (16) and (17) of Hansen and Rattray (1965) in our notation,

$$s_1 = s_0 \frac{\nu}{M} \left[\tfrac{1}{48}\nu Ra (\tfrac{2}{5}n^5 - \tfrac{3}{4}n^4 + \tfrac{1}{2}n^2 - \tfrac{1}{12}) \right]$$

(5-187)

where M is defined by

$$M = \frac{K_z K_{x0}}{\langle \bar{v}_x \rangle^2 h^2}$$

(5-188)

and Ra is defined by (4-129) and the other terms from the discussion following (4-129). We then have from (4-130), (4-132), and (5-168),

$$s_0 \frac{\nu}{M} (\tfrac{1}{48}\nu Ra) = \frac{\langle \bar{v}_x \rangle h^2}{K_z} \frac{\partial s}{\partial x} \left[\frac{v_s}{\langle \bar{v}_x \rangle} \right] = \frac{v_s h^2}{K_z} \frac{\partial s}{\partial x}$$

(5-189)

the same as the coefficient of (5-173). We note that the expressions (5-187) and (5-173) are the same.

Using (5-173) and (5-168), we are now in a position to calculate the circulation salt flux per unit breadth of (5-96), giving

$$S_3 = \int_0^h v_{x1} s_1 \, dz$$

$$= \frac{v_s^2 h^2}{K_z} h \frac{\partial s}{\partial x} \int_0^1 (8n^3 - 9n^2 + 1)(\tfrac{2}{5}n^5 - \tfrac{3}{4}n^4 + \tfrac{1}{2}n^2 - \tfrac{1}{12}) \, dn$$

$$= -0.030 \frac{v_s^2 h^2}{K_z} h \frac{\partial s}{\partial x}$$

(5-190)

In comparing with (5-134) we see that the relative proportion of the longitudinal dispersive flux due to circulation and tidal diffusion effects is expressed by the ratio of $0.030v_s^2h^2/K_z$ to $C^2T \sin \delta/4\pi$. We also have from a relation similar to (5-135) that the corresponding longitudinal circulation dispersion coefficient is given by

$$K_x = 0.030 \frac{v_s^2h^2}{K_z} \qquad (5\text{-}191)$$

These analytic expressions for v_{x1} and s_1, and in turn for S_3, provide us with an interesting alternative method for comparing the net circulation and tidal diffusion contributions to the longitudinal dispersive flux in terms of simple physically observable quantities. We are interested in obtaining an expression for the fraction ν, which we define as the ratio of the tidal diffusion flux to the total upestuary longitudinal dispersion flux. Under steady state conditions and ignoring the contribution S_4, we have from (5-96),

$$0 = S_1 + S_2 + S_3 \qquad (5\text{-}192)$$

or

$$\nu = \frac{S_2}{-S_1} = 1 + \frac{S_3}{S_1} \qquad (5\text{-}193)$$

From expression (5-173) we have for the tidal averaged salinity difference between the bottom and the surface, $\bar{s}_b - \bar{s}_s$, that

$$\bar{s}_b - \bar{s}_s = s_{1b} - s_{1s} = \frac{3}{20} \frac{v_s h^2}{K_z} \frac{\partial s}{\partial x} \qquad (5\text{-}194)$$

Then from (5-96), (5-190), and (5-194) the quantity ν of (5-193) can be written as

$$\nu = 1 - 0.20 \frac{v_s}{\langle \bar{v}_x \rangle} \frac{\bar{s}_b - \bar{s}_s}{\langle \bar{s} \rangle} \qquad (5\text{-}195)$$

in terms of the observable quantities of $\langle \bar{v}_x \rangle$ obtained from the river runoff flow, the net circulation velocity at the surface v_s, and the fractional change in salinity from the bottom to the surface $(\bar{s}_b - \bar{s}_s)/\langle \bar{s} \rangle$.

Equation (5-195) provides us with a very useful and instructive relation for ascertaining the relative importance of the tidal diffusion and net circulation contributions to the total longitudinal dispersion flux. We see, as expected, that the net circulation contribution increases with the salinity stratification in an estuary as measured by the fraction $(\bar{s}_b - \bar{s}_s)/\langle \bar{s} \rangle$; or, conversely, we see that the net circulation contribution is small for a well mixed estuary. We also see that the net circulation contribution increases with the circulation velocity distribution in an estuary as measured conveniently by the net circulation velocity at the surface v_s. Further, we see that the tidal diffusion contribution increases with the river runoff flow as measured by $\langle \bar{v}_x \rangle$; this relation was previously observed in (5-99) and (5-100) and their applications.

It should also be mentioned that Hansen and Rattray (1965, 1966) and Hansen (1967) obtained a similar relation for the fraction ν from their analysis of the net circulation contribution to the total longitudinal dispersion flux. The reader may wish to make a comparison of (5-195) with the stratification-circulation diagram of Fig. 1-1 from Hansen and Rattray (1966). On their log-log plot of $(\bar{s}_b - \bar{s}_s)/\langle \bar{s} \rangle$ versus $v_s/\langle \bar{v}_x \rangle$ the quantity ν has the same linear dependence and the same numerical values as those predicted by (5-195) over the range of values of $v_s/\langle \bar{v}_x \rangle$ for which (5-195) is applicable.

We may extend this type of analysis for S_3 and ν to the conditions for which relations (5-184) and (5-185) apply. Evaluating the integrals of the four polynomial products for S_3 and using these forms for v_{x1} and s_1 gives

$$S_3 = \frac{h^2}{K_z} h \frac{\partial s}{\partial x} [-0.030 v_s^2 + 0.045 v_s \langle \bar{v}_x \rangle - 0.019 \langle \bar{v}_x \rangle^2] \qquad (5\text{-}196)$$

From (5-185) we also have

$$\bar{s}_b - \bar{s}_s = \frac{h^2}{K_z} \frac{\partial s}{\partial x} [0.150 v_s - 0.100 \langle \bar{v}_x \rangle] \qquad (5\text{-}197)$$

so that ν is then

$$\nu = 1 - \frac{0.030(v_s/\langle \bar{v}_x \rangle)^2 - 0.045(v_s/\langle \bar{v}_x \rangle) + 0.019}{0.150(v_s/\langle \bar{v}_x \rangle) - 0.100} \frac{\bar{s}_b - \bar{s}_s}{\langle \bar{s} \rangle} \qquad (5\text{-}198)$$

Relation (5-198) completes the Hansen and Rattray stratification-circulation diagram for smaller values of the ratio $v_s/\langle \bar{v}_x \rangle$. We see from (5-186) that the ratio $v_s/\langle \bar{v}_x \rangle$ has a limiting value of $\frac{3}{2}$, and from (5-182) and (5-183) that as this limit is approached the fraction ν becomes independent of $v_s/\langle \bar{v}_x \rangle$.

Returning to (5-164), we can express the longitudinal circulation dispersion coefficient in a somewhat more general form through the following considerations. We assume that there are no nonlinear tidal variation terms. We take the longitudinal velocity deviation v_{x1} to be given by the form

$$v_{x1} = v_s f(n) \qquad (5\text{-}199)$$

where v_s is again the surface velocity and the vertical mixing coefficient K_z to be given by

$$K_z = K g(n) \qquad (5\text{-}200)$$

where K is the maximum value of K_z in a vertical section, and $g(n)$

correspondingly varies between 0 and 1. Taking the double integration of (5-164), we have

$$s = \langle \bar{s} \rangle + \frac{v_s h^2}{K} \frac{\partial s}{\partial x} [F(n) - \langle F(n) \rangle]$$

(5-201)

where $F(n)$ is given by

$$F(n) = \int \frac{1}{g(n)} \int f(n) \, dn \, dn$$

(5-202)

For the depth-averaged circulation flux we then have

$$\langle v_{x1} s_1 \rangle = -K_x \frac{\partial s}{\partial x} = \frac{v_s^2 h^2}{K} \frac{\partial s}{\partial x} \langle f(n) F(n) \rangle$$

(5-203)

or

$$K_x = -\frac{v_s^2 h^2}{K} \langle f(n) F(n) \rangle$$

(5-204)

We see again that when K_z is large the circulation contribution to the total longitudinal dispersion coefficient is small. As K_z and the longitudinal tidal diffusion coefficient are usually directly related, we anticipate that, for an estuary with strong tidal motions, the longitudinal dispersion coefficient will be principally the longitudinal tidal diffusion coefficient; and for an estuary with weak tidal motions, we anticipate that the longitudinal dispersion coefficient will be a combination of both the tidal diffusion and circulation effects, tending toward dominance by the circulation effect.

We may take as a final example of the vertical velocity shear effect the case of an oscillatory tidal current with a vertical shear. Here we assume for simplicity that the variation in phase of the tidal current with depth is small and can be neglected, so that v_x and v_{x1} are given by

$$v_x = A(z) \cos \omega t$$

(5-205)

and

$$v_{x1} = A_1(z) \cos \omega t$$

(5-206)

where A_1 is defined by the relation

$$A_1(z) = A(z) - \langle A \rangle$$

(5-207)

and, from (5-158) and (5-164), s_1 satisfies the equation

$$\frac{\partial s_1}{\partial t} + v_{x1} \frac{\partial s}{\partial x} = \frac{\partial}{\partial z} \left(K_z \frac{\partial s_1}{\partial z} \right)$$

(5-208)

Assuming a solution for s_1 in the form

$$s_1 = P(z) \cos \omega t + Q(z) \sin \omega t$$

(5-209)

it follows from (5-208), assuming also that K_z does not vary with time, that P and Q satisfy the relations

$$\frac{\partial}{\partial z}\left(K_z \frac{\partial P}{\partial z}\right) = \omega Q + A_1 \frac{\partial s}{\partial x}$$

and

(5-210)

$$\frac{\partial}{\partial z}\left(K_z \frac{\partial Q}{\partial z}\right) = -\omega P$$

If we next consider that Q may be considered small compared with P, a first approximation to P may be found by setting $Q = 0$ in the first equations in (5-210) and then using this value of P to solve for Q in the second equations in (5-210). Following this procedure, we obtain

$$P = R_1 \frac{\partial s}{\partial x}$$

and

(5-211)

$$Q = R_2 \frac{\partial s}{\partial x}$$

where

$$R_1(z) = \int \frac{1}{K_z} \int A_1 \, dz \, dz$$

and

(5-212)

$$R_2(z) = \int \frac{1}{K_z} \int -\omega R_1 \, dz \, dz$$

The depth and tidal averaged salt flux is then

$$\overline{\langle v_{x1} s_1 \rangle} = \frac{1}{2} \frac{\partial s}{\partial x} \langle A_1 R_1 \rangle$$

(5-213)

and from relations (5-199) through (5-204) we see that the K_x value for an oscillating flow is one half that for a corresponding steady flow with the same vertical velocity shear. Generally, in most estuaries the vertical velocity variation in the tidal component is much smaller than that of the net circulation effect, so that this contribution can usually be ignored.

The analysis given above can also be extended to nonlinear tidal terms where, for instance, the coefficient K_z is also expressed in terms of a tidal time variation.

For further information on the material covered in this section the reader may refer to Jacobsen (1913, 1918), Bowden (1963, 1965), Taylor (1953, 1954), Rattray and Hansen (1962), Hansen and Rattray (1965, 1966), and Hansen (1967).

5-7. LATERAL VELOCITY SHEAR, CIRCULATION, AND MIXING EFFECTS

Up to this point we have assumed that the lateral variations in longitudinal velocity and salinity are small and can be neglected. This is not necessarily the case. We may consider an example of estuarine flow for which there is a lateral difference in the net flow, bringing more saline water upestuary on one side and fresher water downestuary on the other side. Clearly, in this example there is a lateral contribution to the net circulation salt flux which conceivably can be as large or larger than the vertical circulation contribution.

We can look at this effect here in only the most general way by extending the analysis of Section 5-4. Instead of (5-93) and (5-94) we now have

$$v_x = \langle \widetilde{\bar{v}_x} \rangle + v_{xt} + \tilde{v}_{x1} + v_{x2} \tag{5-214}$$

and

$$s = \langle \widetilde{\bar{s}} \rangle + s_t + \tilde{s}_1 + s_2 \tag{5-215}$$

where we have omitted the small scale turbulence fluctuations and where the tilde is used to indicate an average value in the lateral direction. The various quantities in (5-214) and (5-215) are defined by the relations

and

$$\langle \widetilde{\bar{v}_x} \rangle = \frac{1}{A} \int_0^h \int_0^b \bar{v}_x \, dy \, dz$$

$$\langle \widetilde{\bar{s}} \rangle = \frac{1}{A} \int_0^h \int_0^b \bar{s} \, dy \, dz \tag{5-216}$$

and by

$$\tilde{v}_{x1} = \frac{1}{b} \int_0^b v_{x1} \, dy$$

and

$$\tilde{s}_1 = \frac{1}{b} \int_0^b s_1 \, dy \tag{5-217}$$

and by

$$\int_0^b v_{x2} \, dy = \int_0^h \tilde{v}_{x1} \, dz = \int_0^b s_2 \, dy = \int_0^h \tilde{s}_1 \, dz = 0 \tag{5-218}$$

where A is the cross sectional area, b is the breadth, and h is the depth. It is to be noted that we have replaced v_{x1}, previously assumed constant at a given depth, by its lateral average value at the same depth. It should also be noted that we have done our averaging with respect to the lateral dimension first, principally to make the results directly comparable with those of Section 5-4. We might instead have first carried out the averaging with respect to depth and then with respect to the lateral dimension.

Following the same procedure as in Section 5-4, we then have for the mean salt flux over a tidal cycle, using relation (5-218),

$$\bar{S} = A\langle \widetilde{\bar{v}_x} \rangle \langle \widetilde{\bar{s}} \rangle + A\langle \widetilde{\overline{v_{xt}s_t}} \rangle + A\overline{\langle v_{x1}s_1 \rangle} + A\langle \widetilde{\overline{v_{x2}s_2}} \rangle$$

$$= S_1 + S_2 + S_3 + S_4 \tag{5-219}$$

Here we have not considered B and C constants, but functions of the lateral and depth coordinates. We have now a fourth term S_4, representing the *lateral velocity shear* or *circulation* contribution. We could break down the analysis of (5-219) further. We could separate the S_2 term into a cross sectionally averaged term and a vertical and a lateral deviation term. We could also have taken the velocity and salinity averages in a vertical direction first, which would give an S_4 term averaged over breadth and time only and an S_3 term averaged over depth, breadth, and time, the differences between the two sets of corresponding terms being the cross product terms of the averaging residues. We could include additional terms representing the contributions from the tidal variations in the cross sectional area, which are usually small. And, finally, in any physical example we should include the actual geometric configuration in the averaging and flux analyses.

For further information on the material covered in this section the reader may wish to refer to Dyer (1973, 1974) and Fischer (1972).

5-8. DIFFUSION INDUCED CIRCULATION

We digress slightly here to consider a circulation effect caused by an existing stratified water condition at the entrance to an inlet, for which there is no driving force other than the longitudinal density gradients associated with the stratified condition maintained at the entrance. Such a circulation was originally inferred by Cameron and Pritchard (1965) from investigations in Baltimore harbor, an inlet tributary to Chesapeake bay for which there is negligible river inflow. The phenomenon was then discussed theoretically by Hansen and Rattray (1972) and by Hansen and Festa (1974), using similarity solution techniques. We carry out a derivation here using somewhat more physical reasoning and standard solution procedures which lead to the results they obtained.

The physical conditions envisaged are those shown in Fig. 5-22. A vertical stratification is maintained at the entrance to the inlet through the normal estuarine circulation conditions, which have been discussed in Section 5-6. We then anticipate that as one progresses into the inlet the vertical diffusion effects tend to smooth out the initial vertical salinity differences at the entrance to the inlet, leading to vertical salinity profiles

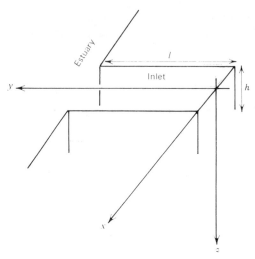

Fig. 5-22

as shown in Fig. 5-23. Taking coordinate directions as shown in Fig. 5-22, we see that the lateral salinity gradient $\partial s/\partial y$ is negative in the upper portion of the water column and positive in the lower portion of the water column. From relation (4-120) we have that the lateral density gradient force increases from zero at the surface to a maximum near middepth and decreases back to zero at the bottom, and that this force leads to a circulation in the direction from the head of the inlet toward its entrance. To balance this and to maintain a condition of no net circulation we must then have a surface slope to produce a force and a circulation in the

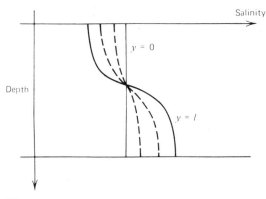

Fig. 5-23

opposite direction. As the surface slope force is constant as a function of depth, we anticipate that the resultant circulation will be a flow into the inlet near the surface, out of the inlet in the middepth region, and into the inlet near the bottom.

We wish to obtain a more quantitative notion of the characteristics of this circulation. From relations (4-120) and (5-33) we have for the lateral pressure gradient,

$$\frac{\partial p}{\partial y} = -g\rho_s \frac{\partial \zeta}{\partial y} + ga\rho_0 \int_0^z \frac{\partial s}{\partial y} \, dz \qquad (5\text{-}220)$$

Substituting this into the lateral relation similar to (4-86), defining i by a relation similar to (4-89), and taking ρ, ρ_s, and ρ_0 to be essentially the same in the resultant equation, we have

$$\frac{\partial^2 v_y}{\partial z^2} = -\frac{gi}{N_z} + \frac{ga}{N_z} \int_0^z \frac{\partial s}{\partial y} \, dz \qquad (5\text{-}221)$$

along with the lateral salt continuity equation corresponding to (5-164),

$$v_y \frac{\partial s}{\partial y} = K_z \frac{\partial^2 s}{\partial z^2} \qquad (5\text{-}222)$$

where we have also assumed that N_z and K_z are constants. We are interested in looking for a special solution which is separable in the variables y and z. We may then presume that v_y can be approximated by

$$v_y = \frac{y}{l} \varphi(z) \qquad (5\text{-}223)$$

varying in a linear manner from $v_y = 0$ at the head of the inlet to a maximum at its entrance. A separable solution can then be obtained if s is of a similar form given by

$$s = s_0 + b\left(\frac{y}{l}\right)^2 \theta(z) \qquad (5\text{-}224)$$

and also that the surface slope is given by

$$i = \frac{y}{l} i_0 \qquad (5\text{-}225)$$

where s_0 and b are defined by

$$s_0 = \frac{s(l,\ h) + s(l,\ 0)}{2} \qquad (5\text{-}226)$$

and

$$b = \frac{s(l,\ h) - s(l,\ 0)}{2} \qquad (5\text{-}227)$$

Equations (5-221) and (5-222) then reduce to the ordinary differential equations

$$\frac{d^2\varphi}{dz^2} = -\frac{gi_0}{N_z} + \frac{2abg}{N_z l} \int_0^z \theta \, dz \tag{5-228}$$

and

$$\frac{d^2\theta}{dz^2} = \frac{2b}{K_z l} \varphi\theta \tag{5-229}$$

It is difficult to obtain an exact solution to these coupled nonlinear differential equations that is readily interpretable. However, we may look for approximate solutions in terms of a series solution in the small quantity a, which also leads to separable equations. We then approximate φ and θ by

$$\varphi = \varphi_0 + a\varphi_1 + a^2\varphi_2 + \cdots$$

and   (5-230)

$$\theta = \theta_0 + a\theta_1 + a^2\theta_2 + \cdots$$

For the zero order terms we have from (5-228) that there is no zero order lateral density gradient term. To maintain a zero net flow condition there is necessarily no zero order surface slope term, and φ_0 is zero,

$$\varphi_0 = 0 \tag{5-231}$$

From (5-229) we see that, if φ_0 is a constant to the order a, that θ_0, meeting the zero flux conditions at the surface and the bottom, is given by

$$\theta_0 = -\cos\frac{\pi z}{h} \tag{5-232}$$

We further see from (5-229) that, as φ_0 is of the order a, $d^2\theta_0/dz^2$ must also be of the order a, so that we can relax the conditions for θ_0 from those given by (5-232) to simply a form of θ_0 that meets the zero flux conditions at the surface and the bottom.

Using the form of θ_0 given by (5-232), we then obtain the following equation for φ_1 to terms of the order a,

$$\frac{d^2\varphi_1}{dz^2} = -\frac{gi_0}{aN_z} - \frac{2bg}{N_z l} \int_0^z \cos\frac{\pi z}{h} \, dz \tag{5-233}$$

which gives on integration:

$$\varphi_1 = -\frac{gi_0}{2aN_z} z^2 + \frac{2bgh^3}{\pi^3 N_z l} \sin\frac{\pi z}{h} + C_1 z + C_2 \tag{5-234}$$

We use the same boundary conditions (4-93) and (4-94) as we did in

Section 4-5, and the continuity condition

$$\int_0^h \varphi_1 \, dz = 0 \qquad (5\text{-}235)$$

which gives, after a small amount of algebra, for φ_1 and then from (5-230) for φ, where n is as defined by (4-99),

$$\varphi = \frac{2}{\pi^3} \frac{abgh^3}{N_z l} \left[\pi(1-n) - \frac{3(\pi^2+4)}{4\pi}(1-n^2) + \sin \pi n \right] \qquad (5\text{-}236)$$

and the relation for i_0,

$$i_0 = -\frac{3(\pi^2+4)}{\pi^4} \frac{abh}{l} \qquad (5\text{-}237)$$

which is the same as the result obtained by Hansen and Rattray (1972) and Hansen and Festa (1974). A plot of (5-236) in terms of v_y/v_{ys} versus n, where v_{ys} is the surface value of v_y, is given in Fig. 5-24. It is to be

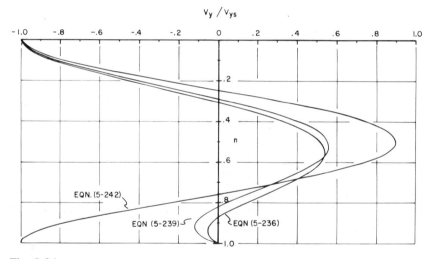

Fig. 5-24

noted that the bottom boundary condition of no motion leads to only a small near bottom motion into the inlet. As we see in the discussion that follows, relaxing this bottom flow condition leads to a more substantial near bottom motion.

Rather than taking a form for θ_0 as given by relation (5-232), we can choose a salinity distribution as given by (5-173) as being more representative of the normal estuarine salinity distribution at the entrance to the inlet. Then, from (5-194) and (5-227), we have for b of (5-228) and

(5-229),

$$b = \frac{3}{40} \frac{v_s h^2}{K_z} \frac{\partial s}{\partial x} \qquad (5\text{-}238)$$

Following the same procedures as before, we then obtain

$$\varphi = \frac{1}{1344} \frac{abgh^3}{N_z l} (\tfrac{16}{5}n^8 - \tfrac{48}{5}n^7 + \tfrac{336}{15}n^5 - \tfrac{112}{3}n^3 + \tfrac{67}{2}n^2 - 1) \qquad (5\text{-}239)$$

and the relation for i_0,

$$i_0 = -\frac{67}{2016} \frac{abh}{l} \qquad (5\text{-}240)$$

The corresponding plot for (5-239) is also included in Fig. 5-24 and is seen to be not too much different than that for (5-236). We may obtain some index of the magnitude of the inlet circulation velocities through the following consideration. Using successively the value for b as given by (5-238) and the values for v_s and λ as given by (4-100), (4-90), and (5-33), we obtain for the inlet circulation velocity at the surface and at the entrance to the inlet,

$$\begin{aligned}
v_y(l,\,0) &= \frac{-1}{1344} \frac{abgh^3}{N_z l} \\[2mm]
&= \frac{-1}{1344} \frac{agh^3}{N_z l} \left(\frac{3}{40} \frac{v_s h^2}{K_z} \frac{\partial s}{\partial x} \right) \\[2mm]
&= \frac{-1}{1344} \left(\frac{3}{40} \frac{v_s h^2}{K_z l} \right) 48 v_s \\[2mm]
&= \frac{-3}{1120} \frac{v_s h^2}{K_z l} v_s \qquad (5\text{-}241)
\end{aligned}$$

Taking typical values for an inlet of $l = 10$ km and $h = 10$ m, and values of $\ddot{v}_s = 10$ cm sec^{-1} and $K_z = 10$ cm^2 sec^{-1}, gives a value of unity for the second factor in (5-241), so that $v_y(l,\,0)$ is a small fraction of v_s.

For completeness and as some index of the importance of bottom friction in this type of circulation, we may wish to consider the limiting case in which there is no bottom friction, so that our bottom boundary condition is given by (4-114) and the salinity distribution at the entrance by (5-175). Following the same procedures again, we obtain in this case,

$$\varphi = \frac{11}{15,120} \frac{abgh^3}{N_z l} (\tfrac{18}{11}n^8 - \tfrac{72}{11}n^7 + \tfrac{252}{11}n^5 - \tfrac{504}{11}n^3 + \tfrac{306}{11}n^2 - 1) \qquad (5\text{-}242)$$

A plot of (5-242) is also included in Fig. 5-24. We see that in this case we have a substantial near bottom flow into the inlet and that the circulation is symmetric about $n = \tfrac{1}{2}$.

5-9. VERTICAL STABILITY AND VERTICAL MIXING

We examine here what is one of the more interesting and one of the more complex problems in the physical oceanography of estuaries. We are concerned with vertical mixing and vertical stability, in particular with the variation in the eddy coefficients N_z and K_z with such stability and with the effects of such variations on the water structure.

We have from the discussion of Section 2-8 the general condition related to vertical stability that the potential energy increase due to mixing must be less than the available kinetic energy of the turbulent motion, which from (2-140) we can state mathematically as

$$gK_z \frac{\partial \rho}{\partial z} < \rho N_z \left(\frac{\partial v_x}{\partial z} \right)^2 \tag{5-243}$$

We can then in a general way think of relation (5-243) as defining three regimes—one in which the left hand side is greater than the right hand side, a second in which the left hand side is much less than the right hand side, and a third which is a transition regime between the first two. The first is a *stable* regime in which there is negligible eddy diffusion across the sharp density gradient; we can treat the problem, as in the previous sections, as a stratified condition with entrainment mixing across the internal interface. The second is a *well mixed* regime in which the eddy coefficients are not variable with the internal water properties; we can treat the problem, again as in the previous sections, as one in which N_z and K_z are constants as a function of depth.

The transition regime is the one we wish to examine here. It is quite difficult to go beyond the defining relation of (5-243), and we can do so at best only in a rather subjective and approximate manner. There are three dimensionless ratios that are helpful in these considerations. One is the Richardson number Ri defined by (2-142) as

$$Ri = \frac{g \frac{\partial \rho}{\partial z}}{\rho \left(\frac{\partial v_x}{\partial z} \right)^2} \tag{5-244}$$

The second is simply the ratio of the eddy diffusivity to the eddy viscosity coefficient K_z/N_z. And the third is the *Richardson flux number Rf*, defined as the ratio of the potential energy increase due to mixing to the kinetic energy of the turbulent motion,

$$Rf = \frac{gK_z \frac{\partial \rho}{\partial z}}{\rho N_z \left(\frac{\partial v_x}{\partial z} \right)^2} = \frac{K_z}{N_z} Ri \tag{5-245}$$

In many ways the Richardson flux number is the more logical ratio to consider as the ratio of the two defining energies with which we are concerned. The Richardson number, however, has the advantage of including only physically observable quantities and not the two derived quantities N_z and K_z.

To obtain relations for N_z and K_z in terms of Rf or Ri we use the general line of argument given by Munk and Anderson (1948) with some changes. In particular we use a different argument to obtain a form for N_z, although both their argument and that given here are conjectural. The principal advantage of the forms developed here is that there are no undetermined coefficients in the equations for N_z and K_z, and that N_z and K_z are given in simple analytic forms expressible as a function of either Rf or Ri. In other words, the forms obtained here are directly comparable with experimental results without the inclusion of an adjustable parameter, and they are sufficiently simple that they permit consideration of inclusion in the defining motion and mixing equations.

To start, several observations should be included in the resultant forms for N_z and K_z. We have under a well mixed condition in which the potential energy term is negligible that N_z and K_z have essentially the same value, which we may express by the condition

$$A_0 = N_z = K_z \qquad Ri = Rf = 0 \qquad (5\text{-}246)$$

For increasing values of Ri or Rf we note that both N_z and K_z are less than A_0, such that their ratio is less than unity and becomes increasingly smaller with increasing Ri or Rf, which we may express in part by the condition

$$K_z < A_0$$
$$N_z < A_0 \qquad Ri, Rf > 0 \qquad (5\text{-}247)$$
$$\frac{K_z}{N_z} < 1$$

We have further that Rf is restricted to values between zero and unity and that Ri may be unbounded in a positive sense,

$$0 < Rf < 1$$
and
$$0 < Ri < \infty \qquad (5\text{-}248)$$

We further impose the condition that Rf increases in a regular manner as a function of Ri, which seems reasonable, and in particular that there are no maxima for Rf as a function of Ri,

$$\frac{dRf}{dRi} > 0 \qquad (5\text{-}249)$$

For the well mixed case the turbulent kinetic energy is entirely dissipated by the small scale turbulence of the fluid in the form of heat H_0, which we may express by

$$\rho A_0 \left(\frac{\partial v_x}{\partial z}\right)^2 = H_0 \tag{5-250}$$

As Rf increases from zero toward unity, the vertical turbulence decreases, reaching zero when Rf reaches unity. Let us then make the assumption that H varies linearly with Rf, giving

$$H = H_0(1 - Rf) \tag{5-251}$$

For the transition regime we then have for the energy relation, equating available turbulent kinetic energy to the sum of the dissipation loss plus the potential energy increase due to mixing,

$$\rho N_z \left(\frac{\partial v_x}{\partial z}\right)^2 = H + gK_z \frac{\partial \rho}{\partial z} \tag{5-252}$$

We note that the potential energy term is proportional to the vertical salt flux. Thus, if we have a change in the independent variable Rf such that the velocity shear remains unchanged, we will also have from the mixing equation, such as (5-164), that the longitudinal velocity and the salt flux at any given depth remain unchanged. In (5-252) we then have that the $(\partial v_x/\partial z)^2$ and $K_z(\partial \rho/\partial z)$ terms as well as the H_0 term are invariant with Rf and, from (5-250) and (5-251), that N_z/A_0 is consequently given by

$$\frac{N_z}{A_0} = 1 - Rf \tag{5-253}$$

As this expression includes only the independent variable Rf, we may expect that it will hold for all cases in which (5-252) applies.

In order to obtain a similar relation for K_z/A_0 or K_z/N_z, we have the following considerations as stated by Jacobsen (1930) for oceanic turbulence and diffusion and as discussed more fully in Frenkiel and Sheppard (1959) for atmospheric turbulence and diffusion. A priori there is no particular reason to consider that the details of momentum transfer of turbulence as related to the coefficient N_z and the material transfer of salt or heat by diffusion as related to the coefficient K_z are the same, even though the generating eddy scale is the same in both cases. On an elemental scale we can consider that the momentum transfer is rapid by collisions of turbulent mass particles with mass particles of the intruded medium, or alternatively by the frictional drag of the intruding turbulent mass on the new surroundings, whereas the heat, and presumably the salt, transfer occurs more slowly, ultimately by conduction. Stated another way, the spectra of the heat or salt transfer should be of lower frequency than the spectra associated with the momentum transfer, which is the case. Then, for a vertically stratified medium as contrasted with a homogeneous water mass we may anticipate that the turbulent elements will be moved to new surroundings by gravitational forces before the equalization can be completed, and consequently that there will be an additional effect such that K_z becomes smaller than N_z and the ratio K_z/N_z is in some way dependent on Rf. This is indeed the case, and we have in general that K_z/N_z decreases with increasing Rf. As in the case of (5-251), let us then make the assumption, for lack of any better reasoning, that K_z/N_z also varies linearly with Rf, or

$$\frac{K_z}{N_z} = 1 - Rf \tag{5-254}$$

We then have in summary from (5-253), (5-254), and (5-245), for the various parameters in terms of Rf,

$$\frac{N_z}{A_0} = 1 - Rf$$

$$\frac{K_z}{A_0} = (1 - Rf)^2$$

$$\frac{K_z}{N_z} = 1 - Rf \tag{5-255}$$

$$Ri = \frac{Rf}{1 - Rf}$$

From the fourth equation in (5-255) we also have for the various parameters in terms of Ri,

$$\frac{N_z}{A_0} = \frac{1}{1 + Ri}$$

$$\frac{K_z}{A_0} = \frac{1}{(1 + Ri)^2}$$

$$\frac{K_z}{N_z} = \frac{1}{1 + Ri} \tag{5-256}$$

$$Rf = \frac{Ri}{1 + Ri}$$

We see then that relations (5-255) and (5-256) do indeed satisfy our observational conditions of (5-246) through (5-249).

It is of interest to compare these results with those given by Munk and Anderson (1948) whose defining equations for N_z/A_0 and K_z/A_0 in terms

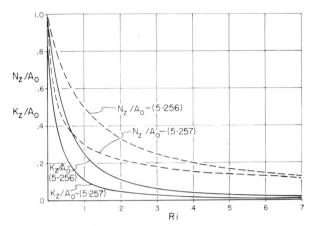

Fig. 5-25

of Ri are

$$\frac{N_z}{A_0} = (1 + 10Ri)^{-1/2}$$

(5-257)

$$\frac{K_z}{A_0} = (1 + \tfrac{10}{3}Ri)^{-3/2}$$

Figure 5-25 is a graph of the two functions for the two cases. We see that the equations in the two situations have a similar variation, the Munk and Anderson curves showing somewhat lower values for any given Ri.

It is of further interest to compare these results with experimental data. As may be anticipated there is a paucity of sufficiently detailed measurements to permit calculations of N_z, K_z, and Ri. In cases in which measurements are that detailed, the only justifiable comparison appears to be that of K_z/N_z versus Ri. Two such sets of measurements appear to be appropriate, one by Jacobsen (1913, 1918) off the Danish coast and the other by Bowden and Gilligan (1971) for the Mersey estuary, England. These results are plotted in Fig. 5-26 with the curve for K_z/N_z as given by the third equation in (5-256). The Jacaobsen data are also summarized in Taylor (1931). We see that the curve gives a surprisingly good fit to the experimental data, showing somewhat lower values than those measured for small Ri, and somewhat higher values than those measured for large Ri.

From the graphs of Figs. 5-25 and 5-26 we can think of the transition regime as being defined, at the outside, by values of Ri within the range defined by $0.1 < Ri < 10$. For values of $Ri < 0.1$ we may consider the relations of (5-246) to apply, and for values of $Ri > 10$ we may consider

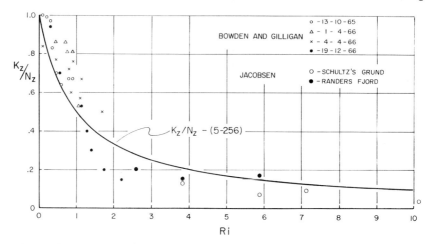

Fig. 5-26

the fluid in this region to be stratified with an internal discontinuity interface across which there is exchange by entrainment only.

We are now in a position to apply these results to the longitudinal motion and vertical mixing equations to ascertain the resultant circulation velocity and vertical salinity structures. In Sections 4-5 and 5-6 we solved these equations for assumed constant values of N_z and K_z. We now have five equations, (2-146), (5-164), (5-244), and the first two equations in (5-256), in the five dependent variables v_x, s, N_z, K_z, and Ri to solve. This appears difficult to do even in terms of the variables $\partial v_x / \partial z$ and $\partial s / \partial z$ rather than v_x and s, because of the apparently conflicting nature of the boundary conditions imposed by the motion and mixing equations and by the Richardson number relation.

Rather than seeking a general solution, then, we concentrate on examining the Richardson number effects in the middepth halocline region. For this purpose we arbitrarily choose for the Richardson number as a function of depth an expression of the form

$$Ri = E(n - n^2) \tag{5-258}$$

which has the desired characteristic of having a maximum at $n = \frac{1}{2}$ and decreasing to zero, or to the anticipated normal conditions, at the surface and the bottom. From the motion equation (2-146) in conjunction with the right hand side of (4-88) and boundary conditions (4-93) and (4-114), we have

$$N_z \frac{\partial v_x}{\partial z} = -\frac{gah^2}{2\rho} \frac{\partial s}{\partial n} (n - n^2) \tag{5-259}$$

where a is the constant of proportionality of (5-33) between a salinity increase and the corresponding density increase, and where the surface slope i is again given by relation (4-115). Substituting for N_z from the first equation in (5-256), with (5-258), we then obtain

$$v_x = -\frac{gah^3}{2\rho A_0}\frac{\partial s}{\partial x}\left[\frac{n^2}{2}-\frac{n^3}{3}+E\left(\frac{n^3}{3}-\frac{n^4}{2}+\frac{n^5}{5}\right)\right]+C$$

$$= \frac{1}{24}\frac{gah^3}{\rho A_0}\frac{\partial s}{\partial x}\left[(1-6n^2+4n^3)+\frac{E}{5}(1-20n^3+30n^4-12n^5)\right]$$

$$(5\text{-}260)$$

where the constant of integration has been evaluated from the continuity relation (4-96) with $r=0$. We could have alternatively used the bottom boundary conditions (4-94) or (4-107) with somewhat more cumbersome algebra, without essentially changing the effect of the Richardson number in the middepth region. We see from relation (5-260) that for large values of the Richardson number the circulation velocity increases linearly with E. A plot of v_x/v_{s0}, where v_{s0} is the surface value of v_x for $Ri=0$, as given by (4-117), is shown in Fig. 5-27A for various values of E corresponding to middepth values of $Ri=0$, 0.25, 0.50, 1.25, 2.50, and 5.00.

From the mixing equation (5-164) we then have, using (5-260) and taking one integration with respect to z,

$$K_z\frac{\partial s_1}{\partial z}=\frac{gah^4}{\rho A_0}\left(\frac{\partial s}{\partial x}\right)^2\left[\frac{1}{24}(n-2n^3+n^4)+\frac{E}{120}(n-5n^4+6n^5-2n^6)\right] \quad (5\text{-}261)$$

where the constant of integration disappears in consideration of the boundary conditions (5-170). Substituting for K_z from the second equation in (5-256), with (5-258), we have

$$\frac{\partial s_1}{\partial z}=\frac{1}{24}\frac{gah^4}{\rho A_0^2}\left(\frac{\partial s}{\partial x}\right)^2\left[(n-2n^3+n^4)+\frac{E}{5}(n-5n^4+6n^5-2n^6)\right]$$

$$\times[1+E(n-n^2)]^2 \quad (5\text{-}262)$$

We see from this relation that, for large values of the Richardson number, the vertical salinity deviation increases as E^3. A plot of $s_1/v_{s0}h^2(\partial s/\partial x)/A_0$ versus n for the same values of E is given in Fig. 5-27B. Taking the horizontal scale as an index of salinity, which is not unreasonable, we see that the salinity structure changes rather rapidly from a condition that would be described as well mixed to one that would be described as noticeably stratified. Beyond values of about $E=10$ there is a still further rapid increase in the salinity gradient in the middepth region, and in this case the salinity difference from the surface to the bottom is determined

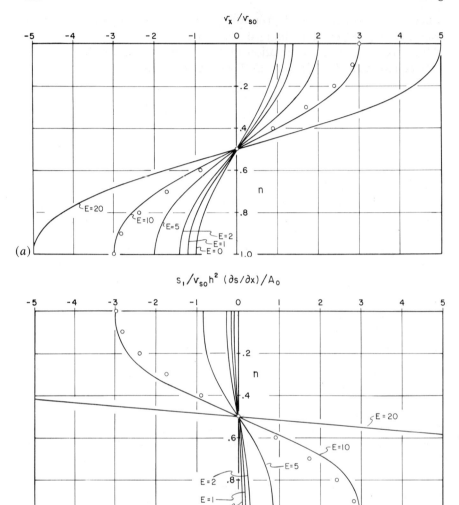

Fig. 5-27

by other longitudinal and entrainment factors and not simply by relation
(5-262).

It is of interest to make a comparison of these results with what would
have been obtained had we used a constant value for the coefficients N_z
and K_z. We have included such a comparison for $E = 10$ in Figs. 5-27A
and 5-27B, using equations (4-116) and (5-175) and choosing values of
N_z and K_z, respectively, to agree with the surface and bottom values of

TABLE 5-2

	$E = 0$	$E = 1$	$E = 2$	$E = 5$	$E = 10$	$E = 20$
N_z/A_0, depth-averaged	1.00	0.86	0.76	0.58	0.43	0.31
N_z/A_0, minimum	1.00	0.80	0.67	0.44	0.29	0.17
N_z/A_0, constant value	1.00	0.83	0.71	0.50	0.33	0.20
K_z/A_0, depth-averaged	1.00	0.75	0.59	0.37	0.23	0.15
K_z/A_0, minimum	1.00	0.64	0.45	0.20	0.08	0.03
K_z/A_0, constant value	1.00	0.68	0.50	0.24	0.10	0.04

the circulation velocity and the salinity deviation. These calculations are shown as the open circles in the two figures, and it is seen that they are not too dissimilar from the plotted curve. We can conclude that, for well mixed and stratified estuaries with a vertical stability or Richardson number effect, the simple, constant coefficient equations can still apply with a judicious choice of the coefficient values used. In Table 5-2 are summarized the depth-averaged and minimum values of N_z and K_z determined from relations (5-256) along with (5-258), and the constant values of N_z and K_z required to fit the resultant curves. We see that the constant values are less than the depth-averaged values, approaching the minimum values as the value of E increases.

In conclusion it is of some interest to go back and look at the form of Ri in terms of the relations (5-256). From the first two of the relations (5-256) we have that $N_z^2 = A_0 K_z$ so that equation (5-244) may be rewritten as

$$Ri = \frac{ga\rho A_0}{f_{zx}^2} K_z \frac{\partial s}{\partial z} \qquad (5\text{-}263)$$

In this expression the Richardson number is directly proportional to the vertical diffusion salt flux and inversely proportional to the square of the internal frictional stress. In general, in estuaries the f_{zx}^2 term is determined by tidal velocity shear. In other words, the available kinetic energy of the turbulent motion is derived from tidal motion, particularly as determined by tidal bottom frictional stress. Taking this as an index of f_{zx}^2, we have from (3-59),

$$\overline{f_b^2} = \frac{1}{2} \left(\frac{8}{3\pi} k\rho C_b^2 \right)^2 \qquad (5\text{-}264)$$

We then have that Ri varies as C_b^{-4}, so that the transition regime is abbreviated in terms of the amplitude of the tidal velocity variation. We may then expect, as discussed before, that there will be two preferred situations, either well mixed or noticeably stratified.

For further information on the material covered in this section the reader may refer to Munk and Anderson (1948), Taylor (1931), Jacobsen (1918, 1930), Bowden and Gilligan (1971), Richardson (1920), and Rossby and Montgomery (1935).

5-10. SEASONAL THERMOCLINE AND HEAT EFFECTS

The *seasonal thermocline* is a thermal gradient in the upper portions of the water column, which is associated with the annual heat cycle. It is a typical feature of all areas with large yearly variations in surface temperature. It is directly related to the net downward heat flux across the ocean surface during the late spring, summer, and early fall months.

As surface waters gradually grow warmer with the advance of spring, a negative temperature gradient develops as a function of the vertical, or depth, coordinate. This then corresponds to a positive density gradient. As we shall see later in this section, for an annual heat cycle with a constant eddy diffusion coefficient K_z, the temperature gradient is a maximum at the surface. Depending then on the magnitude of $(\partial v_x/\partial z)^2$, or f_{zx}^2, near the surface, the Richardson number of (5-244) may become sufficiently large so that the amount of vertical mixing is curtailed or limited, with a consequent substantial decrease in K_z. If these conditions are met, a thermocline will then develop, and this thermocline will act as a deterrent to the flow of heat to the cooler waters beneath. As the temperature of the layer above the thermocline continues to rise, the thermocline itself grows steeper and becomes more of a barrier to downward mixing. As we see later in this section, the thermocline progresses downward with a velocity dependent in part on the lowered value of K_z associated with it.

The conditions for the development of a seasonal thermocline are of course dependent on the denominator in (5-244). For a vertically extended body of water with negligible tidal effects, such as the deep ocean, wind-induced internal frictional effects decrease rapidly with depth. We may then anticipate a seasonal thermocline to develop under such conditions. For a coastal or estuarine locality with tidal currents the tidal induced internal frictional effects decrease upward in a nearly linear manner according to (3-64). Depending on the magnitude of this combined effect, the Richardson number condition may or may not be reached near the surface. In the latter case the normal value of K_z is effective throughout the entire water column; the vertical mixing remains good and the observed temperature gradients are slight. In the former case a thermocline develops, which progresses downward toward the bottom or to a depth at which the Richardson number is further reduced by the tidal frictional effects. For a stratified estuary with a permanent halocline we may anticipate that this density stratification will act as a barrier to further descent of the thermocline.

One of the difficulties in giving an adequate theoretical description of the development of the seasonal thermocline, particularly in deep ocean waters, is in the definition of the stress function f_{zx}^2. Presumably, from a wind-induced circulation such as that described in Section 4-10 the wind stress decreases in some exponential fashion with depth. Then, below the thermocline we should have stress terms less than those in the vicinity of the thermocline. It is also an observed fact that the temperature is nearly isothermal beneath the thermocline. Thus, since Ri has returned to a normal value, we have from relation (5-263), where $K_z(\partial s/\partial z)$ now represents the vertical heat flux, that either $K_z(\partial s/\partial z)$ must decrease substantially or f_{zx}^2 increase substantially over their corresponding values above the thermocline. The latter condition implies a minimum for f_{zx}^2 at the thermocline and is in violation of what we normally consider the variation in f_{zx}^2 to be. The former condition seems the more likely, and it does imply that for the deep ocean the temperature beneath the thermocline should show only a relatively slight seasonal dependence. For coastal and estuarine waters we have a different situation for the f_{zx}^2 dependence in that, from the combined wind and tidal effects, a minimum for f_{zx}^2 occurs near the surface. Condition (5-263) can now be relaxed with regard to the flux dependence. There can be a substantial heat flux in the lower layer, and the temperature beneath the thermocline can show a substantial seasonal dependence. These difference considerations for the waters beneath the thermocline in the deep and shallow water cases are indeed in accordance with the observations.

Another consideration associated with this problem should be mentioned. There is a substantial difference between this heat flux problem and the salt flux problems we previously considered, in that there is now a flux across one of the bounding surfaces. Previously our condition was that of no flux across the boundary, so that the vertical temperature or salinity gradient approached zero near the boundary. We now have the opposite condition that the gradient is a maximum, or near maximum, at the surface.

A good example of the seasonal thermocline for an unrestricted deep ocean area is shown in Figs. 5-28 and 5-29. Figure 5-28 is a seasonal plot of average monthly temperature versus depth curves and Fig. 5-29 of average isotherms. The negative temperature gradient at the surface is first observed in April and soon develops into a thermocline. The thermocline itself is most pronounced during the summer months. It progresses downward at a relatively uniform rate, retaining the same sharpness but less of a total temperature decrease into the fall and winter months.

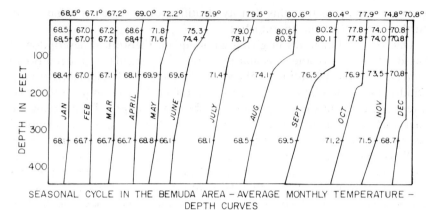

SEASONAL CYCLE IN THE BEMUDA AREA – AVERAGE MONTHLY TEMPERATURE –
DEPTH CURVES

Fig. 5-28   Reproduced with permission from C. O'D. Iselin, 1946, Summary
Technical Report 6A, National Defense Research Committee, Washington D.C.,
Fig. 35.

It is understandably difficult to derive meaningful expressions for the
temperature as a function of depth in the transition regime of the
Richardson number. We can look only at the temperature distribution in
the ocean for a periodic seasonal heat source for the case in which K_z
remains constant as a function of depth. We correctly take the heat flux to
be uniform over the ocean surface, so that the problem reduces to a one

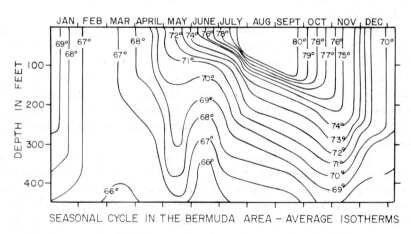

SEASONAL CYCLE IN THE BERMUDA AREA – AVERAGE ISOTHERMS

Fig. 5-29   Reproduced with permission from C. O'D. Iselin, 1946, Summary
Technical Report 6A, National Defense Research Committee, Washington D.C.,
Fig. 36.

dimensional vertical heat flow problem. The derivation of the defining equation is the same as that given in Section 2-4 for salt flux, for which we have from (2-57),

$$\frac{\partial T}{\partial t} = K_z \frac{\partial^2 T}{\partial z^2} \tag{5-265}$$

where K_z is considered a constant. The corresponding heat flux is given by

$$F_T = -c\rho K_z \frac{\partial T}{\partial t} \tag{5-266}$$

where c is the specific heat. Equation (5-265) is the one dimensional heat conduction equation whose solutions are given in detail in such standard texts as Carslaw and Jaeger (1959). For a periodic heat flux at the surface of the form $e^{i\omega t}$, the general solution to (5-265) for the temperature variations T about the yearly ambient temperature is

$$T = A e^{i(\omega t - \epsilon)} e^{\pm(1+i)kz} \tag{5-267}$$

as can be seen by substitution back into (5-265), where

$$k = \sqrt{\frac{\omega}{2K_z}} \tag{5-268}$$

We see that solution (5-267) defines in part a wave motion propagated with a velocity V given by

$$V = \frac{\omega}{k} = \sqrt{2\omega K_z} = \sqrt{4\pi \frac{K_z}{P}} \tag{5-269}$$

where P is the period. Applying (5-269) to the descent of the seasonal thermocline as defined in Fig. 5-29, we obtain a value of $K_z = 0.05$ for the eddy diffusion coefficient in the vicinity of the thermocline; we note that this value is two orders of magnitude or more less than the values normally associated with estuarine waters, presuming that such a simplified calculation has any meaning at all.

For the example of an unbounded ocean in the vertical direction and for a constant K_z and with a seasonal heat flux at the surface given by

$$F_T = F_0 \cos \omega t \tag{5-270}$$

the form of solution for T is

$$T = A e^{-kz} \cos (\omega t - kz - \epsilon) \tag{5-271}$$

from (5-267). Using (5-270) to determine A and ϵ, we have

$$F_T = F_0 \cos \omega t = -c\rho K_z \left(\frac{\partial T}{\partial z}\right)_{z=0}$$

$$= c\rho k K_z A [e^{-kz} \cos(\omega t - kz - \epsilon) - e^{-kz} \sin(\omega t - kz - \epsilon)]_{z=0}$$

$$= \sqrt{2} c\rho k K_z A \cos\left(\omega t - \epsilon + \frac{\pi}{4}\right) \qquad (5\text{-}272)$$

from which T is then given by

$$T = \frac{F_0}{\sqrt{2} c\rho k K_z} e^{-kz} \cos\left(\omega t - kz - \frac{\pi}{4}\right) \qquad (5\text{-}273)$$

We see from (5-272) and (5-273) that the maximum temperature gradient without regard to phase occurs at the surface, and that its maximum at the surface with respect to t occurs at midsummer. We also see that there is a phase lag of $\pi/4$ for the maximum temperature at the surface with respect to the maximum for the heat flux. Now considering (5-273) a function of the dimensionless quantities ξ and τ defined by

$$\xi = kz \qquad \tau = \omega t - \frac{\pi}{4} \qquad (5\text{-}274)$$

a graph of the function $f(\xi, \tau)$ defined by

$$f(\xi, \tau) = e^{-\xi} \cos(\tau - \xi) \qquad (5\text{-}275)$$

as a function of ξ is shown in Fig. 5-30 for various values of τ. We also see from (5-272) and (5-273) that Fig. 5-30 is in addition a graph of $\partial T/\partial z$, where τ is now given by $\tau = \omega t$. For the normal depths and values of K_z associated with estuarine waters a graph of T versus z is crowded into the region from $\xi = 0$ to $\xi = 0.1$, so that without the development of a thermocline the temperature of estuarine waters shows the seasonal temperature cycle with modest and variable temperature gradients from the surface to the bottom.

A somewhat different and more instructive approach to the development and movement of the seasonal thermocline near the ocean surface has been taken in terms of the wind energy and heat input at the ocean surface related to the formation and existence of the isothermal, or nearly isothermal, layer overlying the thermocline. This formulation has been given and applied to the deep ocean by Turner and Kraus (1967) and Kraus and Turner (1967) based in part on some observations and considerations by Stommel and Woodcock (1951), Francis and Stommel (1953), and Tully (1964). It has also been extended and applied to the seasonal thermocline in lakes by Darbyshire and Edwards (1972) and Darbyshire and Jones (1974).

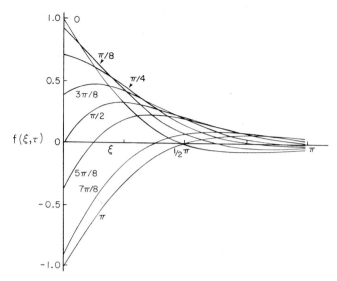

Fig. 5-30 Reproduced with permission from H. S. Carslaw and J. C. Jaeger, 1959, *Conduction of Heat in Solids*, Oxford University Press, London, Fig. 7.

There are two principal considerations in this formulation, which are justified by the observations in nature. First, it is assumed that the heat inputs, principally from solar radiation and beginning in early spring, as measured over time periods of a few days to a few weeks, are confined to the isothermal layer overlying the thermocline. In other words, it is assumed that the temperature of the water beneath the thermocline remains unchanged, which is indeed in agreement with observations in the deep ocean and in lakes. Second, it is assumed that the total energy input of the wind stress at the surface, or a constant fraction thereof, goes into the work that has to be done against gravity, or potential energy decrease, of replacing the colder and denser water by the warmer and lighter water in the isothermal layer. Then it is possible to relate the two dependent variables, the temperature of the isothermal layer and the depth of the thermocline, to the two independent variables, the heat input and wind stress energy input.

Following Darbyshire and Edwards (1972) and Darbyshire and Jones (1974) we then write in simplified form,

$$Q = c\rho_0(T - T_0)d \tag{5-276}$$

and

$$W = \tfrac{1}{2}g(\rho_0 - \rho)d^2 \tag{2-277}$$

where T_0 and ρ_0 are the temperature and density of the water beneath the thermocline, T and ρ are the temperature and density of the water in the isothermal layer above the thermocline, d is the depth of the thermocline, Q is the total heat input beginning sometime in early spring, and W is the corresponding wind stress energy input. Taking the density to be a quadratic function of the temperature in the form

$$\rho = \gamma + \alpha T - \beta T^2 \tag{5-278}$$

where the third term is dominant over the second term in the sensible range of temperatures observed, (5-276) and (5-277) can be solved for d, giving

$$d = \frac{\dfrac{2c\rho_0}{g} \dfrac{W}{Q} - \dfrac{\beta}{c\rho_0} Q}{2\beta T_0 - \alpha} \tag{5-279}$$

The dominant term in the expression (5-279) is W/Q. For increasing total W or decreasing total Q the thermocline will descend. It is anticipated then that following periods of high winds the thermocline will be at a greater depth and in general during the cooling season from late summer into autumn the thermocline will descend. These predictions are in agreement with observations.

During early spring when the incremental heat inputs are small and variable it is anticipated that for periods of high winds and small heat inputs an incipient thermocline will descend rapidly to the lake bottom and thus form a new reference temperature, T_0, for the lake as a whole. This process may be repeated several times until, with the approach of summer and larger incremental heat inputs, a stable seasonal thermocline is established. Again this is in agreement with observations.

To continue this type of analysis further, let us consider another aspect of what can happen during periods of positive incremental heat inputs, ΔQ. During such periods a new thermocline may be formed above the existing thermocline. Consider an interval of time over which there has been an incremental heat input, ΔQ, and an incremental wind stress energy input, ΔW. Then, following the same procedures as before, we define a quantity d' similar to d of expression (5-279) by

$$d' = \frac{\dfrac{2c\rho_0}{g} \dfrac{\Delta W}{\Delta Q} - \dfrac{\beta}{c\rho_0} \Delta Q}{2\beta T_s - \alpha} \tag{5-280}$$

where T_s is the temperature of the existing isothermal layer. If d' is greater than d we proceed as before, simply using (5-279) with ΔQ and ΔW added to Q and W to determine the new depth of the thermocline. If d' is less than d, a new thermocline is formed above the old thermocline at a depth d'. This new thermocline may then in time descend to the depth of the original thermocline or may have other thermoclines formed above it. Using this procedure, a reasonable picture of the seasonal thermal structure in a lake was obtained by Darbyshire and Jones (1974).

For further information on the material covered in this section the reader may wish to refer to Fjeldstad (1933), Dietrich (1954, 1957), Iselin (1946), and Munk and Anderson (1948) in addition to the references cited above.

CHAPTER 6

Pollutant Dispersion

6-1. INTRODUCTION

One of the more important applications of the physical oceanography of estuaries and associated coastal waters concerns the description of the dispersion of various potential pollutants in these waters. We may consider for most cases that any such problem can be separated into a *near field* and a *far field* study. The near field study deals with the pollutant source outlet configuration, geometry, velocities, etc., and can usually be treated as a problem in engineering hydraulics with little recourse to the physical oceanography of the surrounding waters. We are not concerned with such problems here; they properly form a separate area of study in themselves. In the far field study we are concerned with the distant distribution of a potential pollutant throughout the adjoining waters; this is an application of the physical oceanography of estuaries. Generally, for such applications one takes an idealized distributed point or other source condition to approximate the near field source configuration.

In such applications of physical oceanography one immediately runs into several problems which are usually not part of a corresponding general scientific study of the same waters. The source, with its near field configuration and with any time variations in the pollutant input, becomes important. We may be dealing with a conservative pollutant or with a nonconservative pollutant. We may be dealing with a problem that involves coupled differential equations of the pollutant and derived quantities such as biochemical oxygen demand (BOD) and dissolved oxygen (DO). And, most importantly, we must include the actual geometric configuration of the coastal water body.

These various conditions in general, and particularly that of the physical geometry of the surrounding waters, often prohibit a simple analytic solution to the defining equations. Recourse must then be made to computer solutions of the equations in finite difference form with the actual source and boundary conditions. In turn, the size of the matrices involved in such computer calculations often necessitates the inclusion of only the simplest defining equations, such as the longitudinal dispersion equations of Section 5-4. One might say that for such calculations a compromise has been made in the inclusion of the actual geometry to the exclusion of the definition of the physical oceanographic circulation and mixing involved.

We investigate here some pollutant dispersion problems for which simple analytic solutions can be given. We are interested principally in problems for which the defining equations can be given in one dimensional form. These solutions can then serve as an end in themselves for problems to which they are applicable and as an index to the type of results to be obtained from computer calculations with the actual geometry.

6-2. LONGITUDINAL CONSERVATIVE DISPERSION

For a conservative pollutant a logical beginning is the development by Ketchum (1955) as given in the derivation of (5-22) to (5-24) or, equivalently, in the derivation of (5-105) and (5-106). We repeat these equations here for reference,

$$c_0 = \frac{W}{R} f_0 \tag{6-1}$$

where W is the rate of supply of pollutant in mass of pollutant per unit time, R is the total river runoff flow in volume per unit time, c_0 is the cross sectional average pollutant concentration at the outfall in mass of pollutant per unit volume of sea water, and f_0 is the freshwater fraction at the outfall, and

$$c = c_0 \frac{f}{f_0} \tag{6-2}$$

in a downestuary direction, and

$$c = c_0 \frac{s}{s_0} \tag{6-3}$$

in an upestuary direction, where c, f, and s are the pollutant concentration, freshwater fraction, and salinity, respectively, at any given downestuary or upestuary location. The distribution of a pollutant, then, can be given quite simply from the physical observations of the salinity distribution in an estuary without recourse to any further calculations. When it is applicable, it provides a most useful and direct method.

Let us now investigate the analytic solution for the distribution of a conservative pollutant when the effective longitudinal dispersion coefficient K_x and the cross sectional area A can be considered constant. For reference, we first obtain the solution for the freshwater fraction f under such conditions. We effectively obtained this solution previously in (5-119). The defining relation for this case of one dimensional dispersion is that given by (5-103), which we may rewrite as

$$v_x = v_x f - K_x \frac{df}{dx} \tag{6-4}$$

where v_x and f represent $\langle \bar{v}_x \rangle$ and $\langle \bar{f} \rangle$, as before in (5-98) and (5-99), and v_x is defined by

$$v_x = \frac{R}{A} \tag{6-5}$$

Equation (6-4) is a simple ordinary differential equation whose solution is

$$f = 1 + C_1 e^{(v_x/K_x)x} \tag{6-6}$$

As in solution (5-119) we cannot specify a boundary condition at $x = 0$, the upper end of the estuary, since K_x is not zero there. We choose, as before, our boundary condition at the lower end of the estuary,

$$f = 0 \qquad x = L \tag{6-7}$$

so that the final solution is

$$f = 1 - e^{-(v_x/K_x)(L-x)} \tag{6-8}$$

which is the same as (5-119).

For the conservative pollutant problem it is convenient to change our coordinate system so that $x = 0$ at the pollutant outfall, as shown in Fig. 6-1. The defining equations are then (5-101) and (5-102) in a downestuary and an upestuary direction, respectively, which we may rewrite as

$$v_x \psi = v_x c - K_x \frac{dc}{dx} \tag{6-9}$$

and

$$0 = v_x c - K_x \frac{dc}{dx} \tag{6-10}$$

where c represents $\langle \bar{c} \rangle$, and ψ is defined by

$$\psi = \frac{W}{R} = \frac{W}{v_x A} \tag{6-11}$$

and is simply the average concentration of pollutant over a cross section normal to the flow at the outfall location that would result if there were no longitudinal dispersion. The solutions to (6-9) and (6-10) are simply

$$c_d = \psi + C_2 e^{(v_x/K_x)x} \tag{6-12}$$

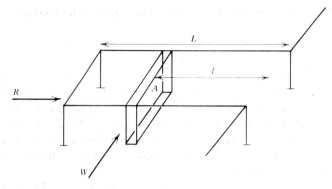

Fig. 6-1

and

$$c_u = C_3 e^{(v_x/K_x)x} \tag{6-13}$$

respectively. To conform with the boundary condition (6-7) we take as our boundary conditions,

$$c_d = 0 \qquad x = l \tag{6-14}$$

and

$$c_d = c_u = c_0 \qquad x = 0 \tag{6-15}$$

Substituting these boundary conditions into (6-12) and (6-13), we obtain for the final solutions

$$c_d = \psi[1 - e^{-(v_x/K_x)(l-x)}] \tag{6-16}$$

and

$$c_u = \psi[1 - e^{-(v_x/K_x)l}]e^{(v_x/K_x)x} \tag{6-17}$$

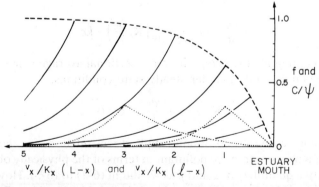

Fig. 6-2 Reproduced with permission from H. Stommel, 1953, *Sewage and Industrial Wastes,* **25,** 1065–1071, Fig. 3.

where the cross sectional average concentration at the source c_0 is given by

$$c_0 = \psi[1 - e^{-(v_x/K_x)l}] \tag{6-18}$$

We see then that, since $L - x$ in (6-8) is the same as $l - x$ in (6-16), a graph of the ratio c_d/ψ follows the freshwater fraction curve, the ratio c_0/ψ is also given by the freshwater fraction curve value, and c_u/ψ decreases in an exponential manner in an upestuary direction. These various relations are shown in Fig. 6-2 for outfall locations at abscissa values of 0.5, 1, 2, 3, 4, and 5. The similarity of this figure and these relations to those given by Fig. 5-4 should be noted.

For further information on the material covered in this section the reader may refer to Ketchum (1955), Stommel (1953), and Dyer (1973).

6-3. LONGITUDINAL NONCONSERVATIVE DISPERSION

Let us now look at the same relations for a nonconservative pollutant. We consider a nonconservative pollutant whose rate of decay is proportional to the amount of pollutant present, so that

$$\frac{\partial c}{\partial t} = -kc \tag{6-19}$$

or

$$c = c_0 e^{-kt} \tag{6-20}$$

in an isolated experiment. For such an experiment the half life of the pollutant decay is given by $0.693/k$ in terms of the decay constant k. We must then go back to our original equation of mass continuity (2-65) in one dimensional form and include this decay, so that the equation now reads,

$$\frac{\partial c}{\partial t} = -v_x \frac{\partial c}{\partial x} + \frac{\partial}{\partial x}\left(K_x \frac{\partial c}{\partial x}\right) - kc \tag{6-21}$$

or, for longitudinal dispersion with cross sectional average values of c and with constant K_x and A under steady state conditions,

$$K_x \frac{d^2 c}{dx^2} - v_x \frac{dc}{dx} - kc = 0 \tag{6-22}$$

We cannot go back and obtain simple direct relations for the concentration of a nonconservative pollutant in terms of the physically observable conservative quantities of s and f as we did in (6-1) to (6-3). However, we can obtain some index of the variation of a nonconservative pollutant in terms of these quantities and the decay constant k through the modified tidal prism method. We previously had the expression (5-20) for the freshwater volume fraction in any tidal prism segment n. When the pollutant outfall is located in the same tidal prism segment, we replace (5-20) for the volume fraction of pollutant V_{cn} in this segment by

$$V_{cn} = V_R[1 + (1 - r_n)e^{-kT} + (1 - r_n)^2 e^{-2kT} + \cdots + (1 - r_n)^m e^{-mkT} + \cdots]$$

$$= \frac{V_R}{1 - (1 - r_n)e^{-kT}} \tag{6-23}$$

where T is the tidal period. We then multiply (6-1) by V_{cn}/V_{fn} to find the average concentration of a nonconservative pollutant in this tidal segment,

$$c_n = \frac{W}{R} f_0 \frac{r_n}{1 - (1 - r_n)e^{-kT}} = c_0 \frac{r_n}{1 - (1 - r_n)e^{-kT}} \tag{6-24}$$

For the next tidal segment seaward the same relations hold, and we multiply again by a similar ratio for this segment to obtain the concentration value there. If r is the same for all segments, we will have for (6-2),

$$c_p = c_0 \frac{f_p}{f_0} \left[\frac{r}{1 - (1 - r)e^{-kT}} \right]^{p+1-n} \tag{6-25}$$

and for (6-3)

$$c_p = c_0 \frac{S_p}{S_0} \left[\frac{r}{1 - (1 - r)e^{-kT}} \right]^{n+1-p} \tag{6-26}$$

To illustrate the relative importance of the tidal exchange ratio r and the natural decay coefficient k, Fig. 6-3 is a graph of c_n/c_0 of (6-24) versus kT for various values of r.

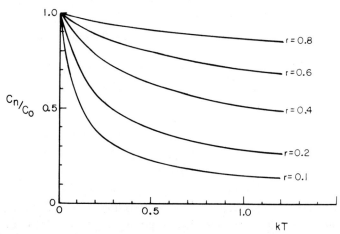

Fig. 6-3   Reproduced with permission from B. H. Ketchum, 1955, *Sewage and Industrial Wastes,* **27,** 1288–1296, Fig. 2.

Returning to the analytic investigation of the problem for constant K_x and A, we have as a first consideration that in this case neither the downestuary flux of pollutant is constant nor the upestuary flux zero, so that we must replace (5-101) and (5-102) by (6-22). Equation (6-22) is a simple ordinary differential equation whose solution is

$$c = C_1 e^{j_1 x} + C_2 e^{j_2 x} \qquad (6\text{-}27)$$

where j_1 and j_2 are given by

$$j_1 = \frac{v_x}{2K_x}\left(1 + \sqrt{1 + \frac{4kK_x}{v_x^2}}\right) \qquad (6\text{-}28)$$

and

$$j_2 = \frac{v_x}{2K_x}\left(1 - \sqrt{1 + \frac{4kK_x}{v_x^2}}\right) \qquad (6\text{-}29)$$

To conform with our previous solution for a conservative pollutant we use the same boundary conditions (6-14) and (6-15) as before. We also include the boundary condition in an upestuary direction,

$$c_u \rightarrow 0 \qquad x \rightarrow -\infty \qquad (6\text{-}30)$$

We must also include an additional source condition at the outfall. If we consider a cross section at the outfall located at $x = 0$ of infinitesimal longitudinal extent Δx, the integrated flux term corresponding to the decay will vanish and, as c is continuous across $x = 0$, the advection term will also vanish. For the flux balance at $x = 0$ we are then left with

$$W = K_x A\left[\left(\frac{dc_u}{dx}\right)_{x=0} - \left(\frac{dc_d}{dx}\right)_{x=0}\right] \qquad (6\text{-}31)$$

Using these four relations to determine the coefficients then gives us for our final solution,

$$c_d = c_0 \frac{e^{-j_2(l-x)} - e^{-j_1(l-x)}}{e^{-j_2 l} - e^{-j_1 l}} \qquad (6\text{-}32)$$

and

$$c_u = c_0 e^{j_1 x} \qquad (6\text{-}33)$$

where

$$c_0 = \frac{\psi}{\sqrt{1 + (4kK_x/v_x^2)}} \frac{e^{-j_2 l} - e^{-j_1 l}}{e^{-j_2 l}} \qquad (6\text{-}34)$$

We see that for $k = 0$ so that $j_1 = v_x/K_x$ and $j_2 = 0$, the solutions (6-32) to (6-34) reduce to (6-16) to (6-18). The dotted lines in Fig. 6-2 are plots of (6-32) and (6-33) for nonconservative pollutant sources located at abscissa values of 1 and 3. We see, as expected, that the concentration levels are further reduced at the outfall and in both a downestuary and an upestuary direction as compared to what they would be for a conservative pollutant.

It is of some further interest to examine in more detail the relative importance of the factors K_x, v_x, and k in the defining equation (6-22) in determining the resultant pollution concentration levels. For this examination we relax our boundary condition at the mouth of the estuary. For steady state with no decay and no diffusion, (6-22) reduces to

$$v_x \frac{dc}{dx} = 0 \qquad (6\text{-}35)$$

whose solution is

$$
\begin{aligned}
c &= c_0 & x > 0 \\
&= 0 & x < 0
\end{aligned}
\qquad (6\text{-}36)
$$

The source flux balance in this case is simply the advection out of a volume of cross sectional area A and extent Δx located at the source,

$$W = c_0 R \qquad (6\text{-}37)$$

so that

$$c_0 = \frac{W}{R} \qquad (6\text{-}38)$$

A graph of (6-36) is shown in Fig. 6-4, corresponding to the distribution of a conservative pollutant in a river with negligible diffusion. For the case of decay and advection but no diffusion, (6-22) reduces to

$$v_x \frac{dc}{dx} + kc = 0 \qquad (6\text{-}39)$$

whose solution is

$$
\begin{aligned}
c &= c_0 e^{-(k/v_x)x} & x > 0 \\
&= 0 & x < 0
\end{aligned}
\qquad (6\text{-}40)
$$

The same source condition applies here, giving as before,

$$c_0 = \frac{W}{R} \qquad (6\text{-}41)$$

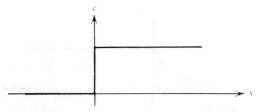

Fig. 6-4

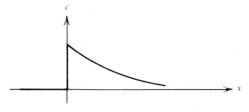

Fig. 6-5

A graph of (6-40) is shown in Fig. 6-5, corresponding in this case to the distribution of a nonconservative pollutant in a river with negligible diffusion. For the case of diffusion and advection with no decay, (6-22) reduces to

$$K_x \frac{d^2c}{dx^2} - v_x \frac{dc}{dx} = 0 \tag{6-42}$$

whose solution is

$$
\begin{aligned}
c &= c_0 & x > 0 \\
&= c_0 e^{(v_x/K_x)x} & x < 0
\end{aligned}
\tag{6-43}
$$

The source condition (6-31) applies here, as the concentration x is continuous across $x = 0$, giving

$$c_0 = \frac{W}{R} \tag{6-44}$$

A graph of (6-43) is shown in Fig. 6-6, corresponding to the distribution of a conservative pollutant in an estuary for which there is no imposed distant boundary condition. For the case of diffusion, advection, and decay, the solution to (6-22) is

$$
\begin{aligned}
c &= c_0 e^{j_2 x} & x > 0 \\
&= c_0 e^{j_1 x} & x < 0
\end{aligned}
\tag{6-45}
$$

The source condition (6-31) again applies, giving

$$c_0 = \frac{W}{R} \frac{1}{\sqrt{1 + (4kK_x/v_x^2)}} \tag{6-46}$$

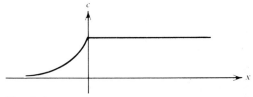

Fig. 6-6

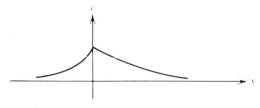

Fig. 6-7

A graph of (6-45) is shown in Fig. 6-7, corresponding to the case of a nonconservative pollutant in an estuary for which there is no imposed distant boundary condition. We also see that the solutions (6-32) to (6-34) reduce to (6-45) and (6-46) for l large as expected. In summary, we see that in the first three cases the source value c_0 is given by W/R, the dilution due to the net flow. In the last example the source value is further reduced by the factor $1/\sqrt{1+(4kK_x/v_x^2)}$. In the last example the effect of the longitudinal dispersion is to reduce the source value and also to reduce the exponential decay rate, that is, increasing the spreading. This is understandable. Diffusion tends to reduce the values of c near the source as compared to what they would otherwise be and to diffuse them out over a broader area. For values of $(4kK_x/v_x^2) \gg 1$, which often can be the case, the last solution reduces further to

$$
\begin{aligned}
c &= c_0 e^{-\sqrt{(k/K_x)}\,x} \qquad x > 0 \\
c &= c_0 e^{\sqrt{(k/K_x)}\,x} \qquad x < 0
\end{aligned}
\tag{6-47}
$$

and

$$
c_0 = \frac{W}{2A\sqrt{kK_x}}
\tag{6-48}
$$

independent of the river runoff or the net longitudinal advection velocity.

For further information on the material covered in this section the reader may refer to Ketchum (1955), Ketchum et al (1952), Stommel (1953), O'Connor (1960, 1965), Thomann (1972), and Dyer (1973).

6-4. VERTICAL DISPERSION EFFECTS

We have up to this point considered only one dimensional effects. We can obtain an index of the vertical variations of a conservative pollutant concentration from the following consideration. We have from (5-164) the defining equation for the vertical salinity deviation s_1 in terms of the longitudinal circulation velocity v_{x1} and the vertical eddy diffusion coefficient K_z, which we repeat here,

$$v_{x1} \frac{\partial s}{\partial x} = \frac{d}{dz}\left(K_z \frac{ds_1}{dz}\right) \tag{6-49}$$

where we assume as before that s_1 is a function of z only. At a sufficient distance from the pollutant outfall that we may consider the vertical distribution of pollutant caused by direct source advection effects to be negligible, the defining equation for the vertical pollutant deviation c_1 taken with respect to the cross sectional mean pollutant concentration is, from (6-49),

$$v_{x1} \frac{\partial c}{\partial x} = \frac{d}{dz}\left(K_z \frac{dc_1}{dz}\right) \tag{6-50}$$

Now, from (6-2) and (6-3), with the use of (5-2), we have simple relations from which we can obtain $\partial c/\partial x$ in terms of $\partial s/\partial x$, giving

$$\frac{\partial c}{\partial x} = -\frac{c_0}{f_0 \sigma} \frac{\partial s}{\partial x} = -\frac{c_0}{\sigma - s_0} \frac{\partial s}{\partial x} \tag{6-51}$$

in a downestuary direction and

$$\frac{\partial c}{\partial x} = \frac{c_0}{s_0} \frac{\partial s}{\partial x} \tag{6-52}$$

in an upestuary direction. Substituting in turn (6-51) and then (6-52) into (6-50) and comparing with (6-49), we then have directly

$$c_1 = -\frac{c_0}{\sigma - s_0} s_1$$
$$= -\frac{W}{R} \frac{s_1}{\sigma} \tag{6-53}$$

for the pollutant concentration deviation in a downestuary direction, and

$$c_1 = \frac{c_0}{s_0} s_1$$
$$= \frac{W}{R} \frac{\sigma - s_0}{\sigma s_0} s_1 \tag{6-54}$$

in an upestuary direction, where for the second of relations (6-53) and (6-54) we have used relation (6-1). We see then that we have a direct correlation between the vertical pollutant concentration deviation and the observed vertical salinity deviation, as we may have anticipated, with the pollutant deviation being the negative of the salinity deviation in a downestuary direction. We also see that the downestuary pollutant deviation remains in a constant ratio to the observed salinity deviation as the outfall location is moved, but that the upestuary ratio decreases as the outfall location is moved toward the mouth of the estuary.

For a nonconservative pollutant we cannot obtain such a simple relation since, because of natural decay, $\partial c/\partial x$ decreases more rapidly than $\partial s/\partial x$. We can, however, write from (6-49) and (6-50) the general expression

$$c_1 = \frac{\partial c/\partial x}{\partial s/\partial x} s_1 \qquad (6\text{-}55)$$

presuming, as previously, that $\partial c/\partial x$ and $\partial s/\partial x$ are independent of z.

6-5. GEOMETRY AND SOURCE CONSIDERATIONS

As mentioned previously, for the dispersion of a pollutant there are two additional variables we must consider in addition to those normally considered in an investigation of an estuarine or coastal water body. These are the geometry of the coastal water body and the temporal and physical characteristics of the source itself. Neither of these variables is particularly amenable to a simple analytic solution approach for somewhat different reasons in each case. As regards the geometry the three dimensional bottom contour configuration of any given coastal water body usually does not lend itself to analytic description, let alone the inclusion of such a description in the defining equations. For the temporal variations in the source we return to considerations of the time dependent characteristics of the tidal, or other eddy, diffusion phenomenon. Here the simple equations we used previously are no longer adequate.

Let us consider briefly, as an example of the geometric variable, longitudinal dispersion of a pollutant in which the cross sectional area A is a function of the longitudinal coordinate x. With reference to (6-21) and (6-22) we now have a dispersion flux term of the form $K_x A (\partial c/\partial x)$, an advection flux term of the form Rc, and a corresponding volume decay term $kcA \, \Delta x$. Considering then the net flux in and out of an infinitesimal cross sectional volume of longitudinal extent Δx, the continuity relation under steady state conditions is

$$\frac{1}{A}\frac{d}{dx}\left(K_x A \frac{dc}{dx}\right) - \frac{1}{A}\frac{d}{dx}(Rc) - kc = 0 \qquad (6\text{-}56)$$

instead of relation (6-22). We have left the quantity R inside the differentiation to accommodate additions to the advection flow from tributaries or from the pollutant source itself. Although analytic solutions can be given for various forms of $A = A(x)$, they are probably not too instructive and are not repeated here in considering the actual variations that occur in A as well as in K_x and R. Computer solutions in finite difference form appear to be a more satisfactory method of approach in such cases.

With regard to temporal variations in the source, we enter an additional realm of complexity. As discussed in particular in Sections 2-5 and 2-6 and also in Sections 5-4 and 5-5, the diffusion phenomenon is a time scale dependent process, varying in the case of estuarine and coastal waters from small scale turbulent times up to that of the tidal period. In this case our equations, such as (6-21), no longer give an adequate description of the processes involved. It is necessary to go beyond the mathematical and physical considerations of (6-21). It appears from the neighbor separation theory of Section 2-6, and from the general considerations that led to the development of that theory, that no differential equation in which position and time are the independent variables, and concentration, expressed as a function of spatial position and time, is the dependent variable can describe the temporal characteristics of a turbulent diffusion process. Further, a solution in terms of the neighbor separation variables, presuming that such a theory is descriptive of the physical processes, does not appear to lend itself to a transformation back into spatial position and spatial concentration. We are indeed left with a difficult problem.

One approach to the resolution of this difficulty has been to express the normal eddy diffusion coefficient K as a function of distance from the source, and in some instances as a function of time as well. If we consider the horizontal, cylindrical spreading away from a restricted source location, such an approach will have some attractiveness in expressing K as an increasing function of the radial distance, which is taken as an index of the neighbor separation distance. Although this suggestion may at times be a useful expedient, it does not appear to meet the original fundamental objections. In the distribution of a dye concentrate, or another observable quantity, the eddy diffusion coefficient envisaged above is instantaneously the same everywhere, for both close and wide pairs of particles. This is contrary to fact. Again this objection may be alleviated to some degree in the above example by expressing K as a function of time as well.

Let us look briefly at some of the solutions that result from this type of approach. Taking the fundamental diffusion equation in cylindrical coordinate form with spatial dependence on the radial distance r only, and with, in this instance, $K = K(r)$, we have

$$\frac{\partial c}{\partial t} = \frac{1}{r}\frac{\partial}{\partial r}\left[K(r)r\frac{\partial c}{\partial r}\right]$$

(6-57)

Following the solution procedures used for (2-66) and normalizing to a unit instantaneous source at $r = 0$ and $t = 0$, we have for the solution to (6-57) for K a constant, corresponding to the simple Fickian diffusion equation,

$$c = \frac{1}{4\pi Kt} e^{-r^2/4Kt} \tag{6-58}$$

We can also take the eddy diffusion coefficient to be proportional to the radial distance,

$$K(r) = Pr \tag{6-59}$$

in which case the solution is

$$c = \frac{1}{2\pi P^2 t^2} e^{-r/Pt} \tag{6-60}$$

We see from the form of either (6-59) or (6-60) that the coefficient P has the units of velocity, and it is sometimes, in this form of solution, referred to as a diffusion velocity. Further, in consideration of the results of Section 2-6, we can take the eddy diffusion coefficient to be proportional to the four thirds power of the radial distance,

$$K(r) = ar^{4/3} \tag{6-61}$$

in which case the solution is

$$c = \frac{1}{6\pi(4a/9)^3 t^3} e^{-9r^{2/3}/4at} \tag{6-62}$$

We see that in these three examples the peak concentration varies, respectively, as t^{-1}, t^{-2}, and t^{-3}, and that the spatial variation in the exponent varies, respectively, as r^2, r, and $r^{2/3}$, with the same time dependence in each case. From diffusion experiments by various investigators involving a dye concentrate or another observable quantity, it has been possible to observe peak concentrate time dependences which can variously be correlated as t^{-1}, t^{-2}, and t^{-3} variations, depending on the particular experimental conditions. From the same experiments it has usually been difficult to sort out which if any of the radial dependences best fit the observed spatial variation.

With regard to the physical characteristics of the source, we have assumed up to this point that the outfall volume flux is small and that the pollutant, on a small scale, is thoroughly admixed with the receiving waters and becomes a part of them without changing their characteristics. This is not always the case as, for example, with a warm water surface discharge of cooling water from an electric power–generating station. As mentioned, the outfall advection flux can be included in the second term of an equation of the form (6-56). For the case of a warm water surface discharge, in the far field, there is usually an interfacial boundary between the warmer surface waters and the underlying cooler receiving waters in which the Richardson number criterion becomes important; and we can then consider entrainment type mixing as we have done previously. Specific examples of these two effects are included in the discussions of applications in the following chapters.

For further information on the material covered briefly in this section the reader may refer to Bowden (1962), Harleman (1966, 1966a, 1969), Joseph and Sendner (1958), Kent (1960), McPherson (1960), Okubo (1962, 1964, 1973), Orlob (1961), and Thomann (1972).

6-6. VELOCITY SHEAR AND POINT SOURCE DISPERSION

We discussed in Section 2-5 some of the characteristics of diffusion from a point source. We now elaborate on this and some of the considerations of the previous section for the case in which there is vertical velocity shear and under the assumption that the Fickian diffusion equation is applicable. We are concerned with the spreading of a pollutant from an instantaneous source in a layer bounded by the surface and the bottom or, as we shall see in Section 8-6, by internal thermocline or halocline features.

To begin we should discuss the physical meaning of the zero, first, and second integral moments of a concentrate distribution defined in general by the relation

$$\theta_{nm}(z, t) = \int_{-\infty}^{\infty} \int_{-\infty}^{\infty} x^n y^m c(x, y, z, t)\, dx\, dy \qquad (6\text{-}63)$$

where $n, m = 0, 1, 2, \ldots$ determines the order. For the zero moment we have

$$\theta_{00} = \int_{-\infty}^{\infty} \int_{-\infty}^{\infty} c\, dx\, dy \qquad (6\text{-}64)$$

which is the total amount of material that has reached a given level z at a time t. Normalizing, as before, the total amount of material released from the instantaneous point source to unity, we have the defining relation for θ_{00},

$$\int_{0}^{h} \theta_{00}\, dz = \int_{-\infty}^{\infty} \int_{-\infty}^{\infty} \int_{0}^{h} c\, dx\, dy\, dz = 1 \qquad (6\text{-}65)$$

where h is the layer thickness. For the first moment in the x coordinate direction we have

$$\theta_{10} = \int_{-\infty}^{\infty} \int_{-\infty}^{\infty} xc\, dx\, dy \qquad (6\text{-}66)$$

which defines the position of the center of mass, or centroid, of the distribution. The x coordinate of the centroid $m_x(z, t)$ is then given by

$$m_x = \frac{\theta_{10}}{\theta_{00}} \tag{6-67}$$

The second moment in the x coordinate direction,

$$\theta_{20} = \int_{-\infty}^{\infty} \int_{-\infty}^{\infty} x^2 c\, dx\, dy \tag{6-68}$$

is related to the mean square value $\overline{x^2}$ of (2-69). The standard deviation $\sigma_x(z, t)$ or the square root of the mean square value about the centroid $\overline{x_1^2}$ is then given by

$$\sigma_x^2 = \frac{\theta_{20}}{\theta_{00}} - m_x^2 \tag{6-69}$$

as can be seen from Fig. 6-8, remembering that the first moment about the centroid is zero.

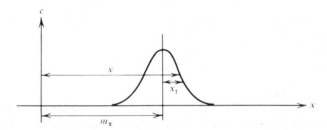

Fig. 6-8

The differential equation that must be satisfied by c is, from (2-65),

$$\frac{\partial c}{\partial t} + v_x \frac{\partial x}{\partial x} = \frac{\partial}{\partial x}\left(K_x \frac{\partial c}{\partial x}\right) + \frac{\partial}{\partial y}\left(K_y \frac{\partial c}{\partial y}\right) + \frac{\partial}{\partial z}\left(K_z \frac{\partial c}{\partial z}\right) \tag{6-70}$$

where v_x, K_x, K_y, and K_z are taken to be functions of z only. For arbitrary distributions of these quantities we cannot find analytical solutions to this equation. However, use can be made of the fact that they are not functions of the x and y coordinates by investigating the integral moments of the concentration distribution. As $x, y \to \pm\infty$, we have the conditions for the instantaneous point source of finite amount that the concentration and the advection, longitudinal diffusion, and lateral diffusion fluxes are zero or equivalently, that the integral moments exist and are finite. Then a differential equation for θ_{00} may be obtained by integrating (6-70) over all x and y, giving

$$\frac{\partial \theta_{00}}{\partial t} = \frac{\partial}{\partial z}\left(K_z \frac{\partial \theta_{00}}{\partial z}\right) \qquad (6\text{-}71)$$

We see then that θ_{00} must satisfy the one dimensional vertical diffusion equation. The differential equation for θ_{10} may be obtained by multiplying (6-70) by x and integrating over all x and y, for which we obtain, after integrating by parts,

$$\frac{\partial \theta_{10}}{\partial t} - \frac{\partial}{\partial z}\left(K_z \frac{\partial \theta_{10}}{\partial z}\right) = v_x \theta_{00} \qquad (6\text{-}72)$$

Again we have the one dimensional diffusion equation but also with a source term $v_x\theta_{00}$. The equation for θ_{20} may be obtained in the same manner by multiplying by x^2 and integrating, giving

$$\frac{\partial \theta_{20}}{\partial t} - \frac{\partial}{\partial z}\left(K_z \frac{\partial \theta_{20}}{\partial z}\right) = 2K_x\theta_{00} + 2v_x\theta_{10} \qquad (6\text{-}73)$$

which again is a one dimensional equation but now has two source terms. For our assumed conditions of parallel longitudinal flow the symmetry of the arrangement requires that the moments $\theta_{01} = \theta_{11} = 0$; and we have for θ_{02} the equation

$$\frac{\partial \theta_{02}}{\partial t} - \frac{\partial}{\partial z}\left(K_z \frac{\partial \theta_{02}}{\partial z}\right) = 2K_y\theta_{00} \qquad (6\text{-}74)$$

corresponding to (6-73).

The solution procedure is then to solve these simpler one dimensional equations first for θ_{00} from (6-71), and then for θ_{10} from (6-72), and then for θ_{20} and θ_{02} from (6-73) and (6-74). The observable quantities m_x, σ_x, and σ_y for a given physical concentration distribution can then be obtained from (6-67), (6-69), and

$$\sigma_y^2 = \frac{\theta_{02}}{\theta_{00}} \qquad (6\text{-}75)$$

remembering that $m_y = 0$.

In the application of this method we must remember that not only are we limited to ascertaining only the gross properties of a distribution but also are limited to conditions for which the Fickian diffusion equation is applicable. We then limit ourselves here to the consideration of conditions a long time t after the initial release. After such a long time, or under such asymptotic conditions, we consider that the concentrate has distributed itself uniformly over the layer of thickness h for which we have the boundary conditions for the vertical flux,

$$K_x \frac{\partial \theta_{00}}{\partial z} = 0 \qquad z = 0, h \qquad (6\text{-}76)$$

From (6-71) we then have

$$\theta_{00} = \frac{1}{h} \qquad (6\text{-}77)$$

We define, as before, vertically averaged quantities,

$$\langle v_x \rangle = \frac{1}{h} \int_0^h v_x \, dz$$

$$\langle \theta_{10} \rangle = \frac{1}{h} \int_0^h \theta_{10} \, dz \qquad (6\text{-}78)$$

and

$$\langle \theta_{20} \rangle = \frac{1}{h} \int_0^h \theta_{20} \, dz$$

From (6-72) and (6-67) we then have

$$\frac{\partial \langle \theta_{10} \rangle}{\partial t} = \frac{\langle v_x \rangle}{h} \qquad (6\text{-}79)$$

or

$$\langle m_x \rangle = \langle v_x \rangle t \qquad (6\text{-}80)$$

which simply states, as expected, that the vertically averaged position of the centroid moves with the mean velocity of the flow.

We next examine the vertical deviation of θ_{10} from its depth-averaged value $\langle \theta_{10} \rangle$. We now have for the equation for the first moment,

$$\frac{\partial \theta_{10}}{\partial t} - \frac{\partial}{\partial z} \left(K_z \frac{\partial \theta_{10}}{\partial z} \right) = \frac{v_x}{h} \qquad (6\text{-}81)$$

where K_z and v_x are functions of z only. We are interested only in the asymptotic approximation to the solution of this linear partial differential equation for large value of t. We see from the form of (6-81) that the complementary function solution can be expressed, by the method of separation of variables, as a product solution of $T(t)$ and $Z(z)$ in which T is expressed as a negative exponential in t. For large values of t this solution can be neglected, and we are left with finding the particular integral solution to (6-81). We can see, by substitution back into (6-81), that the particular integral solution may be written in the form

$$\theta_{10} = \langle v_x \rangle \frac{t}{h} - \frac{1}{h} \int \frac{dz}{K_z} \int (v_x - \langle v_x \rangle) \, dz + C \qquad (6\text{-}82)$$

where the constant of integration C is determined by condition (6-79). We may rewrite this for convenience in the form

$$\theta_{10} = \frac{1}{h} [\langle v_x \rangle t - \varphi(z)] \qquad (6\text{-}83)$$

where the vertical deviation of $\varphi(z)$ is defined by

$$\varphi = \int \frac{dz}{K_z} \int (v_x - \langle v_x \rangle)\, dz \qquad (6\text{-}84)$$

satisfying the relation

$$\int_0^h \varphi(z)\, dz = 0 \qquad (6\text{-}85)$$

We see then that in this asymptotic form the position of the centroid can be expressed as the sum of two terms, one of which is linear in t and the other of which is independent of t. The centroid at any level z moves with the layer-averaged velocity $\langle v_x \rangle$ and is displaced from the mean position $\langle v_x \rangle t$ by an amount given by $\varphi(z)$. In referring back to Section 5-4 we see that this is in the same form taken there for the salinity or pollutant distribution under steady state conditions, in that the distribution is expressed as the sum of a time dependent portion and a time independent vertical deviation.

To continue we then have for the layer-averaged form of the second moment equation (6-73),

$$\frac{\partial \langle \theta_{20} \rangle}{\partial t} = \frac{2 \langle K_x \rangle}{h} + \frac{2}{h} \left[\langle v_x \rangle^2 t - \frac{1}{h} \int_0^h v_x \varphi\, dz \right] \qquad (6\text{-}86)$$

On integrating (6-86) and substituting into (6-69), we then have for the layer-averaged standard deviation $\langle \sigma_x \rangle$,

$$\langle \sigma_x^2 \rangle = \frac{1}{h} \int_0^h \sigma_x^2\, dz = h \langle \theta_{20} \rangle - \langle v_x \rangle^2 t$$

$$= 2t(\langle K_x \rangle + K_x') \qquad (6\text{-}87)$$

where K_x', representing the vertical velocity shear contribution to the net longitudinal dispersion, is given by

$$K_x' = -\frac{1}{h} \int_0^h v_x(z)\varphi(z)\, dz \qquad (6\text{-}88)$$

Referring back to the relations (5-199) through (5-204), we see that this is exactly the same form as obtained previously under steady state conditions for the net circulation, or vertical velocity shear, longitudinal dispersion coefficient, which is indeed what we might have anticipated.

It is possible, as some investigators have done, to go back and consider the complementary function contribution to the solution for shorter times. Such extensions appear, at least to us, to be subject to question in consideration of the applicability of the Fickian diffusion equation under such conditions.

As an example let us take the case in which we have a small layer in which there is a pollutant or dye dispersion and for which we may assume a linear velocity gradient. Taking then, for convenience, the reference velocity to be $v_x = 0$ at $z = 0$ and $v_x = 2U$ at $z = h$, we have for v_x,

$$v_x = 2U\frac{z}{h} \qquad (6\text{-}89)$$

and from (6-78) for $\langle v_x \rangle$,

$$\langle v_x \rangle = U \qquad (6\text{-}90)$$

From (6-84) and (6-85), assuming K_z to be a constant, we can then obtain for φ,

$$\varphi = -\frac{U}{K_z}\left(\frac{z^2}{2} - \frac{z^3}{3h} - \frac{h^2}{12}\right) \qquad (6\text{-}91)$$

and finally from (6-88) for K_x'

$$K_x' = \frac{U^2 h^2}{30 K_z} = \frac{\gamma^2 h^4}{120 K_z} \qquad (6\text{-}92)$$

where the third expression is in terms of the velocity gradient, $\gamma = 2U/h$. We could alternatively have taken an oscillatory vertical velocity shear of the form $v_x = 2V(z/h)\cos\omega t$ which, from the discussion related to (5-213), would give a K_x' value half that of (6-92).

For further information on the material covered in this section the reader may refer to Aris (1956), Bowden (1965), Csanady (1973), Okubo (1967, 1968), and Saffman (1962).

6-7. COUPLED NONCONSERVATIVE SYSTEMS

Often a given chemical or biological decay is related to, or affects, the gain or decay of a second quantity. In turn, the gain or decay of the second quantity may affect that of a third quantity, and so on. In such cases we must then be concerned with the solution of two or more coupled differential equations. We look here at the simplest of such coupled relations, namely, the decay or oxidation of a sewage effluent and its effect on the dissolved oxygen of its associated estuarine or coastal body of water.

If we take a suitable dilution of sewage or other wastewater in an isolated experiment under conditions for which the available dissolved oxygen of the mixture is considerably in excess of the oxygen demand for the decay, we will find that the decay of the carbonaceous sewage material can be approximated by relations of the form (6-19) and (6-20). A convenient measure of this reaction is the amount of dissolved oxygen used in the decay, or the *biochemical oxygen demand* (BOD), usually measured in units such as milligrams of oxygen per liter. We designate this biochemical oxygen demand c_1 and its corresponding decay coefficient k_1.

The dissolved oxygen (DO) in a body of water is then decreased in accordance with the biochemical oxygen demand. It is also increased by reaeration at the water surface. Reaeration itself is a complex phenomenon. For a given estuary condition we may approximate the rate of gain of dissolved oxygen from reaeration by a term of the form $k_2(c_s - c_2)$, corresponding to the right hand side of (6-19), where k_2 is the reaeration coefficient c_2 is the vertically averaged dissolved oxygen content, and c_s is the saturation value of dissolved oxygen for the given water conditions. The coefficient k_2 is inversely proportional to the estuary depth, as reaeration is a surface phenomenon only, and also is in some manner dependent on the vertical mixing and the longitudinal current velocity near the water surface.

To illustrate the nature of the coupling it is instructive to take as a first example the case in which there is no longitudinal dispersion. This corresponds to the case of a nonconservative pollutant in a river or stream with negligible diffusion. Under steady state conditions the equation for the biochemical oxygen demand, assuming no feedback dependent on the amount of dissolved oxygen available, is then simply (6-39), which we repeat here,

$$-v_x \frac{dc_1}{dx} - k_1 c_1 = 0 \qquad (6\text{-}93)$$

The corresponding equation for the dissolved oxygen is

$$-v_x \frac{dc_2}{dx} + k_2(c_s - c_2) - k_1 c_1 = 0 \qquad (6\text{-}94)$$

We may rewrite (6-94), for convenience, as

$$-v_x \frac{dc_3}{dx} - k_2 c_3 + k_1 c_1 = 0 \qquad (6\text{-}95)$$

in terms of the *dissolved oxygen deficit* c_3, defined by

$$c_3 = c_s - c_2 \qquad (6\text{-}96)$$

The solution to (6-93) is simply that given by the relation (6-40), or

$$c_1 = c_{10} e^{-(k_1/v_x)x} \qquad x > 0 \qquad (6\text{-}97)$$

Substituting (6-97) into (6-95), we then have

$$v_x \frac{dc_3}{dx} + k_2 c_3 = k_1 c_{10} e^{-(k_1/v_x)x} \qquad (6\text{-}98)$$

which is a simple, first order, linear ordinary differential equation whose solution is

$$c_3 = C_1 e^{-(k_2/v_x)x} + c_{10}\frac{k_1}{k_2-k_1}e^{-(k_1/v_x)x} \tag{6-99}$$

where the first term is the complementary solution, and the second is the particular integral. In taking a source level for the dissolved oxygen deficit of c_{30} different than zero, or saturation, the solution (6-99) is

$$c_3 = c_{10}\frac{k_1}{k_2-k_1}[e^{-(k_1/v_x)x}-e^{-(k_2/v_x)x}]+c_{30}e^{-(k_2/v_x)x} \qquad x>0 \tag{6-100}$$

For the special case when $k_1 = k_2$, the solution to (6-98) becomes

$$c_3 = c_{10}\frac{k_1 x}{v_x}e^{-(k_1/v_x)x}+c_{30}e^{-(k_1/v_x)x} \qquad x>0 \tag{6-101}$$

The first term of solution (6-100) or (6-101) contains the coupling effect. We see that for $k_2 > k_1$ the first exponential decreases more slowly from unity at $x = 0$ toward zero at $x = \infty$ than the second exponential. The first term then increases from zero at $x = 0$ to a maximum and then decreases more gradually toward zero as x approaches infinity, i.e., the dissolved oxygen shows a minimum, or *sag*, at some position downstream of the sewage outfall. We also see that for $k_1 > k_2$ the same relation still holds. These various relations are shown schematically in Fig. 6-9 for $c_{30} = 0$. The location of the dissolved oxygen minimum is given by setting the derivative $dc_3/dx = 0$, which after a slight amount of algebra gives

$$x_m = \frac{v_x}{k_2-k_1}\log\frac{k_2}{k_1}\left[1-\frac{c_{30}(k_2-k_1)}{c_{10}k_1}\right] \tag{6-102}$$

and the value of c_3 at this minimum is given from (6-95) as

$$c_{3m} = \frac{k_1}{k_2}c_{10}e^{-(k_1/v_x)x_m} \tag{6-103}$$

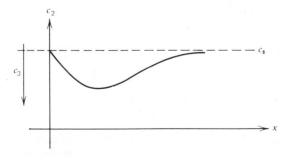

Fig. 6-9

If we now include the longitudinal dispersion coefficient term, the defining equations corresponding to (6-93) and (6-95) will be, from (6-22),

$$K_x \frac{d^2c_1}{dx^2} - v_x \frac{dc_1}{dx} - k_1 c_1 = 0 \qquad (6\text{-}104)$$

and

$$K_x \frac{d^2c_3}{dx^2} - v_x \frac{dc_3}{dx} - k_2 c_3 + k_1 c_1 = 0 \qquad (6\text{-}105)$$

The solution to (6-104) was given previously by (6-45). When this solution is substituted into (6-105), the complementary solution to the

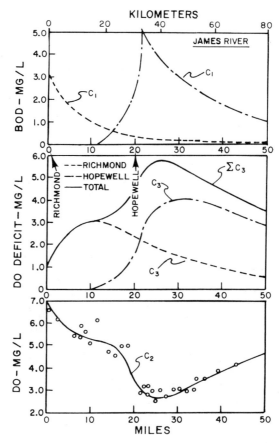

Fig. 6-10 Reproduced with permission from D. J. O'Connor, 1965, *American Society of Civil Engineers, Journal of the Sanitary Engineering Division*, **91**, SA1, 23–42, Fig. 5.

resultant equation is of the same form as (6-45), with k_2 substituted for k in (6-28) and (6-29), and the particular integral is of the same form as the second term in (6-99), with j substituted for k. The final solution for c_3 is then, using the source condition given by (6-31) and assuming that c_{30} is zero,

$$c_3 = \frac{k_1}{k_2 - k_1} \frac{W}{R} \left[\frac{e^{(v_x/2K_x)(1-m_1)x}}{m_1} - \frac{e^{(v_x/2K_x)(1-m_2)x}}{m_2} \right] \qquad x > 0 \qquad (6\text{-}106)$$

and

$$c_3 = \frac{k_1}{k_2 - k_1} \frac{W}{R} \left[\frac{e^{(v_x/2K_x)(1+m_1)x}}{m_1} - \frac{e^{(v_x/2K_x)(1+m_2)x}}{m_2} \right] \qquad x < 0 \qquad (6\text{-}107)$$

where m_1 and m_2 are given by

$$m_1 = \sqrt{1 + (4k_1 K_x/v_x^2)}$$

and

$$m_2 = \sqrt{1 + (4k_2 K_x/v_x^2)}$$

$$(6\text{-}108)$$

The similarity of the solutions for the nondispersive and dispersive cases is readily apparent. Because of the dispersive effect, the location of the dissolved oxygen minimum is nearer the source location in the latter case. Figure 6-10 is a graph of c_1 and c_3 for the combined effect of an outfall located at zero and a second outfall located 20 miles downestuary.

As an example of three stage coupling, we might take the case of the slower decomposition of the nitrogenous material from a sewer outfall. The bacterial nitrification may be thought of as a set of forward reactions from ammonia nitrogen to nitrite nitrogen to nitrate nitrogen. Neglecting any feedback mechanisms and the slower continuation of the nitrogen cycle by the use of these three forms of nitrogen as a nutrient source for algal growth, we have a set of equations of the form,

$$K_x \frac{d^2 n_1}{dx^2} - v_x \frac{dn_1}{dx} - k_1 n_1 = 0$$

$$K_x \frac{d^2 n_2}{dx^2} - v_x \frac{dn_2}{dx} - k_2 n_2 + k_1 n_1 = 0 \qquad (6\text{-}109)$$

$$K_x \frac{d^2 n_3}{dx^2} - v_x \frac{dn_3}{dx} - k_3 n_3 + k_2 n_2 = 0$$

where n_1, n_2, and n_3 are the quantities of ammonia nitrogen, nitrite nitrogen, and nitrate nitrogen, respectively, and k_1, k_2, and k_3 are the corresponding reaction coefficients. These equations may be solved by

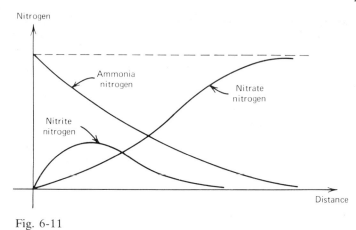

Fig. 6-11

standard methods for given initial outfall contributions W_1, W_2, and W_3. For W_2 and W_3 equal to zero we can see from our previous discussion that the distribution of each form resembles that shown in Fig. 6-11.

For further information on the material covered in this section the reader may refer to Phelps (1944), O'Connor (1960, 1965), and Thomann (1972).

PART II

Application

CHAPTER 7

Great Britain

7-1. INTRODUCTION

In the previous five chapters we have presented a theoretical description of the physical oceanographic processes that occur in estuaries and associated coastal water bodies. In this and the following four chapters we look at the results various investigators have obtained. These results encompass a rather large quantity of scientific and engineering literature. In addition, the quality varies from one report to the next; some just present experimental results, while others include a thorough analysis and interpretation of the results and an extension from the analysis, or considerations related to it, of a theoretical understanding of the relevant physical oceanography.

The following principles have been used in exerpting from this literature the material presented here. A large number of examples from different areas has been included to show in part the wide variation in physical conditions that one can encounter for any given estuary or coastal water body and to show the variations in quality and extent of investigations that have been undertaken. Although many examples are included, coverage is not meant to be all inclusive. Generally only published scientific and engineering articles have been used; and unpublished reports, reports of limited distribution, and company privileged or classified reports have not been included.

Although a general description of each area in terms of its geometry and oceanographic characteristics is given, no strict classification system is used. This is intentional. For example, even for the simplest classification of an estuarine system as a well mixed or a stratified estuary, one can run into difficulties because a given estuary may show either well mixed or stratified conditions as a function of longitudinal distance along the estuary, season of the year, or even in some cases phase of the tidal cycle.

Each area is covered in an abbreviated manner. The detailed physical description of an area is in general curtailed. Little comment is made on the measurement procedures themselves—the instrumentation, their quality, or the care with which they were taken. In any case the reader should go back to the respective references. The analysis procedures and results are, however, discussed in some detail.

In general, the intention in these chapters is to give the reader a sense of what the physical variations may be, the experimental and analysis techniques that have been used, and the results that have been obtained.

7-2. MERSEY ESTUARY AND LIVERPOOL BAY

Bowden (1947) made a series of observations of the fluctuations in the speed of the tidal current at a fixed location in the Mersey estuary. At the same time pressure gauge records were obtained showing the oscillations of the overlying surface wave motion. The location and instrumentation was such that the current measurements could be taken only over the ebb half of the tidal cycle and at a fixed distance above the bottom. The measurements were then repeated at other fixed distances above the bottom on succeeding half tidal cycles. The instrumentation, a Doodson current meter, was sensitive to periods of about 2 sec and greater. The pressure gauge measurements were taken successively at various depths beneath the surface during each measurement period.

Inspection of the current records showed, as the most apparent feature, fluctuations with periods of 4–5 sec, which were of the same order as the periods of the surface wind-generated wave motion. On most records these fluctuations were superimposed on slower fluctuations varying from about 30 sec to several minutes. In turn, both fluctuations were superimposed on the mean current with the tidal period. The pressure records showed similar short period fluctuations but no similar long period fluctuations.

On the assumption that the short period fluctuations were an expression of the water wave motion a Doppler correction for the component of the velocity of the mean tidal current in the direction of the water wave propagation was made for the observed periods. This correction was small, so that the interpretation of the data would be unaffected whether this assumption was correct or not.

The pressure gauge measurements were simply an expression of the overlying surface wave motion. The periods of the pressure fluctuations agreed well with the periods of the surface wave motion. The decrease in amplitude of the pressure fluctuations as a function of depth agreed well with that predicted from the simple theory for small amplitude, short period, single frequency water wave motion in an ocean of finite depth.

The long period current fluctuations showed several interesting characteristics. The periods covered a broader spectrum from about 30 sec up to 7 min or more; the range of the fluctuations increased with the mean tidal current; and the amplitude of the fluctuations was greater near the bottom. On the simplifying assumption that these fluctuations are due to eddies convected with the mean current the scale of the eddies is about 20 m.

The short period current fluctuations showed general characteristics similar to those of the pressure fluctuations, but also characteristics dissimilar to the pressure characteristics. The short period current fluctuations showed mean periods somewhat less than the wave periods, and this discrepancy was more apparent at deeper depths. Further, the amplitude variations did not follow those predicted by the simple theory for water wave motion. The conclusion was reached that these observed fluctuations were a superposition of the water wave motion and of an additional turbulence of unknown origin which could not be separated from the water wave motion for examination and identification.

Bowden and Proudman (1949) continued these measurements on calm days, so that there would be negligible contribution to the observed current fluctuations from the overlying water wave motion. They identified the additional portion of the shorter period current fluctuations as being due to turbulence associated with bottom friction from the tidal current. They argued that near the bottom equations (2-135) and (2-143) apply, so that

$$\overline{v'_x v'_z} = k v_b^2 \tag{7-1}$$

where the overbar indicates a mean value taken over a few minutes, and v_b is the near bottom tidal current velocity. On writing

$$\lambda^2 = \frac{\overline{v'^2_z}}{\overline{v'^2_x}} \tag{7-2}$$

and

$$r = \frac{\overline{v'_x v'_z}}{\left(\overline{v'^2_x}\, \overline{v'^2_z}\right)^{1/2}} \tag{7-3}$$

where λ indicates the ratio of the vertical to the longitudinal turbulence and r is the correlation coefficient between these two components of turbulence, we have

$$\overline{v'^2_x} = \frac{k}{\lambda r}\, v_b^2 \tag{7-4}$$

for the longitudinal current fluctuations in terms of the near bottom tidal current. Taking a characteristic value of $k = 0.0025$ and values of $\lambda = 1$ and $r = 0.3$ from laboratory experiments, we then might expect v'_x to be given in terms of v_b by

$$(\overline{v_x'^2})^{1/2} = 0.09 v_b \qquad (7\text{-}5)$$

at least to an order of magnitude consideration.

The observations of Bowden and Proudman (1949) of the short period current fluctuations showed indeed that the amplitude of the fluctuations was approximately proportional to the tidal current with a mean value near the bottom of $|v_x'|/v_b = 0.117$. If these fluctuations are assumed to be harmonic, (7-5) will give a value of $|v_x'|/v_b = \sqrt{2} \times 0.09 = 0.127$. In a calm sea the fluctuations were small near the surface and increased with depth. If a period of 5 sec and a mean tidal current of 50 cm sec^{-1} are used, the passage of the eddies will correspond to a scale dimension of 2.5 m. A correlogram analysis was also made, computing autocorrelation coefficients of the form (2-75) for the short period fluctuations only. The graph of r_τ versus separation time τ showed no predominant periodicity for the near bottom measurements but resembled that of random variations usually associated with turbulence. A corresponding graph for near surface measurements in the presence of surface waves showed a periodicity corresponding to that of the surface waves.

Observations of the longer period current fluctuations also showed a linear dependence on the tidal current magnitude.

Bowden and Fairbairn (1952) continued these measurements further by using two current meters in a horizontal line transverse to the tidal current and in a vertical line in order to obtain both distance correlation coefficients and autocorrelation coefficients. In addition, they treated the turbulent fluctuations as a continuous spectra rather than making the somewhat arbitrary separation into shorter and longer period fluctuations. The autocorrelation curves they obtained for these longer separation times τ are shown in Fig. 7-1. On the basis of these results they suggested that the current fluctuation observed throughout the whole range of periods should be regarded as similar in character and probably all associated with the turbulent flow of the tidal current.

Bowden and Howe (1963) continued the observations of turbulent fluctuations in the same area with an electromagnetic flowmeter with a flat response to periods greater than 1 sec. Measurements were taken from a fixed platform near the bottom and near the surface. The duration of each record was 5 min, and two variables were measured simultaneously. The observations were variously of v_x' and v_z' at a particular height above the bottom, v_x' at an upper and a lower level, and v_z' at an upper and a lower level. The measurements then permitted the calculation of various correlation coefficients and spectral functions.

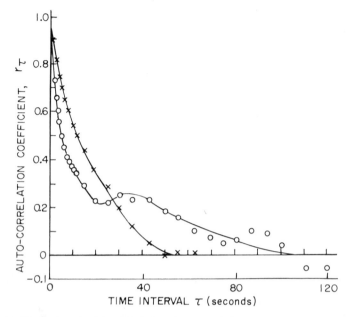

Fig. 7-1 Reproduced with permission from K. F. Bowden and L. A. Fairbairn, 1952, *Philosophical Transactions of the Royal Society of London*, **A244**, 335–356, Fig. 4.

A summary of the results along with similar measurements taken at Red Wharf bay, Anglesey, is given in Table 7-1. In this table r is the correlation coefficient of expression (7-3), which has been evaluated for $\tau = 0$ and for v'_z positive in an upward direction; and $L_{v_x'}$ and $L_{v_z'}$ are the longitudinal characteristic integral scales defined by the relations

$$L_{v_x'} = v_{xt} \int_0^\infty r_\tau(v'_x) \, d\tau$$

and $\qquad\qquad\qquad\qquad\qquad\qquad\qquad\qquad\qquad\qquad\qquad$ (7-6)

$$L_{v_z'} = v_{xt} \int_0^\infty r_\tau(v'_z) \, d\tau$$

If the results near the bottom in the two locations are compared, it will be seen that the mean amplitude of the fluctuations in the Mersey estuary, as given by $(v_x'^2)^{1/2}/v_{xt}$, is approximately half that in Red Wharf bay. The correlation coefficient r is also halved. However, the ratio of the vertical to the longitudinal fluctuations and the integral scale values are approximately the same in the two cases. Comparing the near surface to near bottom data for the Mersey estuary, it will be seen that the amplitude of the longitudinal fluctuations near the surface is about half that near the

TABLE 7-1

Location	r	$\dfrac{(v_x'^2)^{1/2}}{v_{xt}}$	$\dfrac{(v_z'^2)^{1/2}}{(v_x'^2)^{1/2}}$	$L_{v_{x'}}$ (m)	$L_{v_{z'}}$ (m)
Mersey estuary					
4 m below surface	−0.23	0.029	0.55	12.7	3.7
Mersey estuary					
50 cm above bottom	−0.16	0.063	0.41	3.8	1.1
125 cm above bottom	−0.24	0.057	0.52	6.3	0.7
Red Wharf bay					
50–60 cm above bottom	−0.37	0.140	0.46	3.2	0.9
100–125 cm above bottom	−0.39	0.119	0.57	3.3	1.3

bottom, but the ratio of the vertical to the longitudinal fluctuations is nearly the same. In addition, both longitudinal scale values are approximately three times as large near the surface. The difference in turbulent intensity between the two areas is attributed to the difference in bottom type in the two areas, being soft mud for the Mersey estuary and fine sand for Red Wharf bay. The increase in the turbulence scale with increasing height above the bottom is considered a characteristic that is independent of the bottom type.

The spectral functions $F(f)$ of (2-104) were also calculated and compared with a functional relation of the form (2-100). It was found that such a relation could apply with an exponential coefficient of 1.3 for the near surface measurements and of 1.2 for the near bottom measurements rather than $\frac{5}{3}$ over the frequency range 0.02–0.20 cps, corresponding to $k = 10^{-3}$–10^{-2} cm^{-1}, and 0.025–0.25 cps, corresponding to $k = 3 \times 10^{-3}$–3×10^{-2} cm^{-1}, respectively. It should be noted that the spatial wave number of relation (2-100) is proportional to the frequency of the observed fluctuations in this type of measurement through the relation $k = 2\pi f / v_{xt}$, where v_{xt} is the tidal velocity at which a given fluctuation is swept past the sensor.

Hughes (1958) examined tidal mixing in the Narrows of the Mersey estuary. The Narrows is a rectangularly shaped body of water 9.7 km in length with a mean water volume of 147×10^6 m^3 and an intertidal volume of 96×10^6 m^3. The amplitude of the tidal currents is 160–220 cm sec^{-1}, and the river runoff flow velocities are small in comparison with the tidal velocities. The sampling was of temperature and salinity taken at near surface, middepth, and near bottom for a series of cross sections along the Narrows. The sampling utilized Nansen bottles in which the temperature was measured with a glass thermometer on the bottle and the salinity by laboratory conductivity determinations on the water samples. As the measurements over a tidal cycle at each cross section were in turn taken over a sequence of several days to cover all the cross sections, the observed salinity values were adjusted to surface salinity measurements at New Brighton, near the mouth of the Narrows.

The first series of measurements in May 1956 with a lower river runoff flow showed nearly complete vertical mixing at midtide, maximum tidal current velocity, and observable stratification at high and low water. The second series of measurements in March 1957 with a higher river runoff flow showed stratification at all stages of the tide. The surface salinity measurements also showed a slight lateral increase in salinity from right to left across the Narrows looking in a downestuary direction.

By taking mean values for the salinity over a tidal cycle at each cross section, (5-100) was used to calculate longitudinal dispersion coefficients K_x. The results are summarized in Table 7-2. The river runoff flow R was

TABLE 7-2

Date	Distance from New Brighton (km)	Cross sectional area ($\times 10^3$ m^2)	K_x ($\times 10^6$ cm^2 sec^{-1})	R (m^3 sec^{-1})
May 1956	0.0	24.7	1.33	25.7
	−2.6	19.4	1.65	—
	−4.9	17.0	1.86	—
	−8.0	19.1	1.61	—
Mean	—	—	1.61	—
March 1957	−2.4	19.7	3.68	103.
	−7.9	19.2	3.50	—
Mean	—	—	3.59	—

determined from river and stream gauging stations upestuary of the Narrows. The values for the March 1957 period are substantially greater than those for May 1956. During the March 1957 period the river runoff flow was greater, and there was more vertical stratification. We may then anticipate that during this period there is a greater contribution from the circulation flux term S_3 in (5-96) to add to the tidal diffusion flux term S_2 in balancing the net advection flux S_1, thus producing a larger value for the longitudinal dispersion coefficient K_x of (5-100).

By using relation (5-1) the flushing time for the Narrows was also calculated to have a value of 5.3 days for the May 1956 period and a value of 3.8 days for the March 1957 period.

The investigation of circulation and mixing in the Narrows of the Mersey estuary was continued by Bowden (1960). Current meter readings were taken at an anchor station in the Narrows with a Fjeldstad current meter at a sequence of depths from the surface to the bottom every hour over a 50 h period of neap tides during the period 30 June–2 July 1959. The measurements were then repeated over a similar time period from 6 July to 8 July 1959 for spring tides. During the same time period water samples were obtained at stations both upestuary and downestuary from the anchor station for salinity analyses.

From these data the vertical velocity deviations v_{x1} and the tidal averaged salinities $\langle \bar{s} \rangle + s_1$ of (5-93) and (5-94) were determined as a function of depth and are shown in Fig. 7-2. A depth mean flow $\langle \bar{v}_x \rangle$ of

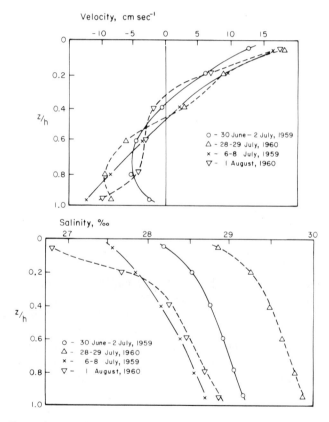

Fig. 7-2   Reproduced with permission from K. F. Bowden, 1963, *International Journal of Air and Water Pollution*, **7**, 343–356, Fig. 2.

4.8 cm sec^{-1} in an upestuary direction was found during the first measurement period and of 1.0 cm sec^{-1} in a downestuary direction during the second period. These values are too large in magnitude, and in one case opposite in direction, to represent the net river runoff flow through a section and are attributed to lateral variations in velocity across the section. The net circulation flow is, as one might anticipate, downestuary in the upper portion of the water column with a surface velocity of about 15 cm sec^{-1}, and upestuary in the lower portion of the water column with a bottom velocity of 5–10 cm sec^{-1}. It is to be noted that the amplitude of the tidal current during these two measurement periods was 93 and 132 cm sec^{-1}, respectively.

From these data for velocity and salinity as a function of depth over several tidal cycles, along with gauging station values for river runoff flow, the salt fluxes S_1, S_2, and S_3 of (5-96) were then calculated. The results are summarized in Table 7-3 along with the values for river discharge R.

TABLE 7-3

Salt flux	30/6–2/7/59	6/7–8/7/59	28/7–29/7/60	1/8/60
$S_1/h(‰\, cm\, sec^{-1})$	6.34	2.63	10.00	6.00
$S_2/h(‰\, cm\, sec^{-1})$	−6.95	−0.44	−12.27	−14.40
$S_3/h(‰\, cm\, sec^{-1})$	−2.36	−3.04	−3.22	−5.15
$(S_1+S_2+S_3)/h$	−2.97	−0.85	−5.49	−13.65
$R(m^3\, sec^{-1})$	41.1	18.0	66.7	42.9

We see that the tidal diffusion flux S_2 and the net circulation flux S_3 are in an upestuary direction, opposing the net river runoff advection flux S_1. Now, if the observations at a given station are typical of those across a complete cross section of the estuary and if the average conditions do not change with time, we expect that $S_1+S_2+S_3 = 0$. We see that this is not the case, and the discrepancies are attributed to probable variations across the estuary, particularly in velocity.

The data also permit calculations of K_z and N_z as a function of depth. From the observed data K_z can be calculated directly from (5-166). The values of N_z can be calculated from (4-123) or (4-127), taking the average values over a tidal cycle. As the coefficient of the first term on the right hand side of (4-127) is not a measured quantity, it must be evaluated by an alternative method. We see that, during the first measurement period, $\partial v_{x1}/\partial z = 0$ at approximately $z = 0.75h$. At this depth we then expect that the first term on the right hand side of (4-127) will be equal to the second, so that the coefficient of the first term can be

TABLE 7-4

z/h	30/6–2/7/59			6/7–8/7/59			28/7–29/7/60	1/8/60
	σ_{zx} (dyne cm^{-2})	N_z (cm^2 sec^{-1})	K_z (cm^2 sec^{-1})	σ_{zx}	N_z	K_z	K_z	K_z
0.1	0.20	9	6	0.34	14	9	6	3
0.3	0.45	27	18	0.86	46	23	27	9
0.5	0.39	40	28	1.03	73	30	71	28
0.7	0.10	43	23	0.86	72	29	41	28
0.9	−0.42	—	8	0.34	25	13	26	15

determined from evaluation of the integral of the second term. By using this approach the values for K_z, N_z, and the stress σ_{zx}, given in Table 7-4, were obtained. It is to be noted that the values of N_z were in every case greater than the corresponding values of K_z by a factor of approximately 2. It should also be noted that the calculated values of N_z are an order of magnitude less than those usually calculated by different methods for tidal currents in an open coastal area.

Bowden (1962a, 1963) continued this type of measurement and analysis at a station downestuary from the previous anchor station on 28–29 July 1960 and at a station upestuary on 1 August 1960. These data and analyses are presented in Fig. 7-2 and Tables 7-3 and 7-4. In addition the Richardson number Ri of expression (5-244), as averaged over a tidal cycle, was calculated for the four measurement series and is summarized in Table 7-5.

Bowden and Sharaf El Din (1966) continued this type of measurement and analysis further. Current measurements were taken at three stations on a section across the Narrows as a function of depth over two tidal cycles with a direct reading Kelvin Hughes current meter recording both magnitude and direction. Temperature and salinity measurements were also taken with *in situ* instrumentation at the same stations and depths

TABLE 7-5

z/h\Ri	30/6–2/7/59	6/7–8/7/59	28/7–29/7/60	1/8/60
0.125	0.92	0.86	0.72	1.36
0.3	1.01	1.32	0.27	1.13
0.5	1.14	1.25	0.25	0.45
0.7	0.48	0.63	0.15	0.47
0.875	0.13	0.12	0.10	0.17

and at corresponding stations on a section upestuary and downestuary from the control section. Field difficulties were encountered with the current measurements, and their reliability may be questioned.

The tidal averaged current velocities v_{x1} showed the same downestuary (north) flow in the upper portion of the water column and upestuary flow in the lower portion of the water column. The corresponding lateral current velocity deviations v_{y1} showed an unusual result—a net flow to the east of about 5–10 cm sec^{-1} at all depths at all three stations across the section. At another station upestuary from this section v_{y1} values taken at a different time were to the west at all depths, with a magnitude of about 18 cm sec^{-1}. As before, the depth and tidal averaged values $\langle \bar{v}_x \rangle$ from the measurements were substantially in excess of those predicted from the river gauging stations R/A. The values for K_z, computed in the same manner as previously, were similar to those given in Table 7-4. The values for N_z, however, again using the $\partial v_{x1}/\partial z = 0$ criterion to evaluate the undetermined coefficient, showed considerable variability, which may reflect uncertainties in the current measurements.

The computed salt flux terms are summarized in Table 7-6. Stations 1, 2, and 3 were across the control section, station 1 being to the east and

TABLE 7-6

Salt flux	1962		1964	
	Station 1	Station 2	Station 3	Station 4
$S_1/h(‰ \text{ cm sec}^{-1})$	4.47	4.53	4.97	5.03
$S_2/h(‰ \text{ cm sec}^{-1})$	14.18	4.46	19.49	−10.30
$S_3/h(‰ \text{ cm sec}^{-1})$	−2.16	−4.20	−1.77	−0.06
$(S_1 + S_2 + S_3)/h$	16.49	4.79	22.69	−5.33
$R(\text{m}^3 \text{ sec}^{-1})$	28.8	29.4	32.7	28.8

station 3 to the west. Station 4 was a midchannel location upestuary from the control section. The values for S_2 for the control section are unusual in being positive and comparatively large, whereas the values in Table 7-3 are negative. The result is indeed surprising, since its physical interpretation is that on the average the water flowing out on the ebb tide had a higher salinity than the water flowing in on the flood tide.

In addition, measurements were taken of salinity and temperature across two sections once a month over the period of a year. The results showed, as might be expected, that there is a strong negative correlation between the salinity at each section and the river discharge, and that there is a tendency for the horizontal and vertical differences in salinity to increase with increasing river flow.

Bowden and Sharaf El Din (1966a) extended these measurements with the same techniques to Liverpool bay, the coastal water body adjacent to the Mersey estuary. Five current stations were occupied a little seaward of the 5 fm (9 m) contour about 6–8 mi (10–13 km) from the coast. Temperature and salinity measurements were taken at these stations and also at stations seaward and landward from them. The same analysis procedures were used as before, with the mixing and motion equations now in finite difference form and the Coriolis term included. The results for K_z, N_z, and the salt fluxes showed an extraordinarily wide range of variation. It appears that there was not sufficient control in the locations and time sequences of the measurements to permit such two dimensional calculations to be made, and the calculated results may or may not be of significance.

The most detailed, and probably the most reliable, sequence of measurements on the circulation and mixing characteristics of the Narrows of the Mersey estuary is that by Bowden and Gilligan (1971). Four transverse sections with five stations per section were occupied over numerous tidal cycles from November 1964 through December 1966. The same instrumentation was used as before, the measurements being taken from an anchored vessel at each location. The section and station locations are shown in Fig. 7-3.

The residual or mean velocity over a tidal cycle was considered to consist of four components—a longitudinal velocity deviation v_{x1} taken to be a function of z only, a net outflow $\langle \bar{v}_x \rangle$ given by R/A, a lateral velocity deviation v_{x2} taken to be a function of y only, and a component due to the diurnal inequality of the mean tide level taken to be independent of both depth and lateral position. The zero position for the v_{x1} values was then adjusted to give an integrated value over each cross section of R.

The salt fluxes were also treated in a slightly different manner than previously. Only the salt fluxes S_1 and S_3 of (5-219) were computed directly. The steady state condition of no net flux, or simply

$$0 = S_1 + S_2 + S_3 + S_4 \qquad (7\text{-}7)$$

was then imposed for the determination of $S_2 + S_4$, where S_1, is numerically the negative of and equal to the sum of the other three fluxes. The coefficient v defined by

$$v = \frac{S_2 + S_4}{-S_1} \qquad (7\text{-}8)$$

was then determined. We see that v represents the fractional contribution

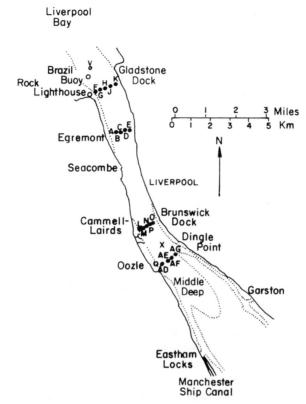

Fig. 7-3 Reproduced with permission from K. F. Bowden and R. M. Gilligan, 1971, *Limnology and Oceanography*, **16**, 490–502, Fig. 1.

to the total upestuary flux of the tidal diffusion and lateral circulation contributions. A summary of these determinations is given in Table 7-7. We see that the fraction v is least in the Cammell Laird section, where the density current circulation is the most highly developed. It is highest at the Rock Light section, where the exchange between the Narrows and Liverpool bay probably takes place largely through tidal action.

Also included in Table 7-7 are values for the fraction v determined by Bowden and Gilligan (1971) from the stratification-circulation diagrams of Hansen and Rattray (1966) and values of v determined from their data using (5-195), which as explained in Section 5-6 is essentially the same as the stratification-circulation diagram. These computed values of v show good agreement with those obtained from the salt flux calculations, except for Cammell Laird section. For this section the values of v computed from (5-195) are quite variable, being negative for three of the

TABLE 7-7

Section	Number of observations	ν			
		Mean	Standard deviation	Hansen and Rattray	Eqn. (5-195)
Rock Light	8	0.85	0.04	0.7–0.9	0.87
Egremont	11	0.51	0.16	0.3–0.55	0.45
Cammell Lairds	5	0.30	0.11	0.1–0.25	0.00
Oozle Wreck	6	0.62	0.15	0.3–0.6	0.56

six cases in which data are given and having a value of $\nu = -0.95$ for one of these three. If the negative values are eliminated as being unrepresentative of conditions to which either (5-195) or the stratification-circulation diagram applies, probably due to transverse circulation effects, an average value of $\nu = 0.25$ is obtained from (5-195).

Calculations for K_z, N_z, and Ri were also made. They show good internal consistency and are probably quite reliable. They have been summarized in Fig. 5-26, showing a progressive decrease in the ratio of K_z/N_z from unity to about 0.2 over the range of Ri values from 0 to 2.0.

Price and Kendirick (1963) made a thorough and exhaustive physical model study of the Mersey estuary with model verification from field current observations. The study was aimed at studying the reasons for siltation in the estuary from source materials in Liverpool bay. The importance of salinity, or density, net circulation currents is emphasized, confirmation being provided by experiments on a large tidal model with a moveable bed in which the natural salinity distribution was reproduced. In addition, Abbott (1960a) applied condition (4-142), using observations of the salinity in the estuary to obtain $\lambda = \partial\rho/\partial x$ and the tide gauge observations along the estuary to obtain $i = \partial\zeta/\partial x$. The longitudinal location at which the equality given by (4-142) is obtained agrees satisfactorily with the site of sediment accumulation.

7-3. SEVERN ESTUARY AND BRISTOL CHANNEL

Bassindale (1943) gives a rather thorough description of the physical characteristics of the Severn estuary and Bristol channel. The Severn estuary extends from Upton-upon-Severn, the highest point at which tidal effects are observed, a distance of 70 mi (112 km) to the Bristol channel, which in turn extends a distance of 66 mi (106 km) to the open coast of the Irish sea. From Upton to Gloucester the estuary presents the normal

TABLE 7-8

Distance (km)	K_x ($\times 10^6$ cm^2 sec^{-1})	Distance (km)	K_x ($\times 10^6$ cm^2 sec^{-1})	Distance (km)	K_x ($\times 10^6$ cm^2 sec^{-1})
6	0.17	32	0.26	58	0.53
10	0.26	35	0.47	61	0.26
13	0.27	39	0.53	64	0.27
16	0.46	42	0.25	68	0.22
19	0.89	45	0.18	71	0.17
23	0.97	48	0.20	74	0.10
26	0.93	51	0.18	77	0.09
29	0.36	55	0.69	80	0.07

appearance of a large river. From Gloucester to the Bristol channel it forms a long waterway, gradually widening in breadth.

From Bassindale's data Stommel (1953) calculated the longitudinal dispersion coefficient K_x, using (5-100), which was developed by him, for summer conditions. These values are summarized in Table 7-8 with distances measured downestuary from Gloucester. It is interesting to note that the largest values of K_x occur at distances of 19–26 km, which coincide with the location of the Severn bore. Bowden (1963) recalculated the values for K_x, including the freshwater contributions to the Severn estuary from its tributary rivers. These results are given in Table 7-9. They show, as one might normally expect, generally larger values for K_x downestuary than those given in Table 7-8. They also illustrate the importance of accurate knowledge of freshwater inputs in such calculations.

From the calculated values of the longitudinal dispersion coefficient Stommel (1953) used the procedures of Sections 6-2 and 6-3, which were also developed by him, for the determination of the distribution of a conservative and a nonconservative pollutant. By using the pollutant

TABLE 7-9

Location	Distance (km)	K_x($\times 10^6$ cm^2 sec^{-1})
Sharpness	39	1.22
Aust	55	1.74
Potishead	71	1.06
Weston-super-Mare	90	0.54

equations in finite difference form, numerical solutions were then obtained using relaxation methods. For the pollutant outfall, chosen at
Sharpness, the results resemble those shown in the numerical example of
Fig. 6-2 with a substantial reduction in the nonconservative pollutant
values for a half life of 4 days as compared to those for a corresponding
conservative pollutant.

De Turville and Jarman (1965) examined the warm water effluent from
an electric power-generating station on the Usk river. The Usk river is a
tributary of the Bristol channel. From the observed temperature field of
the cooling water discharge and the known river flow and cooling water
discharges rates, and with various approximations as to the source distribution at the outfall, they determined a value for the lateral eddy
diffusion coefficient K_y of $2 \times 10^4 \, cm^2 \, sec^{-1}$, which they related to the
lateral dimensions of the river.

Grace (1936) used standard tide height and current observations to obtain evaluations of the bottom frictional coefficient k at various locations
in the Bristol channel, using the method discussed briefly in Section 3-4.
It is of some interest to go over this methodology in more detail here.
First, it is presumed that the motion is essentially longitudinal and
sinusoidal. Then, the defining equations for the finite difference calculations are the relations (3-54), (3-44), and (3-67) repeated here as

$$\frac{\partial \langle v_x \rangle}{\partial t} = -g\frac{\partial \eta}{\partial x} - \frac{f_b}{\rho h} \tag{7-9}$$

$$2\omega_0 \langle v_x \rangle \sin \varphi = g\frac{\partial \eta}{\partial y} \tag{7-10}$$

$$\frac{\partial}{\partial x}(A\langle v_x \rangle) = -b\frac{\partial \eta}{\partial t} \tag{7-11}$$

where b is the channel breadth, and A is the cross sectional area in the
continuity equation (7-11). Figure 7-4 shows the general area of the Bristol
channel, the section lines used for the calculations, the locations of the
tide gauge stations, and the locations of the two light vessels, Scarweather
and Breaksea, for which tidal current measurements were available. From
the coastal tide gauge station data for tidal height and phase, the heights
and phases of the semidiurnal tide were inferred along the medial line
midway between the section lines. Then, the incremental values of the
parentheses of the left hand side of (7-11) from section 0 to each
subsequent section were calculated and adjusted in absolute value to the
observations at the two light vessels. Equation (7-10) was next used to
obtain the correction to be made to the values of η at a tide gauge station
to obtain the corresponding value on the medial line; and these values

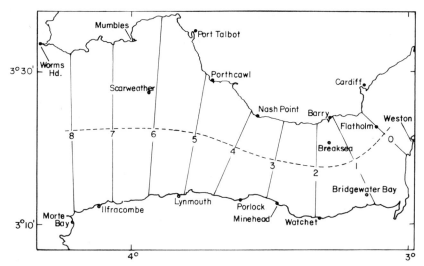

Fig. 7-4   Reproduced with permission from S. F. Grace, 1936, *Monthly Notices of the Royal Astronomical Society, Geophysics Supplement*, **3**, 388–395, Fig. 1.

were then used in turn to find the revised values of $\langle v_x \rangle$ from (7-11). Finally, these revised values of η and $\langle v_x \rangle$ were used in (7-9) to obtain f_b, from which the values of k were obtained from (2-144).

The values determined for k for each section line are given in Table 7-10. The phase for the friction f_b was roughly the same as that for the tidal current $\langle v_x \rangle$, but the values of k varied irregularly, with a mean of 2.4×10^{-3}.

Prior to this Taylor (1921a) made a simple, straightforward, interesting investigation of the variation in tidal amplitude in Bristol channel and the lower portion of Severn estuary. In a general way the amplitudes of the tide in a tidal channel increase as the depth shoals and the breadth narrows. This phenomenon can be and often is the principal reason for tide variations in a tidal channel.

TABLE 7-10

Section	$k(\times 10^{-3})$	Section	$k(\times 10^{-3})$
1	3.0	5	3.5
2	1.4	6	2.5
3	2.4	7	1.4
4	4.1		

By taking the simple equation of motion (3-2) without friction and the continuity equation in the form (7-11) to accommodate variations in the breadth b and the depth h, the defining equation, corresponding to (3-4), in terms of η is

$$\frac{\partial^2 \eta}{\partial t^2} = \frac{g}{b} \frac{\partial}{\partial x}\left(hb \frac{\partial \eta}{\partial x}\right) \qquad (7\text{-}12)$$

Then, taking η to be in the form $\eta = \eta(x)e^{i\omega t}$, (7-12) reduces to

$$\frac{g}{b} \frac{d}{dx}\left(hb \frac{d\eta}{dx}\right) + \omega^2 \eta = 0 \qquad (7\text{-}13)$$

Presuming next that h and b can be represented by linear functions of the form

$$h = h_0 \frac{x}{x_0} \qquad b = b_0 \frac{x}{x_0} \qquad (7\text{-}14)$$

(7-13) reduces further to

$$\frac{d}{dx}\left(x^2 \frac{d\eta}{dx}\right) + \kappa x \eta = 0 \qquad (7\text{-}15)$$

where κ is

$$\kappa = \frac{\omega^2 x_0}{g h_0} \qquad (7\text{-}16)$$

which is a form of Bessel's differential equation whose solution is

$$\eta = \frac{A}{\sqrt{\kappa x}} J_1(2\sqrt{\kappa x}) \qquad (7\text{-}17)$$

where A is a constant, and J_1 is the first order Bessel function.

By representing the breadth and the depth of the deep channel portions of the lower Severn estuary and Bristol channel from Potishead to section 8 of Fig. 7-4 by linear functions of the form (7-14), which can be done satisfactorily, and evaluating the constant A from the observed tide heights at section 8, the calculated curve shown in Fig. 7-5 was obtained. Also shown in Fig. 7-5 are the observed tide heights at various locations. The amplitude agreement is very good. However, the omission of friction makes the tides simultaneous and leaves unexplained the time differences along the channel.

Kirby and Parker (1973, 1974) made a most interesting investigation of the *flocculent layers* that occur in the Severn estuary and Bristol channel area.

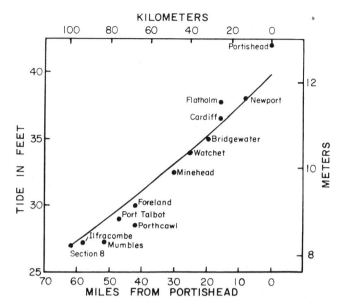

Fig. 7-5 Reproduced with permission from G. I. Taylor, 1921, *Proceedings of the Cambridge Philosophical Society*, **20,** 320–325, Fig. 4.

Before discussing their investigations let us describe, following Einstein and Krone (1962) and Dyer (1972), the flocculation process itself and the formation of flocculent layers. We do this in an abbreviated manner. Most of the sediment in motion in estuaries is carried in suspension and is of small size. Particles less than 2 μm in diameter are mainly composed of the clay minerals illite, kaolinite, and montmorillonite. They normally have a negative surface charge arising from broken intermolecular bonds at the edges of the clay mineral lattice and cationic substitution within the lattice. In fresh waters containing small amounts of salts, cations accumulate by hydration on these surfaces. In such waters the electric field of the accumulated cations is large compared with the dimensions of clay particles and prevents the particles from coming close together. As the anion concentration of the surrounding waters increases to values of salinity of 1‰ in estuarine waters, the effect of the cation field is repressed by the abundance of anions in solution, and the particles do not repel one another at the greater distances. They can then approach one another freely; and an attractive force that is effective only at the shorter distances predominates and mutually attracts and binds the particles, forming through successive collisions a multiparticle aggregation or floc. The shapes of the clay particles cause the structure of a floc to be open and largely to contain water.

Then, for a sufficient anion concentration in the surrounding waters the probability of flocculation occurring depends on the dispersed particle concentration. Collisions can occur by Brownian motion and are substantially promoted by turbulent mixing and by vertical velocity shear of the surrounding waters. The relative importance of turbulent mixing and velocity shear increases with floc size. Laboratory experiments by Einstein and Krone (1962) show that little flocculation occurs for suspended sediment concentrations less than 200–300 ppm in the absence of turbulent mixing and velocity shear. With adequate concentrations flocculation of illite and kaolinite is complete above a salinity of about 4‰. The size, and consequently the fall velocity, of montmorillonite floccules, however, varies over the entire salinity range up to 35‰.

If suspended sediment is present in sufficiently high concentrations, flocculation will occur rapidly. The floccules will not fall as separate units but will form a bonded, three-dimensional flocculent structure, again largely consisting of water. This will then form a distinct layer above the bottom with a well defined interface between its upper surface and the overlying relatively clear water. This interface shows as a clear and distinct acoustic reflection on conventional, high frequency echo sounder records, and in conjunction with the normal bottom reflection return defines the thickness of the flocculent layer. Flocculent layers have been observed in several localities and have been variously referred to as a fluid mud layer or fluff. The sediment concentrations in flocculent layers

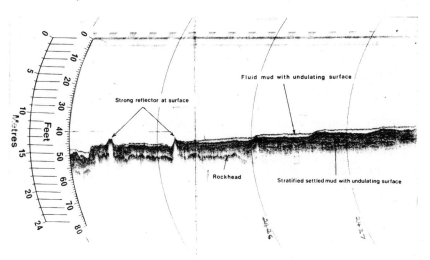

Fig. 7-6   Reproduced with permission from R. Kirby and W. R. Parker, 1974, *Dock and Harbor Authority*, **54,** 423–424, Fig. 1.

are in the range 100,000–300,000 ppm and have corresponding densities of 1.1–1.4 gm cm^{-3}.

Kirby and Parker (1973, 1974) found that flocculent layers existed in small pockets near the confluence of the Severn estuary and Bristol channel, with dimensions of the order of 5–10 km length, 1–2 km breadth, and a fraction of a meter to 1 m thickness. They also found that the existence of the flocculent layers was quite variable with the neap to spring tidal cycle. Figure 7-6 is an example of an echo sounder record over a flocculent layer. The layer is readily distinguishable and can be easily mapped. Figure 7-7 is a comparison of an echo sounder record and the trace from a gamma density probe lowered through the layer. The correlation of the sharp density contrasts, not only at the upper surface of the flocculent

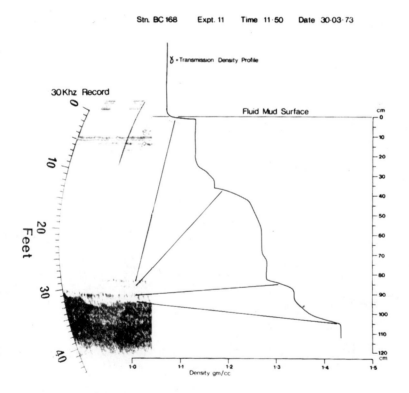

Gamma attenuation profile compared with simultaneously recorded 30kHz echo sounder record, Severn Estuary.

Fig. 7-7   Reproduced with permission from R. Kirby and W. R. Parker, 1974, *Dock and Harbor Authority,* **54,** 423–424, Fig. 2.

layer but also within the layer, with the acoustic reflection returns is excellent.

Kirby and Parker (1973) are the only investigators who have sampled and analyzed chemically the materials of a flocculent layer. The sampling of the flocculent layer was done with a conventional water sampler. To ensure that the sampler was in the layer when tripped, the sampler lowering was followed on the echo sounder. They found that the flocculent layers had high heavy metal concentrations as compared with the normal estuarine waters of Bristol channel and that these layers provide a reservoir for such heavy metal pollutants. The mean concentration values they found for the heavy metals investigated were 133 ppm for lead, 344 ppm for zinc, 44 ppm for copper, and 2.5 ppm for cadmium.

7-4. SOUTHAMPTON ESTUARY AND THE SOLENT

The Southampton estuary is formed by the drowned lower portions of the Test and Itchen rivers. Dyer (1973) made a study of the circulation and mixing characteristics of this estuary. Measurements were taken over a tidal cycle on several different dates of temperature, salinity, and current speed and direction with *in situ* instrumentation at the locations shown in Fig. 7-8. The tide height and tidal current variations are more complex

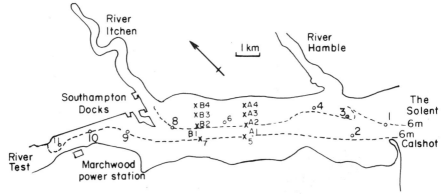

Fig. 7-8 Reproduced with permission from K. R. Dyer, 1973, *Estuaries: A physical introduction,* John Wiley, London, Fig. 4.23.

than those usually encountered, showing overtide components. The measured salinity distribution showed a considerable gradient across the estuary, with fresher water on the left hand side looking downestuary. This situation is opposite that expected under the influence of the Coriolis force and is probably related to topographic and river flow effects. The patterns of salinity showed that at high water the lower part of the estuary was fairly homogeneous vertically; at low water the estuary was stratified throughout. The lateral salinity gradient increased with river discharge and with decreasing tidal range, the maximum lateral gradient being about three times the corresponding longitudinal gradient. One might then anticipate that the circulation and mixing characteristics of the Southampton estuary are complex and, from the limited observational data usually available in any estuary, that some of the simplified analysis techniques used in other estuaries are not applicable here.

The mean salinities $\bar{s}$ and longitudinal velocities, $\bar{v}_x$ taken over a tidal cycle on four different occasions for station 5 are shown in Fig. 7-9. Only

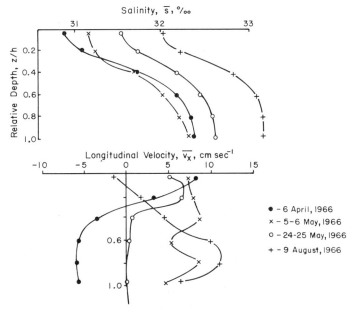

Fig. 7-9 Reproduced with permission from K. R. Dyer, 1973, *Estuaries: A physical introduction*, John Wiley, London, Fig. 4.27.

one station shows what one would normally think of for the longitudinal circulation velocity distribution. Undoubtedly there must be considerable lateral variation in the longitudinal velocity distribution. Measurements taken on 9 August 1966 at the four stations A1 to A4 across a portion of the estuary showed depth mean longitudinal velocities of 4.3, −0.2, −8.5, and −10.1 cm sec^{-1}, respectively. By using the simplified continuity equation

$$\frac{\partial \bar{v}_x}{\partial x} + \frac{\partial \bar{v}_y}{\partial y} + \frac{\partial \bar{v}_z}{\partial z} = 0 \qquad (7\text{-}18)$$

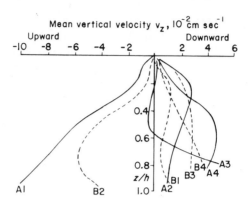

Fig. 7-10 Reproduced with permission from K. R. Dyer, 1973, *Estuaries: A physical introduction*, John Wiley, London, Fig. 4.29.

the gradient values $\partial\bar{v}_z/\partial z$ were calculated from the observations of $\bar{v}_x$ and $\bar{v}_y$. Then, imposing the condition of zero vertical velocity at the surface, the vertical velocity distributions shown in Fig. 7-10 were determined for sections A and B. It is to be noted that the calculated vertical velocities do not return to zero at the bottom. This could be attributed to observational inadequacies or inaccuracies in extending them to such calculations as this, or possibly to a mean longitudinal current of different sign and magnitude at the various locations over a sloping bottom. The measurements seem to indicate that there is a tendency for upward flow in the central portion of the estuary and downward flow on the shallower eastern side.

A somewhat different approach was taken for the salt balance analysis than that used for the Mersey estuary. The differential advection terms in the salt continuity equation (2-65) were calculated in the form $\partial(\bar{v}_x\bar{s})/\partial x$, $\partial(\bar{v}_y\bar{s})/\partial y$, and $\partial(\bar{v}_z\bar{s})/\partial z$, with the inclusion of the volume continuity relation (2-60). All three terms were comparable, with substantial variation in sign and magnitude as a function of depth and station location. Their sums at each depth were small, leaving only a small portion to be balanced by the differential diffusion terms. The quantity $\partial(\overline{v_{xt}s_t})/\partial x$ was also calculated and found to be small, presumably as in other similar calculations elsewhere, the quantity $\overline{v_{xt}s_t}$ being large but its longitudinal gradient small. In addition, the longitudinal dispersion coefficient K_x of (5-100) was calculated to have a value of 1.58×10^6 cm^2 sec^{-1}.

Dyer (1974) analyzed the salt balance further along the lines of (5-219) with the inclusion of the cross product terms and terms representing the variation in the cross sectional area during the tidal cycle. The net result of this analysis was that the lateral circulation flux terms were found to be of magnitude comparable to that of the longitudinal circulation flux terms.

Dyer and King (1975) investigated the residual water flow through West Solent. West Solent is an elongated body of water 20 km in length, 3 km in breadth, and 15 m in depth located between the Isle of Wight and the mainland adjacent to and west of the mouth of the Southampton estuary. The investigations were made using the vertically integrated tidal equation of motion (3-54). Tide guage readings at both ends of West Solent were used to determine finite difference values of $\partial\eta/\partial x$ over the distance between the two gauges and thence to determine $\langle v_x \rangle$, taking v_b to be the same as $\langle v_x \rangle$, at the midpoint between the two gauges from (3-54) through a computer calculation program. A value for the bottom friction coefficient of $k = 5.0 \times 10^{-3}$, as determined from previous investigations in the same area, was used in the calculations. The velocity determinations were carried out at 10 min intervals for an 8 month period of tide gauge readings from 16 February to 6 October 1972.

The 10 min velocities were then integrated over $12\frac{1}{2}$ hr tidal cycles from slack water to the following corresponding slack water to give the residual velocity. Completing this integrating twice with a time lag of about $6\frac{1}{4}$ hr from the intermediate slack to its following corresponding slack water permitted values of the residual velocity at intervals of half a semidiurnal tidal cycle to be determined. There was considerable variation in the magnitude of the residual velocity which reached almost 30 cm sec^{-1} in both eastward and westward directions. Westward residual velocities occurred generally at spring tides with low barometric pressure and southwesterly winds. Eastward residual velocities occurred generally at neap tides with high pressure and northeasterly winds. When averaged over the entire 8 month period, the residual velocity gave a flow of the order of 1 cm sec^{-1}, which was considered too small to be significant compared with the possible errors involved.

Residual water transports were also calculated. In this case, because the current turns before high water and low water, a westward flowing current is present during the high water stand and an eastward current over low water in West Solent. This has the effect of causing a total transport of water toward the west during a tidal cycle, which is independent of the sense of the residual current. The residual transport over a tidal cycle $\bar{Q}$ was then calculated from the relation

$$\bar{Q} = \overline{(\langle \overline{v_x} \rangle + v_{xt})(\bar{A} + A_t)} = \langle v_x \rangle \bar{A} + \overline{v_{xt} A_t} \qquad (7\text{-}19)$$

where $\bar{A}$ is the tidal mean cross sectional area, and A_t is its tidal varying component. The resulting residual transports reached values of 10,000 m^3 sec^{-1}, with a long term average transport of 1000 m^3 sec^{-1}

toward the west rather than toward the east, and with the last term of (7-19) giving values varying from $3000 \, \text{m}^3 \, \text{sec}^{-1}$ for spring tides to $1000 \, \text{m}^3 \, \text{sec}^{-1}$ for neap tides toward the west.

7-5. THAMES ESTUARY

Preddy (1954) developed and applied a rather interesting method for examining the mixing characteristics of the Thames estuary. The method is similar in principle to that of the horizontal dispersion equations (5-98) and (5-99), but with the inclusion of the tidal diffusion concepts of (5-139) through (5-151) the formulation is entirely different. The method is dependent, at least in part, on the assumption that the tidal excursion ξ_0 of Fig. 5-16 remains essentially constant along the length of the estuary, which appears to be the case for the major portion of the Thames estuary. Although the method is interesting and instructive in itself, it appears, at least to us, that it would be simpler to use relations (5-105) and (5-106) to determine the steady state distribution of a pollutant from the observed salinity distribution or, alternatively, to use relation (5-100) to determine the longitudinal dispersion coefficient K_x as a function of distance along the estuary and then to use the time dependent, one dimensional Fickian diffusion equation to determine the temporal changes in salinity or a pollutant where the time scale, necessarily, must be long compared with the tidal period.

In brief, the method is as follows. With reference to the discussion related to (5-139) through (5-151) it should be apparent that after several tidal cycles the distribution approaches that of a Gaussian distribution predicted by the one dimensional Fickian diffusion equation. If we are then looking at time periods long compared with that of a single tidal cycle, it makes little difference what particular form of mixing distribution we take for a single tidal cycle; and for simplicity we assume a uniform distribution over a single tidal cycle as indicated in Fig. 5-16. However, because of the changing geometry of the estuary and the differing effects that may occur on an ebb tide as compared with a flood tide, and because not all the water originally at $x = 0$ is mixed over a tidal cycle, we assume that the proportion of water that is uniformly distributed over a distance ξ_0 toward the ocean P_1 is different than that distributed over the same distance P_2 in the opposite direction, and also that a portion $1 - P_1 - P_2$ is left in its original position. For comparison the assumption in Fig. 5-16 is of course that $P_1 = P_2 = \frac{1}{2}$. In addition, the numerical values of P_1 and P_2, determined from observations of the salinity distribution, also depend on the numerical value chosen for ξ_0.

With reference to Fig. 7-11, for salt at position B_1 with a salinity s

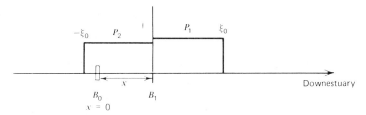

Fig. 7-11

there is a distribution upestuary with a salinity per unit length of sP_2/ξ_0, so that an amount of salt $(sAP_2/\xi_0)(\xi_0-x)$ passes B_0 in an upestuary direction by mixing during a tidal period, where A is a mean cross sectional area. Then, the total amount of salt passing upestuary past B_0 from points downestuary of B_0 is the integral of this expression from 0 to ξ_0. There is a similar expression for points upestuary of B_0, and their difference should be equal to that net advection river runoff salt flux during a tidal period, or

$$\int_0^{\xi_0} sAP_2\frac{\xi_0-x}{\xi_0}\,dx + \int_0^{-\xi_0} sAP_1\frac{\xi_0+x}{\xi_0}\,dx = RTs \qquad (7\text{-}20)$$

where R is the river runoff flow, and T is the tidal period. In addition, since P_1 is not equal to P_2, there is a corresponding volume continuity relation

$$\int_0^{\xi_0} AP_2\frac{\xi_0-x}{\xi_0}\,dx + \int_0^{-\xi_0} AP_1\frac{\xi_0+x}{\xi_0}\,dx = 0 \qquad (7\text{-}21)$$

From the observed values of the vertically averaged salinity, (7-20) and (7-21) can then be used to determine P_1 and P_2 sequentially down the estuary. It is of interest to ascertain what relations exist between the coefficients P_1 and P_2 and the longitudinal dispersion coefficient K_x of (5-99). Considering A to be a constant and P_1 to be equal to P_2 and to be constant also over the limits of integration, and taking s in the form $s = s_0 + (ds/dx)x$, where ds/dx can also be considered a constant over the limits of integration, (7-20) can then be integrated directly, giving, on comparison with (5-99),

$$K_x = \tfrac{1}{3}P_1 \frac{\xi_0^2}{T} \qquad (7\text{-}22)$$

We see then that P_1 and P_2 essentially determine the coefficient values in a longitudinal dispersion coefficient expressed in the usual form of the square of a scale distance divided by a scale time. We also see, as stated previously, that this method probably has application only in cases for which the scale distance ξ_0 can be considered a constant along the estuary.

By using the numerous observations of river runoff flow and salinity, where the salinity is essentially constant in a vertical cross section for the Thames, values for P_1 and P_2 were determined using a relaxation procedure for mean conditions in 1948. As verification, salinity distributions at the end of a 2 week period during which the river runoff flow increased and at the end of a second 2 week period during which the river runoff flow decreased were calculated sequentially from one tidal period to the next using the salinity distribution at the beginning of each 2 week period as input data and equational relations derived in a manner similar to the derivation of (7-20). The results are shown in Fig. 7-12, where curve 1 is plotted through the salinity values on the beginning date and curve 2 derived from it. The open circle values are the observed salinities on the ending date; the agreement is indeed excellent.

Gameson et al. (1957) applied the same methodology as Preddy (1954) to the consideration of the temperature distribution in the Thames. In this case the temperature T is a nonconservative quantity, and we must consider a generalized longitudinal dispersion equation in the form (6-21). Heat is lost at the estuary water surface, and to a first order we can consider the temperature dependence related to the surface heat loss to be of the form (6-19), or

$$\frac{\partial T}{\partial t} = -\frac{f}{h} T \qquad (7\text{-}23)$$

where f is the effective surface heat exchange coefficient and where, as the heat loss is a surface phenomenon only, the multiplying coefficient is also inversely proportional to the depth, expressing the ratio of the surface area to the volume, similar to the relation for the oxygen reaeration coefficient of (6-94). From historical records of the temperature and river runoff for the Thames, and by using the values of P_1 and P_2 obtained previously, average values of the coefficient f were determined. Predictions were then made for the effect on the temperature distribution in the Thames from the warm water effluent from an electric power generating–station as a function of site location along the estuary and net flow conditions.

Preddy and Webber (1963) applied the same methods to consideration of the longitudinal distribution of dissolved oxygen, ammonial nitrogen, and oxidized nitrogen using defining equational relations of the form (6-104), (6-105), and (6-109). Average reaction coefficients were determined from the observed data for the estuary as a whole, and the curves

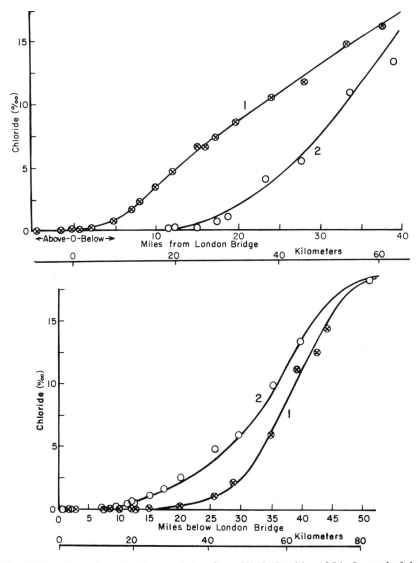

Fig. 7-12   Reproduced with permission from W. S. Preddy, 1954, *Journal of the Marine Biological Association of the United Kingdom,* **33,** 645–662, Figs. 6 and 7.

calculated with these determined coefficients were then compared with the observational data as a function of longitudinal distance. Using estimated polluting loads Gameson et al. (1964) extended the analysis to prediction of the dissolved oxygen distribution and variation for the changing river runoff flow conditions of 1964.

Bowden (1963) calculated a few selected values of the longitudinal dispersion coefficient K_x of (5-100), obtaining 0.53×10^6 cm^2 sec^{-1} at 10 mi (16 km) and 0.84×10^6 cm^2 sec^{-1} at 25 mi (40 km) below London bridge during low river flow conditions, and 3.38×10^6 cm^2 sec^{-1} at 30 mi (48 km) below London bridge during high river flow conditions.

Dyer and Taylor (1973) suggested an alteration in the modified tidal prism concept of Section 5-1 to account for incomplete mixing in each segment of each tidal cycle and applied their method to the Thames estuary. As they have stated, the modified tidal prism method as well as their alteration are probably more of pedagogical interest in understanding tidal mixing than in direct application to the determination of salinity or pollutant distribution in an estuary.

Hunt (1964) made a rather thorough examination of the tidal oscillations in the Thames estuary using the tidal equations of Section 3-4 with a linearized bottom friction term. Analytic solutions are given for an estuary of uniform depth and of an exponentially increasing breadth, which are a good approximation for the Thames from the vicinity of London bridge downestuary. A comparison is given in Fig. 7-13 of the observed and calculated times of high water for 11 February 1948, where the observed values of the tide at the mouth of the estuary have been

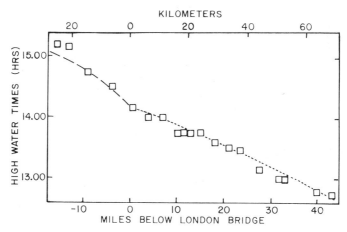

Fig. 7-13 Reproduced with permission from J. N. Hunt, 1964, *Geophysical Journal of the Royal Astronomical Society*, **8**, 440–455, Fig. 4.

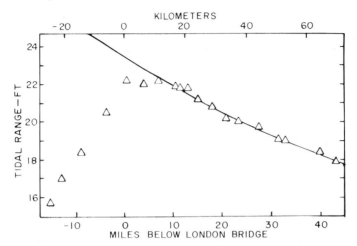

Fig. 7-14 Reproduced with permission from J. N. Hunt, 1964, *Geophysical Journal of the Royal Astronomical Society*, **8**, 440–455, Fig. 5.

used as input data for the calculations and where a value for the bottom friction coefficient of $k = 2.4 \times 10^{-3}$ has been used. The dotted curve is for a mean depth of 26 ft (7.9 m) downestuary from London bridge, and the dashed curve is for a depth of 15 ft (4.6 m) for the portion upestuary of London bridge where the depth gradually shoals. The agreement is quite good. Figure 7-14 is a comparison of the observed and calculated tidal ranges, where the solid curve is for a mean depth of 26 ft (7.9 m). The agreement is again quite good up to London bridge, upestuary of which the bottom shoals and the semidiurnal constituent of the tide becomes gradually less dominant. It is emphasized that, in such a situation of changing geometry and changing phase due to bottom friction, it is incorrect to think of the tide as either a simple progressive or a simple standing wave.

Inglis and Allen (1957) made a thorough and definitive field and physical model study of the Thames estuary with emphasis on near bottom motion, suspended sediment load, and siltation. They found that there is a net upestuary, near bottom flow in the middle to lower reaches of the Thames and that dredged material dumped near the mouth of the estuary reenters the estuary and is carried upestuary to the observed areas of sediment deposition, accounting for the bulk of the sediment deposited in these areas of accumulation. They emphasize the importance of density currents related to the longitudinal salinity gradient.

They also investigated flocculent layer distribution and transport in the Thames estuary. In the Mud reaches of the Thames, where the suspended sediment distribution is the highest, they found that at low water the water turbulence decayed and some of the suspended solids settled out; this settlement led to higher concentrations in the lower portions of the water column of the order of 1000 ppm. Immediately above the estuary bed, however, a layer formed in which the concentrations were as high as 100,000–150,000 ppm, which they called a fluid mud layer. The layer had a clear-cut upper boundary, which showed as a distinct reflection on echo sounder records and was easily distinguished from the normal estuary bottom reflection. They state that the fluid mud tends to consolidate slowly under its own weight at the interface between the fluid mud and the already consolidated muds, thus gradually raising the bed level. During neap tides large pools of this mud persist throughout the tidal cycle, so that consolidation is continuous; but during tides of larger range the layer is disturbed soon after low water and then either flows with the flood current near the bed, maintaining its very high concentrations, or more often is partially mixed with the overlying water and travels 3–4 ft (0.9–1.2 m) above the bed in concentrations ranging from 2000 to 20,000 ppm or more. As might be expected, concentrations varied greatly with time, the material sometimes appearing to move in clouds of high concentration separated by relatively clear water.

Abbott (1960) applied the concepts of boundary layer theory to ascertain the relations between the net flow near the bottom and the tidal observables of amplitude C and phase ϵ as measured outside the bottom boundary layer. For the tidal equation of motion in the bottom boundary layer he used an equation with an effective viscosity coefficient N'_z, appropriate to the boundary layer rather than the simplified bottom boundary condition (2-144) used in the derivations of Section 3-4. For the water column outside the boundary layer he used the usual equation of motion, including the inertial term but excluding any frictional term. These relations are consistent with those used in Chapter 3 for the consideration of vertically averaged tidal motions and are in essence an extension of them to a more detailed consideration of the boundary layer effect and to second order considerations in the net tidal flow. He obtained a relation in terms of the measurable quantities C and ϵ,

$$\frac{dC}{dx} + C\frac{d\epsilon}{dx} \gtrless 0 \qquad (7\text{-}24)$$

for seaward or landward net bottom flow, respectively, where x is measured in a downestuary direction. When this criterion is applied to the tides of the Thames, the location of the principal depositional area of the Mud reaches is correctly predicted.

In a companion paper Abbott (1960a) derives relations (4-142) and

(4-145) of Section 4-7. In applying the relation (4-142), he obtains a net seaward drift near the bottom at all locations in the Thames. It appears from the data of Abbott (1960) that the application of (4-145) would correctly predict a landward drift in the lower reaches of the estuary and perhaps also the location of the depositional area of the Mud reaches. It appears, at least to us, that the two methods, Abbott (1960) and Abbott (1960a), are compatible—one being derived from boundary layer theory in terms of tidal observables and the other from the usual estuary equations with a bottom boundary condition (2-144) in terms of the estuary parameters and tidal observables. Mention is also made that the Thames is probably unusual in the dominance of the nonlinear tidal term in (4-145), and that for most estuaries the simpler relation (4-142) would apply.

7-6. TAY ESTUARY

West and Williams (1972) evaluated the longitudinal dispersion coefficient K_x, using equation (5-100), for the Firth of Tay, or Tay estuary, as a function of longitudinal distance along the estuary under various river flow conditions. The estuary extends as a tidal body through 50 km to the North sea. The Tay river and its main tributary, the Earn river, both located at the upper end of the estuary, are the principal sources of fresh water with mean discharges of $160 \text{ m}^3 \text{ sec}^{-1}$ and $20 \text{ m}^3 \text{ sec}^{-1}$, respectively, and combined monthly mean discharges ranging from 60 to $250 \text{ m}^3 \text{ sec}^{-1}$.

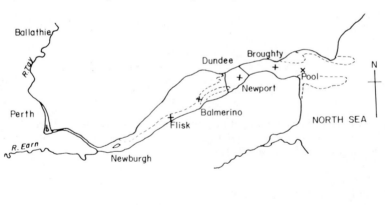

Fig. 7-15   Reproduced with permission from J. R. West and D. J. A. Williams, 1972, *American Society of Civil Engineers, Proceedings of the Thirteenth Conference on Coastal Engineering,* 2153–2169, Fig. 1.

Salinity measurements were obtained from water samples and laboratory conductivity determinations at the five midchannel stations shown in
Fig. 7-15 over several tidal cycles and for various river flow conditions.
Vertical profiles were also taken on several occasions and demonstrated
that the cross sectional mean salinity could usually be given to 1‰ or
better by the midchannel data, indicative of well mixed conditions. The
calculated values of K_x are shown in Fig. 7-16 for the five stations for

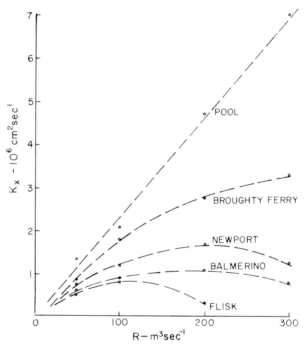

Fig. 7-16   Reproduced with permission from J. R. West and D. J. A. Williams,
1972, *American Society of Civil Engineers, Proceedings of the Thirteenth Conference on Coastal Engineering,* 2153–2169, Fig. 6.

four different river flow conditions, along with the curves fitted to the
data at each station expressing K_x as a function of R. The determined
coefficient K_x is clearly sensitive to the effects of the variation in river
flow. At the three upestuary stations the observed data indicate that the
coefficient passes through a maximum value as the river flow increases.
Near the mouth of the estuary there is a marked variation in K_x with river
flow; elsewhere there is a comparatively more gradual variation.

7-7. RED WHARF BAY, ANGLESEY

Red Wharf bay is a relatively open coastal area off Anglesey, North
Wales. The investigations here closely parallel the similar investigations
into turbulence and internal friction for the Mersey estuary, which have
been discussed in Section 7-2.

Bowden and Fairbairn (1952a) made a determination of the frictional
forces in the tidal current from measurements of tide height at
two locations and the tidal current as a function of depth at one location.
The two tide gauge stations were located in the direction of the
mean tidal current a distance of 9.45 km apart. The Favé tide gauges
used measured hydrostatic pressure at a bottom location with a pair of
Bourdon tubes. The use of the method described below required an
accuracy in the tide measurements of 0.5 cm in amplitude and 0.1° in
phase. Current measurements were taken as a function of depth over the
24 hr observational period. The instrumentation was a Doodson current
meter, and readings of magnitude and direction were taken by eye as
averaged over a few minutes. Five sets of simultaneous records of
elevations at both stations and current at one station for a 24 hr period
were obtained. From three of these, significant values of the friction at
the bottom and the internal friction at various depths were determined;
for the other two sets of records, taken at neap tide, the measurements
led to frictional effects which were apparently so small that they probably
did not exceed the experimental errors in the other terms.

We have from (3-51) for the two dimensional tidal equation of motion
with the inclusion of the Coriolis term of (2-119),

$$\frac{\partial v_x}{\partial t} + 2\omega_0 v_y \sin \varphi = -g\frac{\partial \eta}{\partial x} - \frac{1}{\rho}\frac{\partial f_{zx}}{\partial z} \tag{7-25}$$

and

$$\frac{\partial v_y}{\partial t} - 2\omega_0 v_x \sin \varphi = -g\frac{\partial \eta}{\partial y} - \frac{1}{\rho}\frac{\partial f_{zy}}{\partial z} \tag{7-26}$$

where we have assumed, as before in (3-51), that the density is constant
and the tide-generating force of the moon and the inertial terms can be
neglected. Integrating (7-25) from the surface to the bottom we then
have, similar to (3-54),

$$\frac{\partial \langle v_x \rangle}{\partial t} + 2\omega_0 \langle v_y \rangle \sin \varphi = -g\frac{\partial \eta}{\partial x} - \frac{f_{bx}}{\rho h} \tag{7-27}$$

It should then be possible from observations of $\langle v_x \rangle$ and $\langle v_y \rangle$ as a function
of longitudinal distance to calculate the f_{bx} variations through the tidal
cycle. Similarly, from (7-25) and v_x, v_y, and η at a particular depth the

quantity $\partial f_{zx}/\partial z$ can be determined; and with the assumption that $f_{zx} = 0$ at the surface, f_{zx} can then be calculated sequentially as a function of depth.

The various observed quantities were expressed in harmonic form such that $\eta = \eta(x) \cos[\omega t - \delta(x)]$ and, similarly for v_x, v_y, and f_{zx} with the inclusion of phase lags α, β, and γ for each of the latter with respect to η. The appropriate derivatives were then taken for v_x, or $\langle v_x \rangle$, and η in (7-25), or (7-27), and the resultant equations expressed in finite difference form for the calculation of $\partial f_{zx}/\partial z$, or f_b.

The calculated values of $f_{zx} \sin \gamma$ and $f_{zx} \cos \gamma$, where f_{zx} is the amplitude of the internal friction expressed in its harmonic form, are shown in Fig. 7-17, the $f_{zx} \cos \gamma$ component showing an essentially linear variation

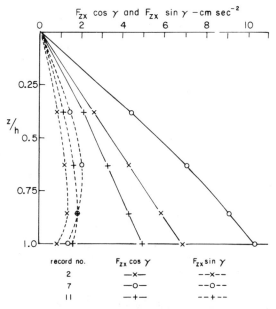

Fig. 7-17   Reproduced with permission from K. F. Bowden and L. A. Fairbairn, 1952, *Proceedings of the Royal Society of London*, **A214**, 371–392, Fig. 4.

with depth. Values of a frictional coefficient k' were also determined from expression (2-143), where f_b and $\langle v_x \rangle$ were considered as magnitudes only, with an average for the three stations of $k' = 1.8 \times 10^{-3}$ and with a phase lag between the depth-averaged velocity and bottom friction of about $13°$ for v_b values of about $65 \, \text{cm sec}^{-1}$. Values of N_z' were also determined from expression (2-135), again using magnitudes for the determining quantities with a range of 200–$500 \, \text{cm}^2 \, \text{sec}^{-1}$.

Bowden and Fairbairn (1956) and Bowden (1962b), using an electromagnetic flowmeter, made an investigation of turbulent fluctuations in

Red Wharf bay. Three stations were occupied in open coastal waters with low water depths of 12, 16, and 20 m, and a tidal range of 6 m. A fourth station was located in a strait to the southeast of the previous measurements with a low water depth of 12 m. The bottom was flat and composed of hard sand; and the tidal currents were nearly linear, reaching surface velocities of 100 cm sec^{-1} during spring tides. Measurements were taken from an anchored boat with a bottom-mounted instrument. The duration of each record was 5–10 min, and the observations were variously of v'_x and v'_z at a particular height above the bottom, v'_x at an upper and a lower level, and v'_z at an upper and lower level.

A summary of the resultant calculations, and a comparison with similar measurements for the Mersey estuary was given in Table 7-1. A further summary with additional calculations is given in Table 7-11. The Reynolds frictional stress of expression (2-135) is the quantity $-\rho v'_x v'_z$ from the calculations; and on the assumption that the Reynolds stress at a height 50–75 cm above the bottom may be equated to the stress at the bottom, it follows that the bottom frictional coefficient k is given by $k = -v'_x v'_z / v^2_{xt}$, remembering again that v'_z in these calculations is taken positive in an upward direction. It is gratifying to see that the mean value of k of 2.4×10^{-3} is in agreement with the results obtained by other methods.

The spectral functions $F(f)$ of (2-104) were also calculated and are shown in Fig. 7-18, using again the relation for the wave number k that $k = 2\pi f / v_{xt}$. The v'_x fluctuations are seen to contain considerably more energy at lower wave numbers than the v'_z fluctions, which might have been anticipated from generalized horizontal and vertical scale considerations. It is also seen that most of the shearing stress appears to be due to wave numbers $k < 10^{-2}$ cm^{-1}, with the peak approximately at $k = 6 \times 10^{-4}$ cm^{-1}.

Bowden et al. (1959) examined the stress variations within a tidal cycle at the same location. These investigations were different from the previous ones by Bowden and Fairbairn (1952a) in that no assumption of or reduction to harmonic components was made and in that the vertical distribution of the current velocity was used to determine the acceleration terms in the equation of motion. Estimates of the bottom friction were made independently. In terms of the defining equations, then, (7-27) was subtracted from (7-25) to eliminate the elevation term, giving a resultant equation similar to equation (3-64), with the inclusion of the Coriolis term. The calculations for f_{zx} as a function of depth were then made in finite difference form starting from the bottom with the independently determined values of f_b.

Measurements were taken at three locations with low water depths of

TABLE 7-11

Station	z (cm)	r	$(v_x'^2)^{1/2}$ (cm sec⁻¹)	$(v_z'^2)^{1/2}$ (cm sec⁻¹)	$-\overline{\rho v_x' v_z'}$ (dyn cm⁻²)	v_{xt} (cm sec⁻¹)	$\dfrac{(v_z'^2)^{1/2}}{(v_x'^2)^{1/2}}$	k (×10⁻³)
2	150	−0.42	3.1	2.0	1.4	32	0.65	—
2	75	−0.47	3.6	2.4	2.1	32	0.66	2.0
2	75	−0.38	3.6	2.1	3.0	34	0.58	2.5
2	150	−0.24	3.6	2.6	2.5	38	0.72	—
2	75	−0.42	3.8	2.6	4.1	40	0.68	2.5
1	150	−0.39	3.5	1.9	2.5	28	0.54	—
1	150	−0.33	4.2	2.8	4.0	50	0.67	—
4	150	−0.29	3.5	2.6	2.8	35	0.74	—
4	75	−0.39	3.5	2.1	2.9	35	0.60	2.4

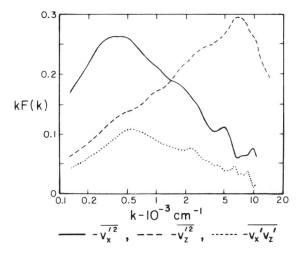

Fig. 7-18 Reproduced with permission from K. F. Bowden, 1962, *The Sea*, Vol. I, 802–825, John Wiley, New York, Fig. 1.

12, 16, and 19 m, respectively. Current measurements as a function of depth were taken with a continuous recording, Doodson current meter with an estimated accuracy in direction of about 5° and in magnitude of 1 cm sec^{-1}. In addition, current magnitudes were measured in the 2 m immediately above the bottom at five fixed distances from the bottom. The instrument consisted of five sets of cup wheels mounted on a vertical tube, the tube being attached to a tripod base designed to give a firm seating on the sea bed.

For the bottom stress determinations it was assumed that the velocity profile near the sea bed, in the bottom boundary layer, could be approximated by a logarithmic relation of the form

$$v_x(z) = \frac{1}{k_0}\left(\frac{f_b}{\rho}\right)^{1/2} \log \frac{z}{z_0} \tag{7-28}$$

where z_0 is a roughness distance and k_0 is an empirical constant with a value of 0.4. The near bottom velocity measurements were then fitted to the logarithmic formula, and the values of f_b so deduced. Because of deficiencies in such vertical current data for all station locations and observing times, calculations were also made according to relation (2-143), using a measured or interpolated v_b at 1 m above the bottom. By comparing these calculations with those from (7-28), at the selected locations for which (7-28) could be applied, a mean value of $k = 3.5 \times 10^{-3}$ was determined. In the previous observations of Bowden and

Fairbairn (1952a) a value k' was determined using the amplitude C_b of $\langle v_{xt} \rangle$ expressed in harmonic form. From (3-59) we then see that k' must be multiplied by $(3\pi/8)(v_b/\langle v_{xt} \rangle)^2$ to compare with k of these determinations. The quantity $v_b/\langle v_{xt} \rangle$ was found to be 0.77, so that the multiplying factor is 2.0, and $k' = 1.8 \times 10^{-3}$ of the previous observations corresponds to $k = 3.6 \times 10^{-3}$.

The calculated variations in f_{zx} as a function of depth and phase of the tidal cycle are shown in Figs. 7-19 and 7-20 along with the corresponding variations in v_x and v_y. The results show that, although near maximum flood and maximum ebb currents the acceleration terms are small and the f_{zx} profile is nearly linear from surface to bottom, at other stages of the

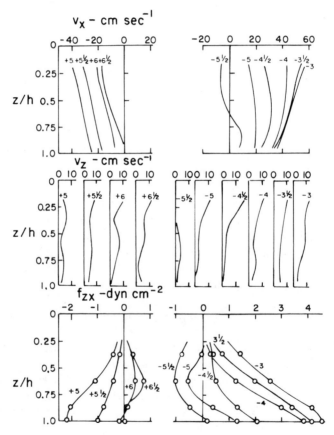

Fig. 7-19   Reproduced with permission from K. F. Bowden, L. A. Fairbairn, and P. Hughes, 1959, *Geophysical Journal of the Royal Astronomical Society,* **2,** 288–305, Fig. 2.

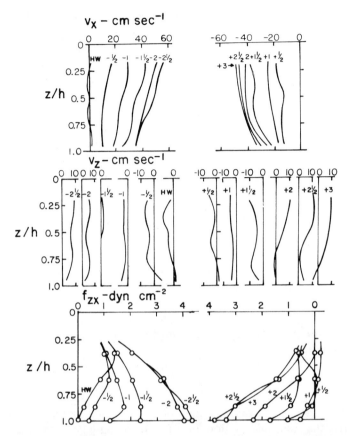

Fig. 7-20   Reproduced with permission from K. F. Bowden, L. A. Fairbairn, and P. Hughes, 1959, *Geophysical Journal of the Astronomical Society*, **2**, 288–305, Fig. 3.

tidal cycle these terms have a considerable effect. The shape of the f_{zx} curve versus z/h depends on whether the current is increasing or decreasing in magnitude, and the curves show a systematic sequence during the tidal cycle. When the current is increasing, the curve is concave upward; and the stress at intermediate depths is less than that corresponding to a linear variation. When the current is decreasing, the effect is reversed, the stress at intermediate depths being greater than a linear variation.

Numerical values of N_z were also calculated for the corresponding values of f_{zx} from relation (2-143). They were of the order of 270 cm^2 sec^{-1} during the flood and 130 cm^2 sec^{-1} during the ebb, corresponding to depth mean currents of 35 and 39 cm sec^{-1}, respectively. It should be noted that the calculations were made only for phases of the tide for which both f_{zx} and $\partial v_x / \partial z$ were large, and that even with this restriction the calculated values show a large range of variation.

It is important to make a small digression at this point. In general, the calculated values of N_z for open coastal tidal areas are in the range 10^2–10^3 cm^2 sec^{-1} and for tidal estuaries 10–10^2 cm^2 sec^{-1}. It is not at all clear, at least to us, why this difference exists. If it is presumed that the difference is real, it could be due, at least in part, to characteristic differences in bottom types in open coastal and estuarine areas as discussed in relation to Table 7-1, and also to differences in the magnitude of the tidal currents. The difference could also be due to the extended horizontal, lateral dimension in open coastal areas as compared with the restricted lateral dimension in estuaries with consequent greater lateral horizontal turbulence and horizontal tidal elliptical motion as opposed to simple longitudinal motion, and the consequent importance of f_{zy} in the formation of vertical turbulence. There is also the possibility that the difference is due to the analysis methods generally employed in each case. For open coastal areas N_z is usually calculated from measurements within the tidal cycle using equations of the form (3-64). For estuaries N_z is usually calculated from measurements averaged over a tidal cycle using equations of the form (4-127).

7-8. IRISH SEA

Taylor (1919) applied the relations of Section 3-5 to the Irish sea. He calculated the tidal energy flow into the Irish sea past a cross section to the south from Arklow to Bardsey and past a cross section to the north from Red bay to the Mull of Cantire using (3-87). The tide heights and phases were obtained from the tide tables for the area. The tidal current magnitudes, phases, and directions were obtained from measurements taken across each section. By fitting these observations to the sinusoidal forms of (3-84) and (3-85) and correcting for the Coriolis variation in tide height across each section from (2-123) and for the angle between the tidal current and the normal to each cross section, values for the tidal energy flow into the Irish sea of 6.4×10^{17} ergs sec^{-1} and 6.2×10^{15} ergs sec^{-1} from the south and from the north, respectively, were obtained. By taking the surface area between the two cross sections this then gave a tidal energy flow per unit surface area of 1640 ergs cm^{-2} sec^{-1}.

From relation (3-93) the work done by the moon on the waters of the Irish sea was also calculated, giving a value of -110 ergs cm^{-2} sec^{-1} and indicating that the tides of the Irish sea do a small amount of work on the moon. This then leaves an amount of $1640 - 110 = 1530$ ergs cm^{-2} sec^{-1} available for energy dissipation by bottom friction or by other effects.

The bottom frictional energy loss was calculated separately from (3-88), using a value for the bottom frictional coefficient of $k = 2.0 \times 10^{-3}$, giving a result of 1300 ergs cm^{-2} sec^{-1}. The conclusion was reached that most of the tidal energy in the Irish sea is dissipated by bottom friction rather than by internal friction or by other effects.

Bowden (1948) investigated the thermal effects of annual heating and cooling across a section of the Irish sea from Kingstown to Holyhead. Assuming that the conditions are sufficiently uniform north and south of the section so that longitudinal heat transfer can be neglected, the heat balance for the entire cross section can be expressed by an equational relation of the form

$$F_S - F_B = (F_E + F_C) - F_T \tag{7-29}$$

where the left hand side of the equation represents the net heat gain by radiation due to the heat absorbed by incoming solar radiation F_S minus the effective back radiation from the sea to the atmosphere F_B, the parentheses represent the net heat transferred to the atmosphere by evaporation F_E and by conduction F_C, and F_T represents the heat used in raising the temperature of the water within the region. The basic ocean temperature data came from surface temperature measurements three times a week on steamer routes from Holyhead to Kingstown over a period of years, along with a few vertical temperature profiles which indicated the temperature was constant as a function of depth. For the calculations of F_T the cross section was divided into three parts, F_{T1} and F_{T3} representing the heat in two shallow sections on either side, and F_{T2} that of the deeper central portion. The other terms were calculated from formulas appropriate to such calculations. A reasonable heat balance resulted throughout the year among the three heat expressions of (7-29).

In addition it was observed that the central portion F_{T2} showed a greater gain of heat in summer than the two coastal sections F_{T1} and F_{T3}, and a greater loss in winter. This is consistent with the idea that there is a net lateral transfer of heat from the coastal areas toward the central portion in the summer and in the reverse direction in the winter. By using relation (5-266) in finite difference form for the lateral transfer of heat between the sections, where K_y is replaced by K_{12} or K_{23}, values for the lateral eddy diffusion coefficient were then calculated, giving results of $K_{12} = 1.1 \times 10^6$ cm^2 sec^{-1} and $K_{23} = 14.8 \times 10^6$ cm^2 sec^{-1}. It is interesting to note that, even under the rather gross simplifications used, fairly reasonable results for the determined coefficients were obtained.

Bowden (1950) and Proudman (1948, 1953) investigated the salinity

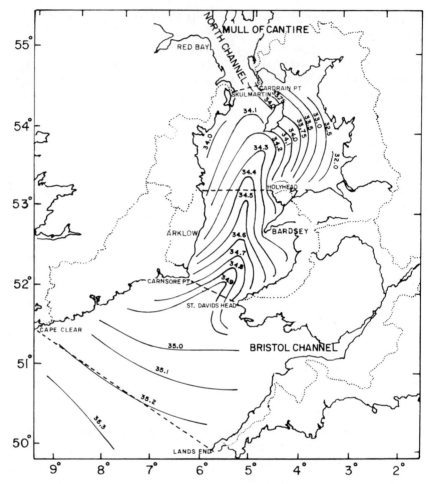

Fig. 7-21   Reproduced with permission from K. F. Bowden, 1950, *Monthly Notices of the Royal Astronomical Society, Geophysics Supplement,* **6,** 63–87, Fig. 1.

distribution in the Irish sea. The mean surface salinity distribution over a year for the Irish sea, taken from numerous measurements over several years is as shown in Fig. 7-21. From selected vertical salinity profiles the salinity is essentially constant as a function of depth north of the section from Carnsore point to St. David's head. Using the simplified continuity relations of (2-39) and (2-40), sometimes referred to as *Knudsen's relations,* and assuming that the net longitudinal dispersive transfer of salt can be neglected in the central portion of the Irish sea, these investigators obtained a mean flow velocity of 0.35 cm sec^{-1} for the section from Kingstown to Holyhead. The corresponding value for the flushing time of (2-41) for the region between the Carnsore point to St. David's Head section and the Kingstown to Holyhead section was then calculated to be 1.4 yr. By using this value for the mean velocity in the net advective salt flux term and arguing further that the salt balance across the southern section from Cape Clear to Land's End would have to be maintained by a net longitudinal dispersive salt flux, a value for the longitudinal dispersion coefficient K_x of 5–$9 \times 10^6 \text{ cm}^2 \text{ sec}^{-1}$ was obtained.

From the salt continuity equation (2-65) we see that, if the longitudinal salinity gradient is a constant and if the vertical terms and the lateral advection term can be neglected, we will be left in the steady state with a balance between the longitudinal advection term and the lateral diffusion term. In other words we anticipate that the salinity distribution will be linear in a longitudinal direction and parabolic in a lateral direction. From Fig. 7-21 we see that this is a reasonable first order approximation to the salinity distribution for the central portion of the Irish sea. From (2-65) we can then calculate K_y/v_x from the observed salinity distribution from the relation

$$\frac{K_y}{v_x} = \frac{\partial s/\partial x}{\partial^2 s/\partial y^2} \tag{7-30}$$

For the central portion of the Irish sea a mean value of $K_y/v_x = 5 \text{ km}$ was obtained; using an approximate value of $v_x = 0.5 \text{ cm sec}^{-1}$, we then have $K_y = 0.25 \times 10^6 \text{ cm}^2 \text{ sec}^{-1}$.

CHAPTER 8

Europe

8-1. HARDANGERFJORD AND NORDFJORD

As mentioned in various preceding chapters, the circulation and mixing in a fjord is quite different from what one encounters in the more usual situation of a stratified or well mixed estuary. For our purposes a fjord is taken to be an elongated indenture of a coastline containing a relatively deep basin with a shallower sill at its mouth. The determining factor is the shallow sill at the mouth of the fjord. This leads to a stratification interface defined by the sill depth and the controlled flow conditions at the estuary mouth. The water below the interface is essentially of oceanic salinity, and the water above the interface increases in salinity from zero at the estuary head to near oceanic salinity at its mouth. There is generally little exchange across the interface either by entrainment or diffusion. Consequently, the circulation of the water below the interface is very slow, measured often in terms of months to years; and the circulation and salinity distribution in the water above the interface are governed largely by the seasonal variation in the river runoff flow conditions.

Saelen (1967) discuses in general the hydrography of the Norwegian fjords and in particular that of Hardangerfjord and Nordfjord. In general the stratification is most pronounced during the summer when the supply of fresh water is greatest and the surface heating strongest. For the fjords in central western Norway the stratification is less pronounced in the winter with its associated sharp decrease in river runoff, but there is little change in the salinity of the deeper layer. Under the more severe conditions in the northern part of the country the stratification may be completely broken up in the winter, with mixing reaching from the surface to the bottom, such as occurs for the fjords in the vicinity of Tromso. Figures 8-1 and 8-2 illustrate these seasonal differences for Hardangerfjord in central western Norway.

The influence of the freshwater supply on the conditions of the upper layer is illustrated in Fig. 8-3. The bar graphs in this figure are the monthly discharges from the four main tributaries of Hardangerfjord, and the open circles are the surface salinities at station H4 in Figs. 8-1 and 8-2. The inverse relation between the two is readily apparent.

298

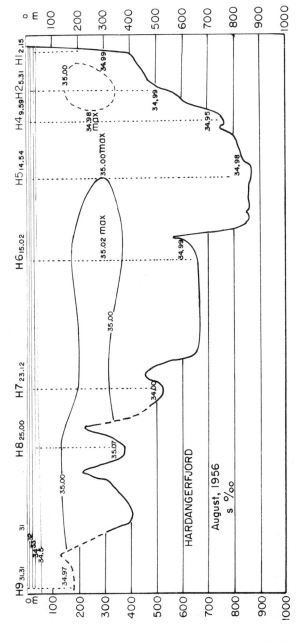

Fig. 8-1 Reproduced with permission from O. H. Saelen, 1967, *Estuaries*, American Association for the Advancement of Science, Washington, D.C., pp. 63–70, Figure 2.

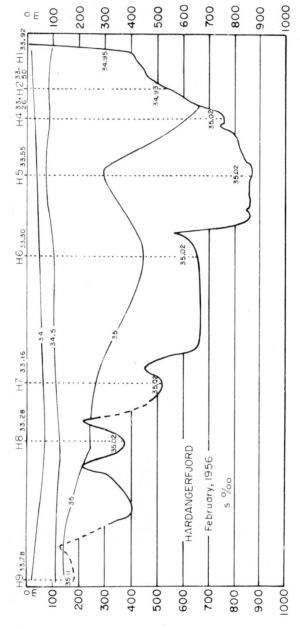

Fig. 8-2 Reproduced with permission from O. H. Saelen, 1967, *Estuaries*, American Association for the Advancement of Science, Washington, D.C., pp. 63–70, Figure 3.

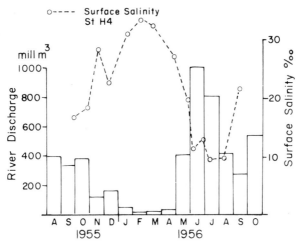

Fig. 8-3 Reproduced with permission from O. H. Saelen, 1967, *Estuaries,* American Association for the Advancement of Science, Washington, D.C., pp. 63–70, Fig. 4.

To gain some insight into the entrainment flow from the lower layer to the upper layer the Knudsen relations of (2-39) and (2-40) were applied at the mouth of Hardangerfjord. With reference to Fig. 2-5 these relations may be written as

$$Q_2 = \frac{s_1}{s_2 - s_1} R \qquad (8\text{-}1)$$

and

$$Q_1 = \frac{s_2}{s_2 - s_1} R \qquad (8\text{-}2)$$

where s_1 and Q_1 are the salinity and net flow out in the upper layer, s_2 and Q_2 are the salinity and net flow in the lower layer, and R is the river runoff flow. From (8-1) we also have for the continuity relation for the entrainment flow,

$$wB = Q_2 = \frac{s_1}{s_2 - s_1} R \qquad (8\text{-}3)$$

where w is the average vertical entrainment velocity, and B is the surface area of the fjord. For the near steady state conditions of midsummer the factor $s_2/(s_2 - s_1)$ varies between an interval from 2 to 6. From (8-2) this means that the surface outflow of low salinity water is between two and six times as large as the freshwater supply. For the same conditions, w of (8-3) is between 2×10^{-4} cm sec^{-1} and 8×10^{-4} cm sec^{-1}, corresponding to a daily vertical ascent of the order of $\frac{1}{2}$ m. It should be remembered that the Knudsen relations ignore any diffusion flux contribution.

It is of further interest to see if the simple concepts of fjord type entrainment mixing involved in the derivation of (5-52) have application here. We can take as an example the conditions shown in Fig. 8-1, where there is a pronounced summer stratification and a correspondingly strong surface layer flow. Using a value for the surface layer thickness of $h_1 = 35$ m from Fig. 8-1 and a value of $C = 3.5 \times 10^{-4}$ from Section 5-3, the theoretical curve calculated from (5-52) is as shown in Fig. 8-4. The

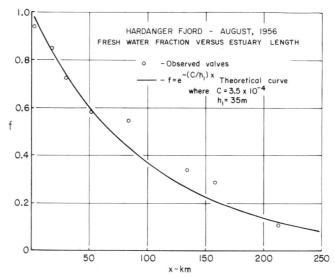

Fig. 8-4

values for the freshwater fraction f of (5-50) are taken from the observed surface salinities of Fig. 8-1 and the observed oceanic salinity of $\sigma = 35.0‰$. The agreement is quite good and appears to indicate that such a simple entrainment mixing concept has applicability.

The second principal feature of fjord circulation is the movement of the deeper layer. It is closely connected with the depth of the sill or, with the possibilities of communication with the coastal and oceanic water masses. For fjords with shallow sills there may be long intervals between inflows, so that stagnant conditions may sometimes result with an absence of oxygen and the development of hydrogen sulfide. Relatively few of the Norwegian fjords apparently reach this condition, most showing oxygen contents of 5–6 ml l^{-1}, which correspond at the ambient temperatures there to saturations of 80–90%.

An example is given of the deeper layer circulation characteristics of Nordfjord, which is divided by a second sill into an inner and an outer basin. The outer basin shows an annual influx of the more saline oceanic water during the summer. In the inner basin the deeper layer is undisturbed for longer periods of time. This condition is shown strikingly by Fig. 8-5, which is a plot of the oxygen content at 400 m at a station in the

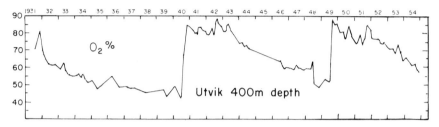

Fig. 8-5 Reproduced with permission from O. H. Saelen, 1967, *Estuaries,* American Association for the Advancement of Science, Washington, D.C., pp. 63–70, Fig. 12.

inner basin over a period of years. From 1932 to 1940 the percentage of oxygen saturation decreased steadily to 43%, but in 1940 it suddenly increased to 90% and was accompanied by a temperature decrease of 1°C. After 1940 the oxygen again decreased steadily until 1947, when it again rose suddenly to 90%, this time accompanied by a marked increase in temperature and salinity. In the following years the oxygen content again decreased steadily. Evidently the inflows that occurred annually in the outer basin have not been strong enough to carry any appreciable quantity of water across the second sill. Apparently, a change to high oxygen content in the inner basin takes place only in years with an exceptionally strong inflow. It is surmised from this and other data that the deep circulation in the fjords is essentially independent of the upper layer circulation.

8-2. FRIERFJORD AND OSLOFJORD

Carstens (1970) examined the turbulent diffusion and entrainment salt fluxes across the stratification interface of the waters of Frierfjord. The data are particularly interesting in themselves; the calculations, as this investigator mentions, are certainly subject to criticism not only because of the simplifying assumptions used but also because of the lack of control data.

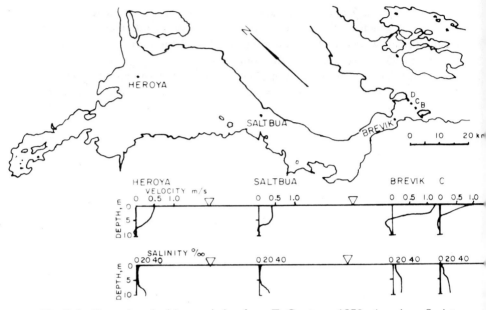

Fig. 8-6 Reproduced with permission from T. Carstens, 1970, *American Society of Civil Engineers, Journal of the Waterways and Harbor Division,* **96,** WW1, 97–104, Fig. 3.

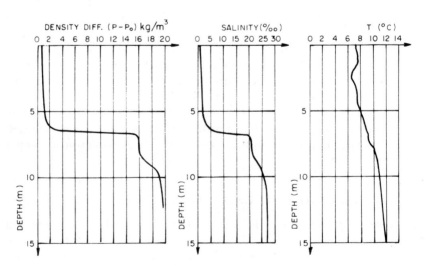

Fig. 8-7 Reproduced with permission from T. Carstens, 1970, *American Society of Civil Engineers, Journal of the Waterways and Harbor Division,* **96,** WW1, 97–104, Fig. 1.

The geography of Frierfjord and the net current velocity and salinity profiles at three locations in the estuary are shown in Fig. 8-6. The differences between the velocity and salinity characteristics of the upper and lower layers is quite apparent. Figure 8-7 shows in more detail the salinity and density stratification at the interface.

If it is assumed that steady state conditions exist and that there is negligible horizontal diffusion flux, neither of which assumptions is necessarily true, the salt flux balance is given by

$$S_2 - S_1 = S_e + S_d \qquad (8\text{-}4)$$

with reference to Fig. 8-8, where S_2 is the advection salt flux out the mouth of the fjord, S_1 is the advection salt flux in at the head of the fjord,

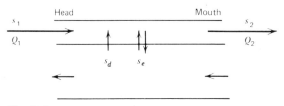

Fig. 8-8

and S_e and S_d are the entrainment and diffusion salt fluxes across the interface. For purposes of calculating S_e and S_d we also have the relations

$$\begin{aligned}
S_1 &= Q_1 s_1 = 0 \\
S_2 &= Q_2 s_2 \\
S_e &= (Q_2 - Q_1)\sigma \\
S_d &= (S_2 - S_1) - S_e = S_2 - S_e
\end{aligned} \qquad (8\text{-}5)$$

where s_1 and Q_1 are the averaged salinity and volume flux at the head of the fjord in the upper layer, s_2 and Q_2 are the corresponding quantities at the mouth of the fjord, and σ is the salinity of the lower layer. From the velocity and salinity distributions at the upper and lower locations of Fig. 8-6 and with certain assumptions as to the lateral distribution of each parameter, which are not necessarily correct, the entrainment salt flux was calculated to be 85% of the total exiting salt flux on one date and only 10% on a second date.

The control relation (4-16) was also applied to the restricted geographic conditions at the mouth of the fjord. The calculated and observed layer thicknesses h show a similar seasonal variation with the freshwater influx Q_1, as shown in Fig. 8-9. The difference between the two curves is attributed to an entrainment volume flux in addition to Q_1.

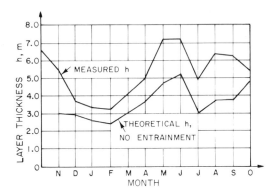

Fig. 8-9   Reproduced with permission from T. Carstens, 1970, *American Society of Civil Engineers, Journal of the Waterways and Harbor Division*, **96**, WW1, 97–104, Fig. 4.

Gade (1963) made a thorough study of the hydrographic conditions for the inner Oslofjord. In a general way we may consider the Oslofjord as consisting of an outer portion with depths of more than 300 m connected to the open coastal waters by a sill at a depth of 120 m and to the central portion, or Breiangen, with depths greater than 200 m by a sill at a depth of 110 m. The central and inner portions are connected by Drobak sound with a sill depth of 20 m. The inner portion in turn consists of two parts, Vestfjord and the innermost Bonnefjord, both with depths of more than 150 m and separated by a sill at 50 m.

Figure 8-10 are temperature, salinity, and oxygen contours for a typical station in Vestfjord during the calendar year 1959. As can be seen from the oxygen diagram and also from the salinity diagram, there was an influx of deeper water into Vestfjord from the Breiangen during the winter of 1958–1959. Then during the calendar year of 1959 there was a progressive decrease in the oxygen concentration of the deeper and more stagnant layer as a result of normal oxidation processes. Also, as can be seen from the temperature diagram, there is a downward progression of the temperature contours, due to the seasonal temperature cycle. By using the relation (5-269) averaged values for K_z were determined for each of the three basins Breiangen, Vestfjord, and Bonnefjord, with results of 0.63, 0.37, and 0.15 $cm^2 sec^{-1}$, respectively.

Figure 8-11 is a seasonal plot of the density difference σ_t versus depth for Vestfjord. It is to be noted that the conditions change from that approaching an interface between the surface, fresher water, runoff layer and the deep, more saline water, stagnent layer in the summer to that approaching well mixed in winter. Similar results are obtained for the other two basins.

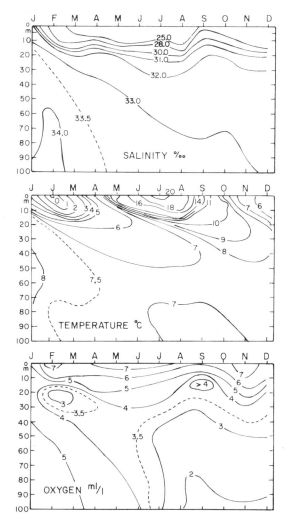

Fig. 8-10 Reproduced with permission from H. G. Gade, 1963, *Det Norske Videnskaps-Akademi I Oslo, Hvalradets Skrifter,* **46,** 1–61, Fig. 5.

8-3. RANDERS FJORD AND SCHULTZ'S GRUND

One of the more fundamental and definitive investigations of eddy viscosity and eddy diffusion in coastal waters was that conducted by Jacobsen (1913) at Schultz's grund in the Kattegat and by Jacobsen (1918) in Randers fjord on the northeast coast of Denmark. The investigations include not only careful and detailed experimental measurements

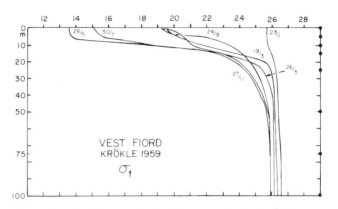

Fig. 8-11 Reproduced with permission from H. G. Gade, 1963, *Det Norske Videnskaps-Akademi I Oslo, Hvalradets Skrifter*, **46**, 1–61, Fig. 9.

but also the development and application of the theoretical relations of (4-123) and (5-164) for the determination of the vertical eddy viscosity and diffusion coefficients N_z and K_z, which have been used by most other subsequent investigators in their determinations.

Both areas have strongly stratified conditions. At Schultz's grund the salinity increases from about 18 to 28‰ from the surface to a depth of 20 m. The vertical salinity gradient increases by about an order of magnitude from the surface to the depth of the halocline and then decreases below. In Randers fjord the salinity increases from 4 to 11‰ from the surface to a depth of 5 m, with a strong vertical salinity gradient at all depths.

TABLE 8-1

Depth (m)	K_z (cm^2 sec^{-1})	N_z (cm^2 sec^{-1})	K_z/N_z	Ri
Schultz's grund				
2.5	0.3	3.1	0.09	7.1
5.0	0.4	3.1	0.13	3.8
7.5	0.18	2.7	0.067	5.9
10.0	0.05	2.2	0.023	10
12.5	0.04	1.9	0.021	28
15.0	0.2	3.8	0.05	125
Randers fjord				
1	0.6	3.5	0.17	5.9
2	0.5	2.5	0.20	2.6
3	0.4	2.6	0.15	3.8

By using tidal averaged values of the longitudinal current velocity and salinity, (4-123) and (5-164) were applied in the determination of N_z and K_z as a function of depth. The coefficient of the first term on the right hand side of equation (4-123) was determined both from the observed values of the surface slope i from tide gauge stations, and from evaluations at the depth for which $\partial v_x / \partial z = 0$. The determined values of N_z and K_z, and also of K_z/N_z and Ri, are given in Table 8-1. They are also summarized in Taylor (1931) and in Fig. 5-26. They give a useful and informative extension at higher Richardson numbers, corresponding to highly stratified conditions of the values of Bowden and Gilligan (1971) at lower Richardson numbers, corresponding to well mixed and weakly stratified conditions.

8-4. ELBE AND EMS ESTUARIES

There has been a substantial number of investigations into the tidal mechanics of the tidal rivers and estuaries of western Europe over the years. These investigations, and their associated calculations, have been particularly involved with the inclusion of the shallow water constituent and inertial terms and with the inclusion in the nonlinear bottom friction term of the river runoff contribution to the tidal motion. Fourier harmonic solutions can be obtained as well as finite difference computer solutions, which also permit the inclusion of varying topography. We do not go into these results here, and the reader may wish to refer to the excellent and detailed treatise by Dronkers (1964).

One of the more recent investigations of the Elbe estuary has been that by Ramming (1972). A numerical computer model is used based on the two dimensional tidal equations of motion with the inclusion of the Coriolis and inertial terms and in which the components of the bottom friction term are given by

$$f_{bx} = \frac{v_{bx}}{v_b} f_b = k \rho v_{bx} \sqrt{v_{bx}^2 + v_{by}^2} \tag{8-6}$$

and similarly for f_{by}. Tide gauge data at various locations were used as controls, and the calculations gave good agreement with observations at other locations. The two dimensional concentrate continuity equation was also applied to the same numerical model of the Elbe estuary. However, in this case the longitudinal dispersion coefficients were taken to be proportional to the product of the longitudinal velocity and the depth, a relation appropriate to near bottom velocity shear effects of small dimension hydrodynamic phenomena, rather than relating them to the anticipated tidal diffusion and circulation contributions.

Dorrestein (1960) and Dorrestein and Otto (1960) investigated the mixing characteristics of the Ems estuary. They used the familiar relation (5-100) to determine the longitudinal dispersion coefficient as a function of distance along the estuary. They then applied the one dimensional, time dependent, conservative concentrate continuity equation, which from (6-21) and (6-56) for an estuary of varying cross sectional area, is

$$A\frac{\partial c}{\partial t} = -\frac{\partial}{\partial x}(Rc) + \frac{\partial}{\partial x}\left(K_x A\frac{\partial c}{\partial x}\right) \tag{8-7}$$

to the distribution of a pollutant located initially at various distances along the estuary after several tidal cycles. Although the method is certainly correct, the results presented appear unusual, perhaps because of the imposed computational boundary conditions at the head and the mouth of the estuary. Eggink (1966) continued the same discussion with application to reaeration and nonconservative pollutants.

8-5. BOSPORUS, DARDANELLES, GIBRALTER, AND BAB EL MANDEB

One of the earliest, and still one of the more interesting, developments in coastal oceanography is the theoretical description and verification of the stratified flow in straits given by Defant (1930), also repeated in Defant (1961). The theory was given in detail in Section 4-4. We are concerned here with its applications to three straits. First, there are the straits between the Black sea and the Aegean sea, namely, the Bosporus, Sea of Marmara, and the Dardanelles. In this case the near surface flow is of lighter water out from the Black sea to the Aegean sea bounded at depth by a strong halocline and underlain by a near bottom flow of heavier water into the Black sea. Second, there is the Strait of Gibraltar between the Mediterranean sea and the Atlantic ocean, with a near surface flow into the Mediterranean and a near bottom flow out of the Mediterranean. And third, there is the Strait of Bab el Mandeb between the Red sea and the Indian ocean, with a near surface flow into the Red sea and a near bottom flow out of the Red sea.

In each case the continuous interchange of water between the enclosed sea and adjacent open ocean, or sea, is controlled by two factors, the depth and width of the passage to the open ocean and the excess, or deficiency, of precipitation and river runoff flow to evaporation. For the Black sea precipitation and river runoff exceed evaporation. For the Mediterranean and Red seas evaporation exceeds precipitation and river runoff.

The physical characteristics of the waters of the straits between the

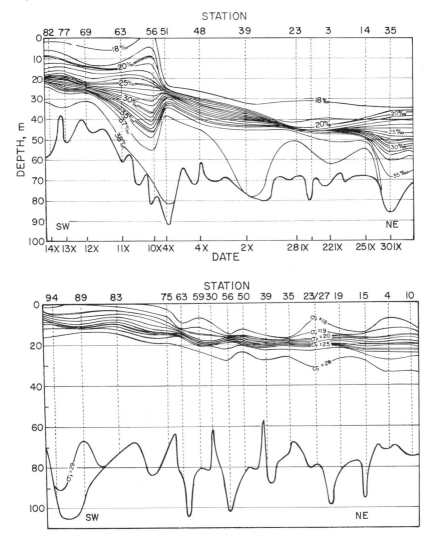

Fig. 8-12 Reproduced with permission from A. Defant, 1961, *Physical Oceanography*, Vol. I, Pergamon Press, New York, Fig. 241.

Black and Aegean seas are shown in Fig. 8-12. The upper portion of the figure is through the Bosporus, with the Black sea on the right and the Sea of Marmara on the left; the contours are for the salinity in parts per thousand. The lower portion of the figure is through the Dardanelles, with the Sea of Marmara on the right and the Aegean sea on the left; the contours are for the density difference σ_t from unity in parts per thousand

at atmospheric pressure. Figure 8-12 clearly illustrates the rather intense
stratification between the upper water layer flowing out from the Black
sea and the lower water layer flowing into the Black sea and shows the
anticipated rise in the stratification interface from the Black sea end of
the straits to the Aegean sea end.

An exact comparison of observed current profiles with the theory of
Section 4-4 is difficult to make. In addition to the observable parameters
of wind stress T, water surface slope i_1, interface slope i_2, and interface
depth h_1, the determining equations include the internal and bottom
frictional coefficients N_z and k. The equations can then reasonably be
applied only to some averaged conditions near an observing station for
which characteristic spatial and temporal values can be used for T, i_1, and
i_2. Further, the theory takes no account of the irregular bottom topography that usually exists in straits, so that a simple bottom boundary
condition with a bottom frictional coefficient k is certainly inadequate.
Nevertheless, the current profiles show the general parabolic shape predicted by the theory; and excellent agreement can be obtained by a
suitable choice of frictional coefficients.

Figure 8-13 shows the observed and predicted current profile for a
portion of the Bosporus and of the Dardanelles. For the Bosporus section
the wind was opposite the direction of the current, sea level fell 10.1 cm
over a distance of 30 km from the Black sea toward the Sea of Marmara,
and the interface rose 36 m over the same distance. For the Dardanelles

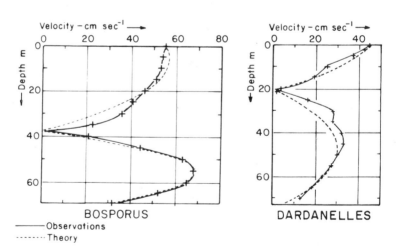

Fig. 8-13   Reproduced from A. Defant, 1930, *Sitzungberichte der Preussischen
Akademie der Wissenschaften, Physikalish-Mathematische Klasse* (Berlin), pp.
191–208, Figs. 3 and 4.

section the wind was in the direction of the current, sea level fell 7.6 cm in 65 km from the Sea of Marmara toward the Aegean sea, and the interface rose 19 m over the same distance. As already stated, the theoretical curves were adjusted to fit the experimental data by a suitable choice of frictional ceofficients.

A similar comparison is shown in Fig. 8-14 for the Strait of Gibraltar and Bab el Mandeb. For the Gibraltar section the wind was in the

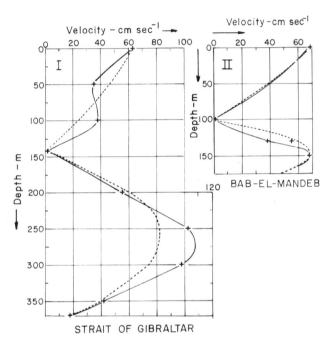

Fig. 8-14 Reproduced from A. Defant, 1930, *Sitzungberichte der Preussischen Akademie der Wissenschaften, Physikalish-Mathematische Klasse* (Berlin), pp. 191–208, Fig. 5.

direction of the current, sea level rose 0.6 cm in 100 km from the Mediterranean sea toward the Atlantic ocean, and the interface fell 15 m over the same distance. For the Bab el Mandeb section the wind was also in the direction of the current, sea level rose 1.1 cm in 100 km from the Red sea toward the Indian ocean, and the interface fell 40 m over the same distance.

The simplified Knudsen relations of Fig. 2-5 and (2-39) to (2-41) can also be applied to the Mediterranean and Black seas to gain some insight into the gross exchange properties of these two bodies. We have from Sverdrup et al (1942), quoting the results of others, that the inflow to the

Mediterranean sea from the Atlantic ocean is computed to be $1.75 \times 10^6 \, \text{m}^3 \, \text{sec}^{-1}$. From relation (2-38) and the observed salinity values one then obtains a value of $1.68 \times 10^6 \, \text{m}^3 \, \text{sec}^{-1}$ for the outflow and a net excess of evaporation over precipitation, river runoff, and net inflow from the Black sea of $7 \times 10^4 \, \text{m}^3 \, \text{sec}^{-1}$. The rapidity of exchange through the Strait of Gibraltar is illustrated by stating that the inflow from the Atlantic is sufficient to provide for a complete renewal of the Mediterranean waters in about 75 yr. The conditions for the Black sea are quite different. There, from current measurements in the Bosporus the outflow to the Aegean sea is estimated to be $12.6 \times 10^3 \, \text{m}^3 \, \text{sec}^{-1}$ and the inflow to the Black sea $6.1 \, \text{m}^3 \, \text{sec}^{-1}$, giving an excess of precipitation and river runoff over evaporation of $6.5 \times 10^3 \, \text{m}^3 \, \text{sec}^{-1}$. The inflow, in this case, is so small in proportion to the total volume of water in the Black sea that complete renewal of the water below a depth of 30 m would take about 2500 yr. For all practical purposes the deep water of the Black sea is stagnant, as is evident from the observation that below a depth of 200 m the water contains no oxygen but large quantities of hydrogen sulfide.

8-6. EUROPEAN COASTAL WATERS

This and the preceding section are, in a certain sense, out of context in comparison to the other sections in this chapter in that they are concerned with the investigation of a specific process rather than of a given geographic area. In the preceding section the investigations of some years ago by Defant on the basic concepts of stratified flow through straits were covered. In this section we are concerned with the more recent investigations by Talbot and Talbot (1974) and by Kullenberg (1971, 1972, 1974) on the details of horizontal mixing and diffusion in coastal waters as determined from dye concentration measurements following instantaneous dye releases.

There are several difficulties in the direct interpretation of such dye measurements. These items have been covered previously in the appropriate sections of Chapters 2, 5, and 6; but it is well to go over them again here as general guidelines for the discussion that follows. When we are concerned with a short time scale, there is a serious question as to the proper applicability of the Fickian diffusion equation. In the case of most measurements following an instantaneous dye release the time scale is of a few to several hours. It is then within the time scale of the semidiurnal tide, and the meaning of a tidal diffusion coefficient becomes abstruse. Presumably the tidal diffusion coefficient increases in some manner from zero up to its steady state value as time increases from zero to a few tidal periods. In other words we have the problem of sorting out tidal advection and tidal diffusion effects that is, of defining an average and a turbulent velocity component depending on the time scale of the investigation. To add to this is the concern in separating the relative tidal diffusion and the net circulation, or velocity shear, contributions to the observed longitudinal and lateral diffusion again as a function of time. For the velocity shear contribution the steady state simplification to a net nontidal circulation effect cannot be made; but the actual oscillatory velocity profile must be used with the correct phase corresponding to that at the time the dye is released. Further, there is the concern of the vertical diffusion contribution and as to what degree, if at all, the dye is limited to horizontal diffusion within a defined vertical layer. It thus appears, at least to us, that although such dye measurements are interesting and instructive in themselves it is difficult to quantify the results.

Talbot and Talbot (1974) made an extensive series of surface dye measurements as a function of time from instantaneous surface rhodamine dye releases in the waters adjacent to the British coast. Sixteen different locations were occupied, some more than once; and they included estuarine, coastal, and open sea situations. The estuarine and coastal measurements usually covered a period of a few to several hours; the open sea measurements covered longer periods measured in days.

For each time after the release time at which the surface dye field was surveyed, longitudinal and transverse dispersion coefficients were determined from the relation

$$K_{x,y} = \frac{\sigma_{x,y}^2}{2t} \tag{8-8}$$

where $\sigma_{x,y}$ is the standard deviation of a Gaussian distribution fitted to the data. We see from (2-69) that this relation strictly holds only for $K_{x,y}$ constant. In general the relation (2-84) applies, in which case the values determined from (8-8) will be time averaged values over the time interval from the release time to the survey time. The longitudinal or K_x value is determined for the direction of maximum extension of the observed dye field, generally in the direction of the tidal current, and the K_y value normal to it. For estuarine and coastal observations K_x values increase in an irregular manner as a function of time, particularly for the values obtained in the time intervals up to 6 hr. For open sea observations at time intervals of 12–190 hr K_x values vary somewhat from one observation to the next but show no apparent trend as a function of time. The K_y values in both cases are of an order of magnitude or more less than the corresponding K_x values.

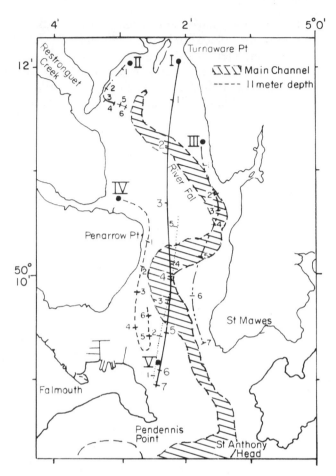

Fig. 8-15   Reproduced with permission from J. W. Talbot and G. A. Talbot,
1974, *International Council for the Exploration of the Sea, Rapports,* **167,** 93–110,
Fig. 84.

Let us look at results for a few of these areas in more detail. Figure
8-15 shows the tidal movements in the Fal estuary as revealed by five dye
releases, the first four being at high water and the fifth at low water. The
arabic numbers in the figure indicate hours after release. The path and
extent of movement following release II show the combined influence of
the water flowing out of Restronguet creek and the decreased tidal
motion in the upper reaches of the estuary. The complex pattern of water
movement in the estuary was often demonstrated by splitting of the dye
patches. Figure 8-16 shows the surface dye distribution 3 hr after the

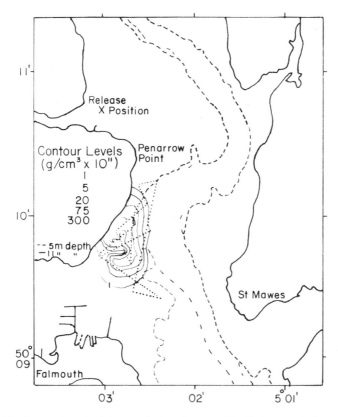

Fig. 8-16   Reproduced with permission from J. W. Talbot and G. A. Talbot, 1974, *International Council for the Exploration of the Sea, Rapports,* **167,** 93–110, Fig. 85B.

release at point IV. It clearly shows the combined effect of tidal advection and horizontal dispersion for measurements taken within this short time interval after release.

Figure 8-17 shows the results for similar measurements in the Blackwater estuary. Survey 1 is 6 hr after release, survey 2 is 12 hr after release, survey 3 is 25 hr after release, and survey 4 is 30 hr after release. Again the combined tidal advection and horizontal dispersion effects are well illustrated. At a time 6 hr after release the dye patch has moved a maximum distance away from its release location. At 12 hr after release the dispersed dye patch has returned to the initial location. The same tidal advection and dispersion process is repeated for the 25 and 30 hr observations.

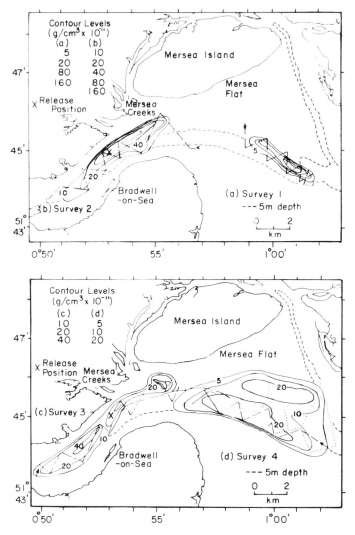

Fig. 8-17 Reproduced with permission from J. W. Talbot and G. A. Talbot, 1974, *International Council for the Exploration of the Sea, Rapports,* **167,** 93–110, Figs. 87B and 87C.

318

TABLE 8-2

Location	Date	Time after release (hr)	K_x $(\times 10^4\,\mathrm{cm^2\,sec^{-1}})$	K_y $(\times 10^3\,\mathrm{cm^2\,sec^{-1}})$
Fal	April–May 1967	2.6	1.6	1.6
		3.5	0.8	—
		3.4	0.4	—
		6.9	3.6	1.4
		2.0	1.4	—
		3.3	—	2.9
Blackwater	April 1966	6.5	15.8	9.0
	June 1968	6	13.4	2.7
		12	27.0	12.5
North sea	October 1969	12	21.7	6.0
		17	14.8	2.7
		23	9.9	2.6
		29	9.6	3.1
		36	11.3	4.3
		48	10.4	2.4
		54	10.7	2.6
		60	11.1	1.4

The values of K_x and K_y determined from (8-8) for these two locations, as well as for the open sea location to be discussed, are given in Table 8-2.

Figure 8-18 shows the dye pattern 36 hr after a release in the southern part of the North sea. Here, as in the other open sea experiments in the same area, the pattern is fairly simple. The observation is made for these open sea areas that, as the distributions spread, they can often be closely represented by constant diffusion coefficients, which is to be expected. They take as a first approximation to this situation the two dimensional Fickian diffusion equation with constant K_x and K_y and with an assumed uniform distribution as a function of depth from the surface to the bottom. As we can see from this defining equation, or more directly from the relation (6-58), the solution is

$$c = \frac{M}{4\pi h \sqrt{K_x K_y}\, t} e^{-[(x^2/4K_x t)+(y^2/4K_y t)]} \tag{8-9}$$

where M is the mass of dye released, and h is the water depth. The observed values can indeed be well fitted to the x and y Gaussian distribution defined by (8-9), which is gratifying in itself but not necessarily conclusive. The decay of the maximum concentration, however, does

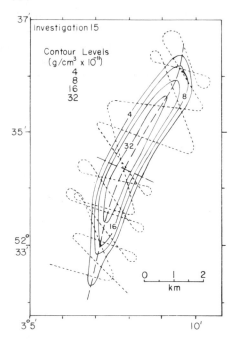

Fig. 8-18 Reproduced with permission from J. W. Talbot and G. A. Talbot, 1974, *International Council for the Exploration of the Sea, Rapports,* **167,** 93–110, Fig. 83D.

not show good agreement in all cases with the defined decay rate of t^{-1}. Undoubtedly this is at least in part related to the actual vertical mixing effects being different than that of an assumed uniform vertical mixing.

Kullenberg (1971, 1974) investigated the vertical diffusion from instantaneous rhodamine dye releases. In this study the dye was injected at depth in a solution whose density was adjusted to that of ambient water at the depth of injection. The tracing was carried out with a specially developed *in situ* fluorometer, which made it possible to detect thin layers of rhodamine and to observe steep variations in the concentration. The output signal, yielding a measure of the volume concentration, was recorded continuously as the ship cruised over the area. Navigation was performed with reference to a parachute drogue placed in the depth interval where the rhodamine was injected and connected to a surface float. When the dye patch grew sufficiently large, standard Decca navigation was used. The tracing period varied between 5 and 15 hr. Most of the measurements were taken in Scandinavian waters. In addition, salinity, temperature, and current profiles were taken before, during, and after the dye measurements.

Two cases are distinguished from the observations. In one there is strong stratification and consequent weak vertical mixing, and in the other there is weak stratification and consequent strong vertical mixing. For the

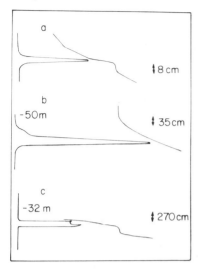

a

↕8 cm

b
-50m

↕ 35cm

c
-32 m

↕ 270cm

Fig. 8-19 Reproduced with permission from G. Kullenberg, 1974, *International-Council for the Exploration of the Sea, Rapports,* **167,** 86–92, Fig. 76.

first case, which was exhaustively examined, the dye was injected in or immediately below a locally sharpened temperature gradient. The dye then spread horizontally in a thin layer associated with this local gradient. Examples are shown in Fig. 8-19.

The simple solution forms of one dimensional, vertical, Fickian diffusion were used to calculate values of K_z. The two layered solution was used for the strong stratification case, and the unbounded solution for the weak stratification case. The time intervals for the calculations were 0.5–1.5 hr after release. The calculated values of K_z were about 10^{-1} cm^2 sec^{-1} for strong stratification cases and about 10 cm^2 sec^{-1} for weak stratification cases. The assumptions in this calculation methodology appear to be subject to question.

Kullenberg (1972, 1974) investigated horizontal diffusion from the same experiments, obtaining values of the standard deviations σ_x and σ_y at times of 1–10 hr after release. Through an extension of the methods of Section 6-6, which include several assumptions as to the determining variables and equational relations, a relation was derived for σ_x^2 and σ_y^2 and compared favorably with the observed values.

CHAPTER 9

Americas, East Coast

9-1. ST. JOHN ESTUARY

The St. John estuary in New Brunswick is one of the most unusual estuaries to be found anywhere. It consists of the St. John river and lake system and St. John harbor, which empties into the upper end of the Bay of Fundy. The St. John river and harbor are separated by a narrow gorge with sill depths less than 5 m, referred to as Reversing falls. The amplitude of the tides in St. John harbor is 7–8 m.

In this gorge the great rise and fall of the tides in the Bay of Fundy produce a phenomenon known as a reversing falls, which has given the name to the gorge itself. At half tide the ordinary current of the river flows out through the narrow gorge in the form of a rapid; but when the water in the harbor falls, the rapid becomes a foaming waterfall. As the tide rises, the level of the water in the harbor increases to a height above that in the river; a strong current then sets inward through the gorge, which becomes a rapid and eventually a waterfall when the tide is at its height.

The physical oceanographic characteristics of the St. John estuary have been examined by Hachey (1935, 1939), Trites (1960), and Neu (1969). The salinity and current velocity distribution through a tidal cycle are illustrated in Fig. 9-1. The salinity distribution in the harbor shows a strong and variable salinity gradient, with intrusion of the more saline waters at depth. The current distribution in the harbor shows a two layer flow for part of the tidal cycle, with a flow out of the estuary in the upper portion of the water column and a flow into the estuary in the lower portion; during the remainder of the tidal cycle there is a single layer flow out of the estuary.

On the supposition that the abrupt widening at the mouth of the harbor acts as a control and that the action of Reversing falls is so efficient that overmixing occurs, Stommel and Farmer (1953) applied relation (5-41) to the observed salinity data of Hachey (1939) at the harbor mouth. The results are shown in Fig. 9-2, where the plotted points are the observed salinity difference ratios for various values of river flow, and the straight line is computed from (5-41), using the values for the breadth and depth

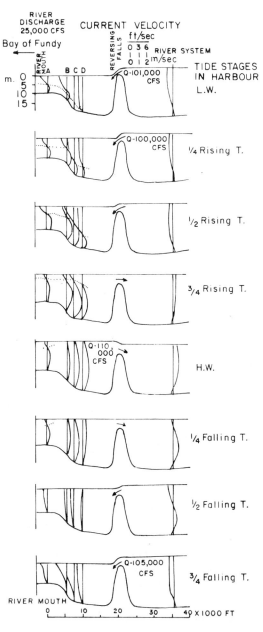

Fig. 9-1 Reproduced with permission from H. A. Neu, 1969, *International Association for Hydraulic Research, Proceedings of the Thirteenth Congress*, **3**, 241–248, Fig. 8.

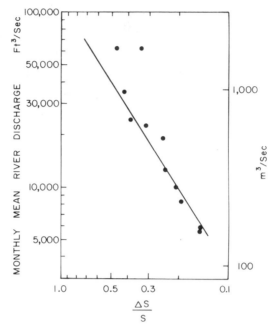

Fig. 9-2   Reproduced with permission from H. Stommel and H. G. Farmer, 1953, *Journal of Marine Research*, **12**, 13–20, Fig. 4.

of the harbor at its mouth. In view of the crude nature of both the data and the theory, the agreement is fairly good; and it was concluded by these investigators that St. John harbor is overmixed.

9-2.   MIRAMICHI, PENOBSCOT, AND KENNEBEC ESTUARIES

Bousfield (1955) has described the general physical oceanographic features of the Miramichi estuary. This estuary consists of an inner, or river, portion with two branches extending up to the tidal limit, and an outer, or bay, portion bounded by the Gulf of St. Lawrence, the overall length being about 50 mi (80 km). The general physical oceanographic features are those of a partially stratified estuary. The tidal cycle is somewhat different than what one normally encounters, varying over the 2 wk tidal period from a regular semidiurnal tide to a nearly diurnal tide.

A feature of particular interest is the effect of the estuary net circulation on the distribution of the larvel stages of a brackish water barnacle *Balanus improvisus* during its 18 day planktonic life period. The first three nauplius stages, found mainly in the upper portion of the water column, are transported successively seaward. The fourth and fifth stages, found near middepths, are concentrated near the mouth of the estuary. And the sixth stage nauplius and the cyprid, found mainly in the lower portion of the water column, are carried progressively landward toward the head of the estuary.

Haefner (1967) has described the physical oceanographic features of the Penobscot estuary. This estuary consists of the lower portion of the Penobscot river from the head of tide to Penobscot bay, extending a distance of 28 n mi (52 km). The Penobscot river is the largest river in Maine and is heavily polluted with industrial and organic wastes. Measurements of temperature, salinity, and dissolved oxygen were taken as a function of depth and tidal cycle during the spring, summer, and fall seasons along the length of the estuary.

The estuary is partially stratified, showing increased stratification in a downestuary direction. During spring runoff conditions the salinity contours are pushed downestuary in the upper portion of the estuary; and a stronger halocline is developed in the lower portion of the estuary, associated with the increased river runoff. During the summer a seasonal thermocline is developed in the lower portion of the estuary.

The dissolved oxygen values show an interesting distribution related to the pollutant oxygen demand and the associated runoff and tidal mixing conditions in the estuary. During the spring runoff conditions of April the dissolved oxygen values are highest throughout the estuary. During the low river runoff conditions of August the upper estuary becomes anoxic, presumably related to the lower river runoff, the decrease in longitudinal dispersion usually associated with lower river flows, and the decrease in tidal mixing usually associated with decreased tidal motions in an upestuary direction.

Francis et al. (1953) made one of the first, if not the first, attempts to measure directly the turbulent transports of momentum and heat. These measurements were made at various depths in the Kennebec estuary. Longitudinal velocity fluctuations were measured with a Von Arx propeller type current meter, and vertical fluctuations were obtained from the angular movements of a vane pivoted on a horizontal axis. Temperature fluctuations were measured with a thermistor. The turbulent stress $\sigma_{zx} = -\rho\overline{v'_x v'_z}$ is then given by $-\rho\overline{v'_x \bar{v}_x \theta'}$, where θ' is the angular fluctuation of the vane, and $\bar{v}_x$ is the mean longitudinal velocity over the measurement time. The vertical turbulent heat flux is given by a similar expression of the form $c\rho\overline{T'\bar{v}_x\theta'}$, where T' is the temperature fluctuation.

As one might expect in such a first attempt, considerable instrumentation difficulties were encountered; and a judicious selection of records had to be made for those suitable for analysis. In addition, the sampling interval for the reading was 3 sec over a record only 2–3 min long. The determined values of the momentum and heat fluxes then permitted the determination of the vertical viscosity and diffusion coefficients N_z and K_z from the defining relations (2-135) and (2-64). The values of K_z for the three determinations given varied from 50 to 600 cm^2 sec^{-1}, and the values for N_z from 500 to 1200 cm^2 sec^{-1}. The corresponding values of K_z/N_z and Ri were 0.14 and 5.2, 0.04 and 19, and 1.01 and 0.14, respectively, which are in general agreement with the values given in Fig. 5-26.

9-3. CONNECTICUT AND MERRIMACK RIVERS

Howard (1940) has briefly reported on and Meade (1966) has analyzed in more detail a series of salinity measurements taken in the Connecticut river during the 5 yr period 1934–1939. The Connecticut is a tidal river which exits through a relatively narrow channel directly into Long Island sound. The coastal waters extend as a salt intrusion up the Connecticut to varying distances depending on river runoff conditions. During the spring runoff the river is virtually free of salt. During the summer low river flow conditions the salt may extend about 13 km up the river. In addition, the salt intrusion shows substantial movement back and forth, corresponding to the tidal excursion.

The seasonal river flow variation of the salt penetration is shown in Fig. 9-3. It is of interest to note the steepening of the salinity gradient with increasing river flow. Figure 9-4 is also of interest in illustrating the details of the salinity gradient, as well as its tidal movements. The measurements were taken on 16 March 1965 during an ebb tide. The tide began to ebb at 1030, the measurements in the upper figure being taken on a downriver traverse and those in the lower figure for a following upriver traverse.

One further important comment should be made on nomenclature. It should be noted from Figs. 9-3 and 9-4 that, although there is a strong salinity gradient, it by no means is a stratification interface. The Connecticut river is classified as a strongly stratified estuary and not as an arrested salt wedge estuary. There is substantial mixing of river and coastal waters, as indicated by the rapid horizontal change in the surface salinity in the vicinity of the salinity gradient. The descriptive term salt wedge is variously used in the literature to indicate an estuary condition in which there is either a stratification interface or a strong salinity gradient. In our theoretical derivations here, however, we restrict the term salt wedge to indicate a stratification interface condition. Care

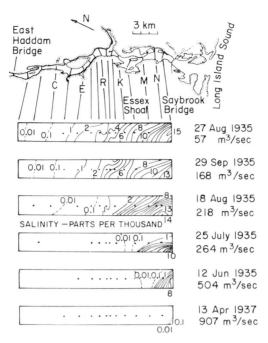

Fig. 9-3 Reproduced with permission from R. H. Meade, 1966, *Water Resources Research*, **2**, 567–579, Fig. 3.

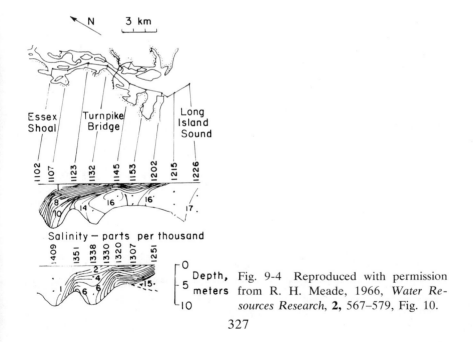

Fig. 9-4 Reproduced with permission from R. H. Meade, 1966, *Water Resources Research*, **2**, 567–579, Fig. 10.

should then be taken by the reader to note which condition is being referred to when the term salt wedge is used in a particular article.

Garvine (1974) investigated the plume of brackish water that extends out from the Connecticut river into Long Island sound. In Long Island sound the principal current structure is that of the tides, parallel to the longitudinal axis of the sound or, essentially parallel to the shoreline near the discharge. Within the sound the river outflow formed a buoyant plume a few meters in depth whose horizontal location was subject to great variation. Tidal phase and brackish water volume discharge were found to control much of this variation. Figure 9-5 gives an indication of this variation for conditions of low slack water and high slack water on 21 April 1972, a time of high river discharge. The contours are of salinity at 0.5 m depth, and the dashed lines are the survey boat's track taken to define the plume.

The plume also formed a transverse front on its offshore side, which was essentially parallel to the shoreline but which during a period of low slack water curved around at its western end to the shore and during a period of high slack water curved around at its eastern end to the shore. Figure 9-6 shows the location of the front during a period of low slack water on 13 April 1973. The contours are again for salinity at a depth of 0.5 m. A transverse section across the plume and across the front along the section line A of Fig. 9-6 is shown in Fig. 9-7. The vertical column of dots in Fig. 9-7 indicates the control points for the measurements. The salinity contours are in parts per thousand and the temperature in degrees Celsius. The density contours are for σ_t, the density difference from unity in parts per thousand at atmospheric pressure, these values being determined from standard tables from the observed salinity and temperature measurements. During the period of these measurements the ambient water of Long Island sound had a characteristic salinity of 25‰, temperature of 5°C, and σ_t of 20. These values were then taken to indicate the boundary of the plume. The front itself was quite clearly a very prominent feature of the plume distribution.

In addition, the front forms a distinctive surface feature. The outflowing river water is yellowish brown in color, whereas the sound water is nearly blue green. The offshore boundary between the two waters masses, which is exceedingly sharp at the front, presents a feature that may be readily photographed from the air. Further, because of the convergent velocities at the surface into the front, it is also often distinguished by the presence of a foam line.

Garvine and Monk (1974) continued these interesting investigations of the frontal features of the Connecticut river plume in more detail. They took several series of measurements of salinity and temperature and of

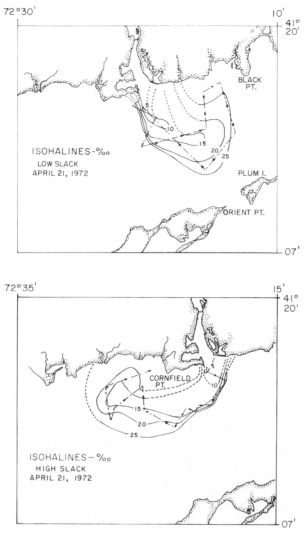

Fig. 9-5 Reproduced with permission from R. W. Garvine, 1974, *Journal of Geophysical Research*, **79**, 831–846, Fig. 12.

horizontal velocity in the immediate vicinity of a front. The results of one such series of measurements are shown in Figs. 9-8 and 9-9. Figure 9-8 is a plot of σ_t contours along a section normal to the front. The vertical column of dots again indicate the control points for the measurements. It is to be noted that the front follows roughly the form of (4-165) from Section 4-8 as shown by the heavy line in Fig. 9-8. Figure 9-9 is a plot of

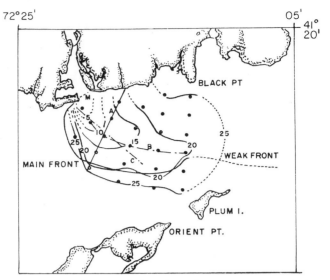

Fig. 9-6 Reproduced with permission from R. W. Garvine, 1974, *Journal of Geophysical Research*, **79**, 831–846, Fig. 3.

horizontal velocity contours normal to the front along a section also normal to the front. It is to be noted that there are strong convergent surface velocities into the front from both sides, and also that the surface velocities on the brackish water side of the front are of the order of half those immediately below the front and in the opposite direction, again as anticipated by the simplified theory of Section 4-8.

To gain some understanding of the evolution of the density fields in the vicinity of a front a sequence of four sections was taken across a front, as shown in Fig. 9-10. The same relative position along the front was maintained by following a marker drogue which progressed with the tidal current. At the conclusion of the fourth run the surface color contrast weakened markedly and soon disappeared. The surface salinity in the general area showed no frontal structure. As determined by an observer moving with the frontal system, the mixing time required to destroy the front following its formation near the mouth was $2\frac{1}{2}$ hr.

From the observations of Fig. 9-10 the following mixing dynamics were postulated. The high level of density contrast associated with a section of freshly formed front at the mouth induces high levels of convergence and sinking. The saline water, driven beneath the lens of brackish water, entrains the lighter water above and mixes it downward. This downward vertical mass flux is supplied by the horizontal inflow of brackish water toward the front. As the mixing continues, the reservoir of brackish water

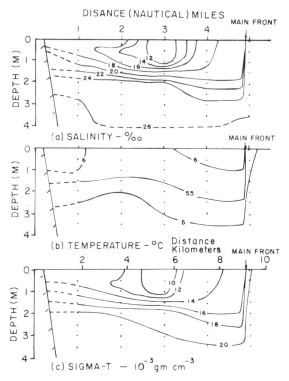

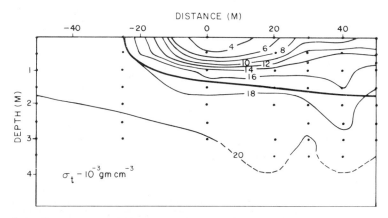

Fig. 9-7 Reproduced with permission from R. W. Garvine, 1974, *Journal of Geophysical Research*, **79**, 831–846, Fig. 4.

Fig. 9-8 Reproduced with permission from R. W. Garvine and J. D. Monk, 1974, *Journal of Geophysical Research*, **79**, 2251–2259, Fig. 6.

331

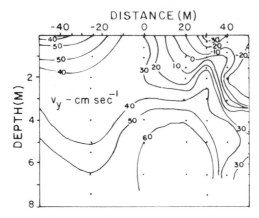

Fig. 9-9   Reproduced with permission from R. W. Garvine and J. D. Monk, 1974, *Journal of Geophysical Research*, **79**, 2251–2259, Fig. 7.

in the plume is gradually depleted; and the downward migration of the σ_t contours is reversed. Finally, the brackish water reservoir is pumped out, the density contrast nearly vanishes, and the cross frontal motion is shut off.

Boyle et al. (1974) made a study of the chemical mass balance for silica and iron in the Merrimack river. Surface samples were taken along the river, and the concentration values for silica and iron compared with the corresponding salinity values. If the source amount for either quantity can be taken as having occurred from continental weathering further upriver from the area of investigation, which is a reasonable assumption in this case, and if there are no tributary sources, which is the case, then if either quantity acts as a conservative substance, we have from relation (5-105) or (5-23) that the concentration c of the substance as averaged over a vertical section should vary linearly with the salinity s or, equivalently, that dc/ds should be a constant. Figure 9-11 is a plot of silica concentration versus salinity for four measurement periods. It follows a linear relation, and we may consider that here silica acts as a conservative substance. The vertical intercepts on the graphs represent source concentrations. The iron versus salinity relation and the corresponding silica versus salinity relation are shown in Fig. 9-12 for another series of measurements. In this case iron clearly shows a nonconservative variation. It might be conjectured that the rapid decrease in iron concentration in the salinity range 0–7‰ represents flocculation and settlement of the clay minerals on which the iron has been adsorbed, which is anticipated to occur in this salinity range, and that the linear decrease for greater

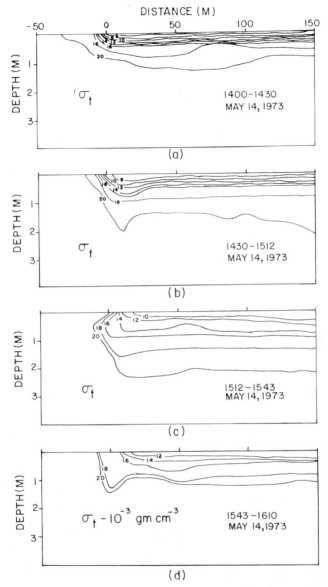

Fig. 9-10   Reproduced with permission from R. W. Garvine and J. D. Monk, 1974, *Journal of Geophysical Research*, **79**, 2251–2259, Fig. 10.

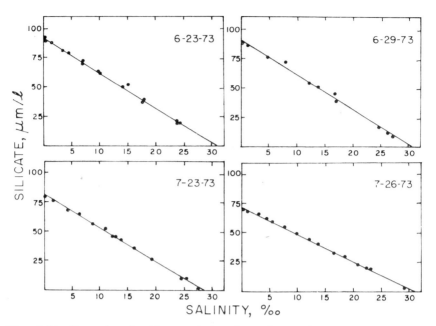

Fig. 9-11 Reproduced with permission from E. Boyle, R. Collier, A. T. Dengler, J. M. Edmond, A. C. Ng, and R. F. Stallard, *Geochimica and Cosmochimica Acta*, **38**, 1719–1728, Fig. 1.

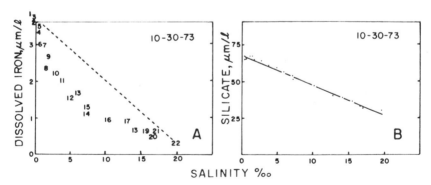

Fig. 9-12 Reproduced with permission from E. Boyle, R. Collier, A. T. Dengler, J. M. Edmond, A. C. Ng, and R. F. Stallard, *Geochimica and Cosmochimica Acta*, **38,** 1719–1728, Fig. 2.

334

salinity values represents the more normal conservative estuarine longitudinal dispersion relation.

9-4. HUDSON RIVER AND ASSOCIATED WATERS

The New York harbor area is a rather intricate waterways system, as shown in Fig. 9-13, consisting of the principal drainage of the Hudson

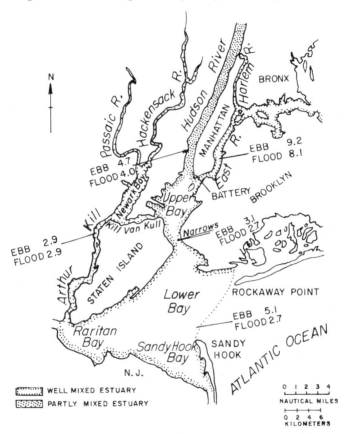

Fig. 9-13 Reproduced with permission from C. M. Duke, 1961, *American Society of Civil Engineers, Journal of the Waterways and Harbor Division,* **87,** WW1, 29–45, Fig. 4.

river into Upper Bay and thence through the Narrows to Lower Bay and the Atlantic ocean, the subsidiary drainage of the Passaic and Hackensack rivers into Newark bay, and the interconnecting waterways of Arthur Kill, Kill Van Kull, and the Harlem and East rivers, which are connected to the east to Long Island sound. Also as shown in Fig. 9-13, the waters of Lower and Upper bays and the interconnecting waterways are described as being well mixed, and those of the Hudson river as being partially stratified.

Duke (1961) and Simmons (1966) examined the sedimentation and shoaling characteristics of New York harbor. According to the latter, and as shown in Fig. 9-14, the net circulation flow at the surface is downstream for the Hudson river and Upper bay from Tappan zee, at the left

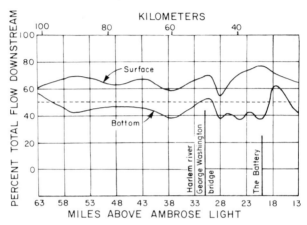

Fig. 9-14   Reproduced with permission from H. B. Simmons, 1966, *Estuary and coastline hydrodynamics*, McGraw-Hill, New York, pp. 673–690, Fig. 16.12.

hand edge of the figure, to the Narrows, at the right hand edge of the figure. In contrast the net circulation flow at the bottom shows a more complex distribution. In the Narrows the net bottom flow is upstream, but in Upper bay it is downstream because of the mixing effects of the East river and Kill Van Kull. In the Hudson, for a freshwater discharge condition of 6000 ft^3 sec^{-1} (170 m^3 sec^{-1}), the net bottom flow is predominantly upstream except for a local area in the vicinity of the George Washington bridge. The cross sectional area of the river channel at the bridge is somewhat reduced, so that local currents and turbulence are above normal. Furthermore, the Harlem river, which is a relatively small waterway connecting the Hudson and East rivers, joins the Hudson just upstream of the bridge. The combination of these two effects is such that the upstream predominance of net bottom flow is interrupted locally at the bridge. The data of Fig. 9-14 refer to conditions in the deep navigation channel of the Hudson; at the shallower depths on the western, or New Jersey, side of the river the net bottom flow is predominantly downstream.

These net circulation conditions produce an interesting sedimentation and shoaling pattern. Available evidence indicates that sediments are moved progressively upstream until they reach the influence of the turbulent region at the George Washington bridge. At this point much of the sediment is resuspended, and some of it is moved downstream along the western side of the river under the influence of the downstream bottom flow at the shallower depths there. Following a maintenance dredging in the navigation channel shoaling always begins first at the upstream end and then progresses downstream on the western side as the upstream portion is filled.

O'Connor (1962) made an interesting investigation of the BOD and the DO distribution in the Hudson river and Upper bay, using the concepts of Section 6-7. The general development and application of these concepts both here and in other estuarine water bodies is largely due to O'Connor.

To start it was observed that the BOD concentration was approximately constant throughout Upper bay due to a combination of tidal mixing and the number and location of the wastewater outlets. The same condition was observed for the Hudson river from the Battery to the city line, somewhat north of the George Washington bridge. This permits a substantial simplification in the equational relations (6-104) and (6-105). With reference to Fig. 9-13 we can then apply the equivalent of Knudsen's relations to the total organic material in Upper bay, presuming that the diffusion contribution out of the bay at the Narrows and into the bay from its various entrances is negligible, giving

$$c_{1B}R_B = W_B + c_{1H}R_H - k_{1B}c_{1B}V_B \qquad (9\text{-}1)$$

where c_{1B} and c_{1H} are the BOD concentrations in Upper bay and the Hudson river, respectively, R_B and R_H are the net flows out of the bay through the Narrows and into the bay from the Hudson river, respectively, W_B is the total mass of BOD discharged to the bay, not including the Hudson river contribution, V_B is the volume of the bay, and k_{1B} is the BOD decay coefficient for the bay. Similarly, for the Hudson river the relation is

$$c_{1H}R_H = W_H - k_{1H}c_{1H}V_H \qquad (9\text{-}2)$$

where W_H is the total mass of BOD discharged to the Hudson, V_H is the volume of the Hudson from the Battery to the city line, and k_{1H} is the BOD decay coefficient for the Hudson. These relations can then be written as

$$c_{1B} = \frac{W_B + c_{1H}R_H}{R_B + k_{1B}V_B} \qquad (9\text{-}3)$$

and

$$c_{1H} = \frac{W_H}{R_H + k_{1H}V_H} \qquad (9\text{-}4)$$

for c_{1B} and c_{1H}.

Under the assumption that the BOD concentration is a constant as a function of longitudinal distance the coupled equations (6-104) and (6-105) reduce to the single equation

$$K_x\frac{d^2c_2}{dx^2} - v_x\frac{dc_2}{dx} + k_2(c_s - c_2) - k_1c_1 = 0 \qquad (9\text{-}5)$$

for the DO concentration c_2, where k_2 is the reaeration coefficient, c_s is the saturation value for the DO concentration, and k_1c_1 is a constant. The further assumption is made that the advection contribution in (9-5) can be neglected for Upper bay. The solutions, then, to (9-5) for Upper bay and the Hudson river are simply

$$c_{2B} = c_s - \frac{k_{1B}c_{1B}}{k_{2B}}\left[1 - e^{-\sqrt{(k_{2B}/K_x)}x}\right] - (c_s - c_{0B})e^{-\sqrt{(k_{2B}/K_x)}x} \qquad (9\text{-}6)$$

and

$$c_{2H} = c_s - \frac{k_{1H}c_{1H}}{k_{2H}}\left[1 - e^{(v_x/2K_x)(1-m_2)x}\right] - (c_s - c_{0H})e^{(v_x/2K_x)(1-m_2)x} \qquad (9\text{-}7)$$

where m_2 is given by

$$m_2 = \sqrt{1 + \frac{4k_{2H}K_x}{v_x^2}} \qquad (9\text{-}8)$$

and where c_{0B} is the dissolved oxygen concentration for Upper bay at $x = 0$, presumably at the Battery, and c_{0H} is the dissolved oxygen concentration for the Hudson river at $x = 0$, presumably at the city line.

The longitudinal dispersion coefficient K_x was determined from the observed values of salinity along the Hudson river using (5-119), under the assumption that K_x could be considered a constant for this stretch of the river and that the variations in cross sectional area were slight. These determinations were made for various observations of salinity over a period of years, and for various flow conditions for the Hudson river. The results are shown in Fig. 9-15 as a function of river flow, the empirical correlation between K_x and R being quite good.

The coefficients k_1 and k_2 were determined from various empirical considerations as a function of the water temperature. The values chosen at a reference temperature of 20°C were $k_{1B} = 0.25 \text{ day}^{-1}$, $k_{1H} = 0.23 \text{ day}^{-1}$, $k_{2B} = 0.23 \text{ day}^{-1}$, and $k_{2H} = 0.09 \text{ day}^{-1}$.

From the observed values of W_B, W_H, R_B, and R_H and using the empirical values of k_{1B} and k_{1H} for mean monthly water temperatures,

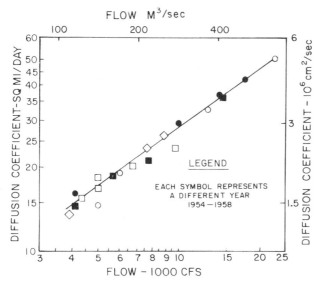

Fig. 9-15 Reproduced with permission from D. J. O'Connor, 1962, *Journal of the Water Pollution Control Federation*, **34,** 905–919, Fig. 2.

values of c_{1B} and c_{1H} were calculated from (9-3) and (9-4) for the mean conditions of each of the four summer months for a series of years. The correlation between the calculated and observed mean monthly values for c_{1B} for Upper bay is shown in Fig. 9-16. A similar correlation was obtained for the Hudson river.

Calculations were also made for the variation in the DO concentration

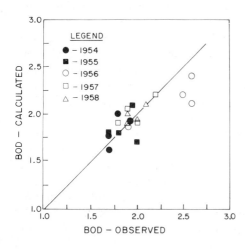

Fig. 9-16 Reproduced with permission from D. J. O'Connor, 1962, *Journal of the Water Pollution Control Federation*, **34,** 905–919, Fig. 7.

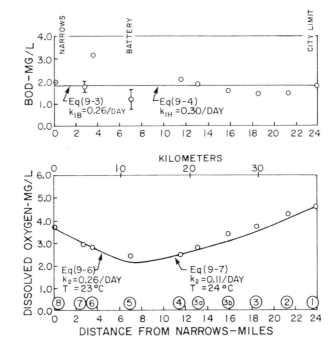

Fig. 9-17   Reproduced with permission from D. J. O'Connor, 1962, *Journal of the Water Pollution Control Federation*, **34**, 905–919, Fig. 9.

as a function of longitudinal distance from (9-6) and (9-7). One such calculation and its comparison with the observed values of c_2 are shown in Fig. 9-17. The initial value c_{0H} in this and the other similar calculations is taken from the observed data, presumably at the city line, and the value of c_{0B} from the resultant calculations of (9-6).

O'Connor (1965, 1966) extended this type of analysis to the East river. In this case the coupled equations of the form (6-104) and (6-105) were used with the inclusion of the cross sectional area variation as a function of longitudinal distance as in (6-56). For the East river the net circulation flow was assumed to be zero, so that the advection terms in (6-104) and (6-105) could be omitted. For the lower portion of the East river from the Battery up to its junction with the Harlem river the cross sectional area was essentially constant; for the upper portion of the East river from the Harlem river to Long Island sound the cross sectional area varied approximately as the square of the longitudinal distance. Solutions were given for c_1 and c_2 for a point and for a distributed source of pollutant for each case. Calculation were then made for the c_1 and c_2 distribution in the East river from the summation of the various pollutant sources, taking the treatment plant contributions as point sources and the untreated sewage and storm overflow contributions as distributed sources, and compared with the observed values of c_1 and c_2.

9-5. RARITAN ESTUARY

Ketchum (1951, 1951a) applied the modified tidal prism concepts, which he developed, to a determination of the longitudinal salinity distribution in the Raritan estuary. The Raritan estuary is composed of the Raritan river and Raritan bay, each approximately 10 mi (16 km) in length. The first step in the application of the method was to determine the cumulative low tide, tidal prism, and high tide volumes as a function of longitudinal distance from bathymetric charts and tide tables. The extent of the zero segment was defined by relation (5-14) for December 1948, when the river flow R was 740 ft^3 sec^{-1} (21 m^3 sec^{-1}). Each succeeding segment was then defined by relation (5-15), dividing the estuary into 14 segments. Relations (5-17), (5-19), (5-3), and (5-2) were then used to determine the fresh-water fraction and the salinity for each segment. The value of the reference salinity of Lower bay into which Raritan bay empties was $\sigma = 27\%$ during this period. A comparison of the calculated and observed values of salinity, also obtained during December 1948, is shown in Fig. 9-18. The

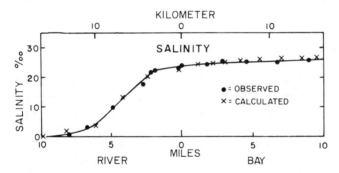

Fig. 9-18   Reproduced with permission from B. H. Ketchum, 1951, *Journal of Marine Research*, **10**, 18–37, Fig. 3.

salinity observations were taken over a tidal cycle, and the high tide values are plotted in the figure. For the bay each observed point is the average from four to seven stations at each cross section, and for the river from a single station.

   Ketchum et al. (1952) made an interesting investigation of the processes contributing to the decrease in coliform bacteria in the Raritan river, using the modified tidal prism concepts of Section 5-1 with an extension to nonconservative quantities as given in Section 6-3. It is an excellent example of the application of sound physical intuition and simplified analyses techniques to the understanding of a rather complex process.

A survey of various physical oceanographic, chemical, and biological parameters was made on 26–27 June 1949 for the Raritan river from New Brunswick near its head to Perth Amboy at its mouth, a distance of approximately 12 mi (19 km). The coliform bacteria counts decreased from maximum values of 110,000 ml^{-1} at New Brunswick, which is taken as the source area of coliform bacteria for subsequent elaborations, to average values of less than 100 ml^{-1} at the river mouth. There was no measurable DO in the water at New Brunswick, but the concentration of oxygen increased downriver so that at Perth Amboy it was about 70% of saturation. Correspondingly, there were no living zooplankton in the waters containing zero oxygen, but near the mouth the zooplankton population was fairly high (about 100 organisms per liter) and was made up of relatively few species, *Acartia tonsa* being predominant.

There are at least three processes that contribute to the decrease in coliform bacteria along an estuary, namely, dilution due to tidal mixing, bactericidal action of sea water, and predation by zooplankton. The purpose of the investigation was to determine the relative importance of each of these effects and to ascertain how well the combination of all three predicted the observed decrease in coliform bacteria along the river from the source area near New Brunswick to its mouth.

For the 2 wk period prior to the survey the river flow was not constant but decreased from a volume of 147×10^6 ft^3 (4.12×10^6 m^3) to a volume of 10.4×10^6 ft^3 (0.29×10^6 m^3) per tidal cycle. The modified tidal prism relation (5-19) was used in conjunction with the observed salinity values, in the reverse sense to the manner used in the previous paragraphs of this section to prove its applicability by comparing predicted and observed salinity values for a constant river flow, to obtain the volumes of fresh river water V_R escaping each segment per tidal cycle. The determined values varied in a regular manner from 14.5×10^6 ft^3 (0.41×10^6 m^3) per tidal cycle for the first segment to 40.0×10^6 ft^3 (1.12×10^6 m^3) for the last segment at the mouth of the river, corresponding in a general way to the decrease in river flow over the same time scale. These determined values of the river flow R were then used in the subsequent calculations for each segment.

The dilution or tidal mixing effect is given directly by relation (5-23). Bactericidal activity is observed, in separate experiments, to follow the first order decay relation of (6-20), with a mortality coefficient of $k = 1.15$ day^{-1}. The combined effect of dilution and mortality is then given by relation (6-24) for the first tidal segment and by relations similar to (6-25) for each succeeding segment, where the appropriate values of r and R are used for each segment.

The predation effect is also observed to follow a first order decay relation of the form

$$c = c_0 e^{-wpt} \qquad (9\text{-}9)$$

where w is related to the volume of water filtered per predator per unit time, and p is the number of predators. From predation measurements with other zooplankton a value for w of 20 ml of water per day was chosen for Raritan river zooplankton. The combined effect for dilution, mortality, and predation can then be calculated from relation (6-24) and the corresponding following relations by substituting $k + wp$ for k, where the observed values for the zooplankton population p are used for each segment.

A comparison of these calculations with the observations is shown in Fig. 9-19. It is to be noted in the calculations that mortality is the

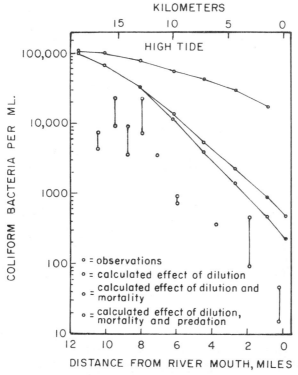

Fig. 9-19 Reproduced with permission from B. H. Ketchum, J. C. Ayers, and R. F. Vaccaro, 1952, *Ecology*, **33**, 247–268, Fig. 4.

dominant contributor, and also that the predation effect accounts for an additional half of the population decrease at the river mouth in excess of the combined effects of dilution and mortality alone. The calculations account for more than 99% of the observed decrease at the river mouth. The decreases, nevertheless, are so large that in spite of this agreement the calculated populations at the river mouth are still several times greater than the observed populations.

9-6. DELAWARE ESTUARY

The Delaware estuary, comprising the Delaware river and Delaware bay, is one of the larger estuaries in the United States and is also located in one of the more heavily industrialized areas. It extends roughly from the fall line at Trenton to Cape May and Cape Henlopen on the Atlantic coast, a distance of 134 mi (216 km). Understandably, it has been the subject of numerous engineering studies related to saline water intrusion and to pollution.

Mason and Pietsch (1940) investigated salinity intrusion in the Delaware estuary. They were interested in determining the variations in the upestuary extent of brackish water as a function of river flow, considering the estuary waters over this reach to be well mixed vertically, which is the case. Their basic data consisted of a rather extensive series of chlorinity, or salinity, measurements taken during 1930 and 1931; and they took as their definition of the fresh to brackish water boundary a salinity of 0.05‰. Their work is of interest not only in itself but also because the simple physical reasoning they used was a forerunner of the much used longitudinal dispersion relations (5-98) to (5-100) and their time dependent counterparts.

They argued that at all times there existed a volume A of fresh water between Trenton and the salinity boundary of 0.05‰, that during each day a volume B at the lower end of this pool of fresh water was assimilated by the brackish water working upestuary through tidal action, and that during each day a river flow volume C replenished the freshwater pool. The equation was then written for the freshwater volume A_1 to start the beginning of the next day as

$$A - B + C = A_1 \qquad (9\text{-}10)$$

and so on for succeeding days.

The volumes B were then ascertained in the following manner. The observed movement of the 0.05‰ salinity boundary over a sequence of days was recorded in units of distance per day. The average cross section area of the estuary, which is directly related in the case of the Delaware to longitudinal distance, at which this movement took place was then plotted against the average daily river flow for the period; and each point was labeled with its corresponding movement per day, some values of course being positive and some negative. The average curve through the positive and negative values was then drawn, corresponding to $A = A_1$, and from (9-10) also $B = C$. The curve then represented the values of B as a function of longitudinal distance or cross sectional area.

By using these values of B, various graphic presentations were then given for the variations in the extent of A as a function of C.

O'Connor (1960, 1965) applied the analytic solutions of the steady state, longitudinal dispersion, coupled BOD-DO equations to the Delaware estuary. These relations were developed by him and were discussed in Section 6-7. The solutions were applied to a section of the Delaware estuary for which the cross sectional area was assumed to be constant. A mean value for the longitudinal dispersion coefficient K_x along this stretch was determined from a log plot of salinity versus distance using (5-119), as was done for the similar investigation for the Hudson river discussed previously. The coefficients k_1 and k_2 were determined from various empirical relations. The calculated values of BOD and DO were compared with slack tide observations with good agreement.

O'Connor (1965) and O'Connor et al. (1968) extended this analytic method. Solutions are given treating the cross sectional area of the Delaware estuary to vary linearly with longitudinal distance, which is a reasonable approximation to the observed variation of the cross sectional area with distance downestuary. The solutions are given in terms of modified Bessel functions of the first and second kind. In addition a distinction is made between the carbonaceous BOD and the nitrogenous BOD, including a time offset for the commencement of the nitrogenous oxidation from its time of introduction into the estuary. The estuary was divided into five sections for the calculations and comparison with observed data.

A word of caution is perhaps in order at this point with regard to the application of these techniques either in an analytic solution form to a portion of an estuary as a whole or in the more usual finite difference, section by section, computer solution form. The latter of course has the distinction of permitting the observed changes in cross sectional area as well as changes in K_x, k_1, and k_2 to be more easily included. These types of solution have received considerable application and have been quite successful and useful in estuary pollution studies. There are, nevertheless, certain vagaries and limitations. There are a substantial number of coefficients and a certain amount of inaccuracy in the determination of each. An exact agreement of calculated and observed results should not then be taken as proof in itself, for there is a certain amount of justifiable variation in the coefficients allowable to permit such an agreement. Further, the defining equations are vertically integrated, or averaged, and tidally averaged, so that only a comparison with such results is permitted. The inputs to the solutions are the various pollutant sources and possible sinks along the course of the estuary and the concentration values at the upper and lower ends of the estuary, the latter values usually being given from observations at these two extreme distances.

Thomann (1963, 1972) applied a finite section computer approach to a stretch of the Delaware estuary extending from Trenton a distance of 84 mi (135 km) downestuary. This stretch was divided into 30 sections for computation purposes. Figure 9-20 shows a comparison of the computed

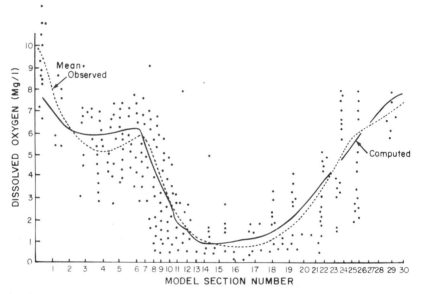

Fig. 9-20 Reproduced with permission from R. V. Thomann, 1972, *Systems Analysis and Water Quality Management*, Environmental Research and Applications, New York, Fig. 7-7.

and observed results for DO for the conditions of June–August 1964. The scatter in the individual observations represents tidal oscillations and day to day fluctuations, among other influences. One of the distinct advantages of such a computer model for an estuary is that it permits estimates to be made for changing pollutant input, river flow, and other system parameters.

This method and its extensions have been discussed further in a series of interesting papers, mentioned only briefly here, as applied to the Delaware estuary. Thomann and Sobel (1964) discuss the method in terms of a systems analysis approach to the forecasting and optimum management of water quality. Thomann (1965) discusses results obtained for variations in the Delaware and tributary river flow and for variations in the determining coefficients. Thomann (1967) discusses the effects of seasonal variations in temperature through a harmonic analysis approach. Using the time dependent forms of (6-104) and (6-105), Pence et al. (1968) include the observed seasonal variations in river flow and the observed daily average water temperature variations for the empirical determination of the reaction coefficients in a computer solution to determine the seasonal variation in the dissolved oxygen content at a series of locations along the Delaware estuary. The calculated and observed results for the calendar year 1964 for a location at Bristol, Pennsylvania, are shown in Fig. 9-21.

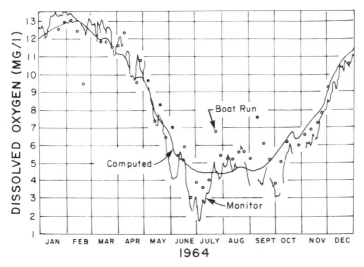

Fig. 9-21   Reproduced with permission from G. D. Pence, J. M. Jeglic, and R. V. Thomann, 1968, *American Society of Civil Engineers, Journal of the Sanitary Engineering Division*, **94**, SA 2, 381–402, Fig. 11.

Thomann et al. (1970) examined numerical modeling of the nitrogen cycle in the Delaware estuary. It should be mentioned that the slower decomposition, or oxidation, of the nitrogenous material had been included in an approximate manner in some of the previous DO deficit examinations. Here, for the nitrogen cycle itself we have a more complex problem. It is of interest to go over the mathematical modeling principles themselves in some detail, if not the specific results obtained. We can consider the feedforward mechanisms of the decay of organic nitrogen and the bacterial nitrification of ammonia nitrogen to nitrite nitrogen and of nitrite nitrogen to nitrate nitrogen. We can also consider the feedback mechanisms of the use of ammonia nitrogen and nitrate nitrogen in the growth of algae and the subsequent release of algal nitrogen on death to an organic nitrogen form. We then have a set of linear equations in the steady state, where we have considered each reaction to be a first order kinetic reaction, similar to (6-109), of the form

$$0 = K_x \frac{d^2 n_1}{dx^2} - v_x \frac{dn_1}{dx} - k_{11}n_1 + k_{21}n_2 + \cdots + k_{m1}n_m$$

$$0 = K_x \frac{d^2 n_2}{dx^2} - v_x \frac{dn_2}{dx} + k_{12}n_1 - k_{22}n_2 + \cdots + k_{m2}n_m \qquad (9\text{-}11)$$

$$\cdots\cdots\cdots\cdots\cdots\cdots\cdots\cdots\cdots\cdots\cdots\cdots$$

$$0 = K_x \frac{d^2 n_m}{dx^2} - v_x \frac{dn_m}{dx} + k_{1m}n_1 + k_{2m}n_2 + \cdots - k_{mm}n_m$$

The quantities $n_1, n_2, \ldots, n_m$ are the various forms of nitrogen, including organic nitrogen, algal nitrogen, ammonia nitrogen, nitrite nitrogen, and nitrate nitrogen. The coefficients k_{ij} are the first order reaction coefficients representing feedforward for $j > i$ and feedback for $j < i$. The coefficient k_{ii} represents the total reaction for substance n_i and in general must satisfy the subsidiary relation

$$k_{ii} = \sum k_{ij} \qquad j \neq i \qquad (9\text{-}12)$$

The equations are in a linear form and are suitable for matrix computer solutions. However, they include numerous coefficients, the values of many of which are only poorly known. In addition, the growth and death of phytoplankton and herbivorous zooplankton populations, e.g., seasonal plankton blooms, together with their utilization of inorganic nitrogen usually cannot be considered a steady state process. Equations (9-11) were applied to the Delaware estuary and a trial and error fit of the mathematical model used to determine the various reaction coefficients.

Ippen and Harleman (1966) examined the tidal oscillations of the Delaware estuary. They considered the estuary to be of uniform depth and to have an exponentially increasing breadth and used essentially the same analytic methods as Hunt (1964) for the Thames, which was discussed previously in Section 7-5. Good agreement was obtained between the observed and calculated values of tidal amplitude and phase.

Szekielda et al. (1972) examined the chemical enrichment associated with convergence into the lateral fronts in Delaware bay. The physical oceanographic features of the fronts examined by them have already been discussed in Section 4-8.

9-7. CHESAPEAKE BAY AND ASSOCIATED WATERS

Chesapeake bay is a somewhat different coastal water body in that it has several sizable tributary estuaries. We examine here some investigations that have been carried out in Chesapeake bay and Baltimore harbor, as well as in the Potomac, Patuxent, Rappahannock, and York estuaries on the western shore of the bay and in the Choptank estuary on the eastern shore. The James estuary at the lower end of Chesapeake bay is covered as a separate item in the next section. The diversity of the different types of investigations that have been carried out in Chesapeake bay and its associated coastal waters should be noted.

Newcombe et al. (1939) discussed some of the general oceanographic characteristics of Chesapeake bay. Pritchard (1952a) reported on a considerably more extensive series of measurements taken in Chesapeake bay and its tributary estuaries, particularly the James estuary, during the period July 1949–January 1951. The measurements were of salinity, temperature, and longitudinal current velocity taken as a function of depth and tidal cycle at several stations over 10 cruises. The salinity measurements were from a specially designed *in situ* conductivity device, as well as from water sample analyses. The temperature measurements were from bathythermograph lowerings. And the current measurements were taken with a modified Jacobsen drag current meter. A characteristic feature of the surface salinity distribution is the obliqueness of the isohalines, resulting in higher salinity on the left side and lower salinity on the right side as viewed in a downestuary direction both for Chesapeake bay and for the tributary estuaries in which measurements were taken. This is attributed to a Coriolis effect.

Pritchard and Carpenter (1960) pioneered the use of dye diffusion techniques for investigation of the mixing and circulation characteristics of estuaries. In principle a dye, such as rhodamine, which can be detected down to concentrations of the order of 10^{-11}–10^{-12}, is used as a tracer in much the same manner that fresh water or salinity in used as a tracer in ascertaining estuary dispersion properties. There are two preferred experimental procedures. In one there is an instantaneous release of dye, and dye concentrations are measured out from the release point as a function of time, longitudinal and lateral distances, and depth. In the other there is a continuous release of dye at a selected location, and measurements are taken as a function of the three spatial coordinates until steady state as measured over a tidal cycle is reached. The method provides a powerful investigative technique both for scientific and engineering purposes. For example, for a proposed outfall of a conservative pollutant that is miscible with the estuary water and for which the outfall flow is small compared with the estuary circulation flows, the pollutant concentration field can be estimated directly from the results of a continuous release, steady state dye diffusion experiment.

In one experiment reported by Pritchard and Carpenter (1960) two instantaneous releases of dye were made in a study of the dispersion and circulation over a potential oyster bed area located in the central portion of Chesapeake bay. In this area a culch bed of clean oyster shells was spread over the bottom just before the oyster larvae, which are free swimming, were ready to set. The purpose of the study was to determine the best location for planting the parent, or brood, stock of oysters so that there would be the greatest possibility of a successful set on the culch bed.

In another experiment the dye diffusion results correctly showed an upestuary near bottom flow in Baltimore harbor, which had been previously predicted. In this case there was an instantaneous release of dye from a diffuser head at a depth of about 10 m, which is in the deeper water layer of the harbor. The longitudinal distribution of the dye 18 hr after introduction is shown in Fig. 9-22. At this time the dye remained in a layer about 1 m thick but spread longitudinally over 3.5 km.

In the Chesapeake bay experiment the maximum concentration of the dye, in the vicinity of the initial release point, multiplied by the thickness of the dye layer was plotted as a function of time over a period of 16–180 hr and found to follow a t^{-2} dependence, in agreement with that given by relation (6-60). As discussed in Section 6-5, it should be mentioned again, however, that the simple Fickian diffusion relation of (6-58) or its extension through relation (6-59) or (6-61) does not appear to be applicable in the immediate vicinity of the source.

Cameron and Pritchard (1965) discuss a diffusion-induced circulation in Baltimore harbor, which was covered previously in Section 5-8. The observed salinity distribution and inferred circulation velocity distribution are shown in Fig. 9-23.

Prichard (1969) made an alteration in the vertically integrated longitudinal dispersion relations by considering the estuary to be divided longitudinally into several segments, each of which in turn was divided vertically into two segments corresponding to the net downestuary circulation flow in the upper layer and the net upestuary circulation flow in the lower layer. The usual mass and salt continuity relations were then applied to determine the defining relations, much the same, in a finite difference form, as the derivation in the second portion of Section 5-3. In this the net longitudinal circulation fluxes were included, but the tidal

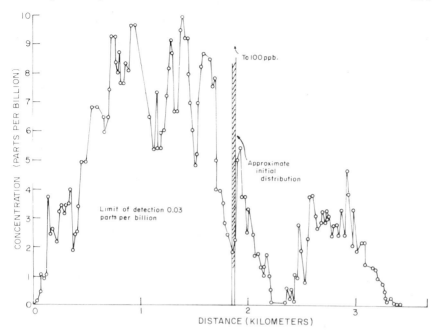

Fig. 9-22 Reproduced with permission from D. W. Pritchard and J. H. Carpenter, 1960, *Bulletin of the International Association of Scientific Hydrology*, **20**, 37–50, Fig. 4.

diffusion fluxes were omitted. The resulting relations were applied to pollutant dispersion consideration in Chesapeake bay.

Schubel (1969, 1971), as also discussed by Dyer (1972), investigated the *turbidity maximum* in the upper reaches of Chesapeake bay. A turbidity maximum zone is characterized by suspended sediment concentrations greater than those found either upstream in the source river or

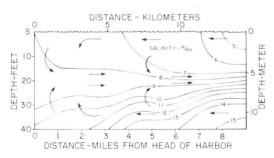

Fig. 9-23 Reproduced with permission from W. M. Cameron and D. W. Pitchard, 1965, *The Sea*, Vol. II, John Wiley, New York, pp. 306–324, Fig. 2.

farther seaward in the estuary. Turbidity maxima are associated with the upper reaches of several estuaries and are related to the change in near bottom circulation from a riverine dominated, downestuary flow to an estuarine dominated, upestuary density gradient flow. Schubel attributes the turbidity maximum to a combination of local resuspension of bottom sediments by tidal action in the accumulation zone and to a *sediment trap*. The sediment trap is described as follows. Particles entering from the river on the mean downestuary near surface flow settle into the near bottom upestuary flow in the lower reaches of the estuary, join particles entering from the ocean, and are carried back upestuary. There the suspended sediment is mixed with the near surface layer in the vicinity of the horizontal salinity gradient zone; and a continuous exchange of particles takes place. This action forms an effective sorting mechanism; and the size range of particles in the turbidity maximum is narrow, as the data indicate, the coarser particles being deposited and the finer ones being swept through.

Hetling and O'Connell (1965) used the familiar relation (5-100) to determine the longitudinal dispersion coefficient K_x sequentially at various locations from 4 to 8 km below Key bridge, Washington, D.C., along the Potomac estuary for a river flow condition of 2000 ft^3 sec^{-1} (56 m^3 sec^{-1}). They then compared these results with a theoretical longitudinal tidal diffusion coefficient having the same functional form as relation (5-135) in terms of the amplitude of the tidal velocity variation and the tidal period. The two show agreement for a value of sin δ in (5-135) of about 0.20. Such a comparison of course is valid only under the assumption that the net longitudinal circulation flux contribution to the total longitudinal dispersion flux can be neglected.

Hetling and O'Connell (1966) extended their investigations of the Potomac estuary with a long term dye diffusion experiment conducted from 10 June to 14 July 1965. Known amounts of a rhodamine dye were injected into a sewer outfall near Washington over a period of 13 days. During the 13 day period of the dye release and for 21 days thereafter dye concentrations were measured both upestuary and downestuary from the release locations over a total distance of 25 mi (40 km). A continuous record of the dye concentrations was made at each high water and low water slack tide during this 36 day period by continuously pumping estuary water from a depth of 1.5 ft (0.5 m) to a recording fluorometer on a moving boat. In addition, the dye distribution in the vertical and lateral directions was determined periodically at 32 selected locations.

The familiar time dependent, longitudinal dispersion equation was used for theoretical comparison purposes and to determine values for the longitudinal dispersion coefficient K_x. A decay term was also included in

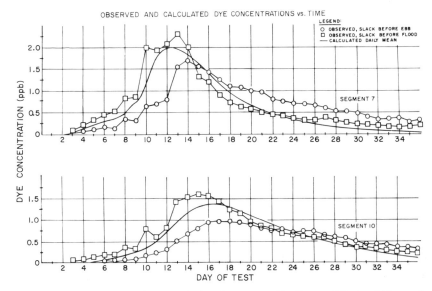

Fig. 9-24 Reproduced with permission from L. J. Hetling and R. L. O'Connell, 1966, *Water Resources Research*, **2**, 825–841, Fig. 7.

the equations for the known rate of decay of the dye. For computer calculation purposes, using the equations in a finite difference form, the 25 mi stretch of the estuary for which the observations were made was divided into 16 segments. The volumes and cross sectional areas for each segment were taken for mean tide conditions from existing navigation charts. The net volume flows along this stretch were taken as the gauged Potomac river inflow, which is the principal inflow, plus the wastewater discharges and an estimate for evaporation. The dye input location was at segment 5, a distance of 7.7 mi (12.4 km) below Chain bridge, Washington, D.C.

A comparison of the observed dye concentrations at high water and low water slack tides with the computer calculations for four segments is shown in Figs. 9-24 and 9-25 as a function of time. Segments 7 and 10 in Fig. 9-24 are located at distances of 12.8 mi (20.6 km) and 15.9 mi (25.6 km) downestuary from Chain bridge, respectively; and segments 13 and 16 in Fig. 9-25 are located at distances of 19.3 mi (31.1 km) and 23.5 mi (37.8 km) downestuary from Chain bridge, respectively. The agreement of both the magnitude of the dye concentrations and their temporal distribution for each segment is excellent and serves as a strong argument for the use of longitudinal dispersion equations for such pollutant distribution investigations.

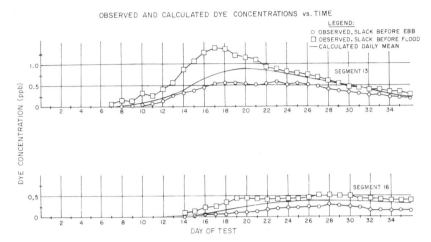

Fig. 9-25 Reproduced with permission from L. J. Hetling and R. L. O'Connell, 1966, *Water Resources Research*, **2**, 825–841, Fig. 8.

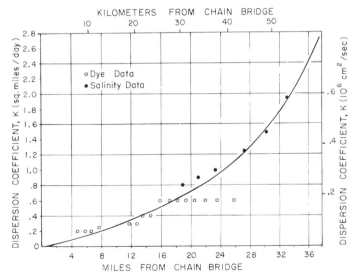

Fig. 9-26 Reproduced with permission from L. J. Hetling and R. L. O'Connell, 1966, *Water Resources Research*, **2**, 825–841, Fig. 11.

Figure 9-26 is a graph of the longitudinal dispersion coefficient K_x as a function of downestuary distance from Chain bridge. The open circles are the values of K_x used in the dye diffusion computer calculations. The solid circles are additional values of K_x calculated from the salinity distribution using (5-100) for the same period. The curve is simply a visual approximation through both sets of data. The figure shows a downestuary increase in K_x, which is in agreement with observations in other estuaries.

Thomann et al. (1970) applied their nitrogen cycle model, discussed in Section 9-6, to a sequence of data collected in the Potomac estuary during July–August 1968. To obtain satisfactory agreement between their computer model and the observed results they ended up, through trial and error substitutions, with the values shown in Table 9-1. Although the

TABLE 9-1

Reaction Coefficients, Potomac Estuary, July–August 1968,
Temperature 28°C

Nitrogen step	Reaction coefficient (day^{-1})	Nitrogen step	Reaction coefficient (day^{-1})
Decay of organic N	0.1	NH_3 N → algal N	0.02
Organic N → NH_3 N	0.1	NO_3 N → algal N	0.1
NH_3 N → NO_3 N	0.28	Algal N → organic N	0.12

exact values of the coefficients may be subject to question, it appears that their relative magnitudes as well as the specific feedforward and feedback nitrogen steps they took are of significance.

Cannon (1971) carried out an interesting experiment in the Patuxent estuary to examine the turbulent fluctuations of the horizontal current velocity. The horizontal velocity measurements were taken with a specially designed propeller type current meter consisting of two propellers mounted one above the other at right angles. Three runs were taken with a bottom-mounted installation; the record duration was 50 hr and the sampling interval 10 sec. The longitudinal current velocity was considered to be given in the form

$$v_x = \bar{v}_x + v_{xt} + v'_x \qquad (9\text{-}13)$$

where $\bar{v}_x$ is the time average over an integral number of tidal cycles, v_{xt} is the tidal velocity component, and v'_x is the turbulent, or residual, velocity component; a similar expression is considered to apply to the lateral current velocity. The amplitude and phase of the tidal velocity components were determined from a Fourier analysis of the records at the frequency of the tidal period. The energy density spectrum of the turbulence was then determined through a computer program, presumably using much the same formulation as that discussed in Sections 7-1 and 7-7. Because of the duration of the records it was possible to extend the turbulence spectra determinations to longer periods, in a range of about 1 min to 1 hr, than had previously been observed in estuaries.

For the first run with four meters located longitudinally along the estuary at a depth of 4 m in a water depth of approximately 12 m the observed turbulence spectrum could be represented as increasing as $f^{-5/3}$ in the period range 1–4 min on a log-log plot and as increasing again on an offset $f^{-5/3}$ line in the period range 8–60 min. For the second run with meters at depths of 2, 4, 7, and 12 m at the same location the spectrum followed much the same dependence for the meters at 7 and 12 m depth. For the meters at 2 and 4 m depth the spectrum showed the same dependence at the shorter periods but increased less rapidly at the longer periods. For the third run with the meters located laterally across the estuary at a depth of 4 m the spectrum increased much the same as before at the shorter periods but increased more rapidly at the longer periods. A comparison was then made of these variations with that predicted by the simple relation (2-100). It is difficult to see, at least to us, how such a comparison can be made directly if we consider that the wave number k is given in terms of the observed turbulence frequency f by the relation $k = 2\pi f/v_{xt}$, where v_{xt} is the tidal velocity at which the turbulence is swept past the sensor; for the duration of these records the magnitude of v_{xt} varies several times between zero and its maximum value C.

The variation in the ratio v'_x/v'_y of the longitudinal to the lateral turbulent velocity component among the three runs was also of interest. For the first run the ratio was approximately $2:1$ over the observed period range. For the second run the ratio was greater, particularly at the longer periods. For the third run the two components were approximately of the same magnitude.

Gordon and Dohne (1973) made an interesting investigation of Reynolds stresses in the Choptank estuary. They employed much the same procedures as Francis et al. (1953), as discussed in Section 9-2, using a suspended pivoted vane type current meter which was self aligning in the direction of the maximum horizontal current. The sampling interval was every 2 sec for a record duration of 10–13 min. The mean velocity $\bar{v}_x$ over the turbulent fluctuations was determined for each record as a function of time by a least squares, third degree polynomial fit to the

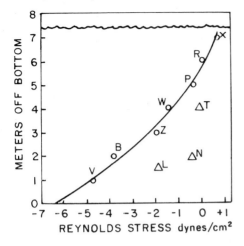

Fig. 9-27 Reproduced with permission from C. M. Gordon and C. F. Dohne, 1973, *Journal of Geophysical Research,* **78,** 1971–1978, Fig. 3.

data. The Reynolds stress $\sigma_{zx} = -\rho \overline{v_x' v_z'}$ was then determined from the observations of v_x' and v_z'. The results are shown in Fig. 9-27; all the measurements were taken with the instrument at the same location near the times of maximum tidal current. The triangular data points are considered less reliable than the circular data points; the curve is simply a least squares, second degree polynomial fit to the circular data points and is included only as a visual convenience. A favorable comparison is then made of these results with those of Bowden et al. (1959), discussed in Section 7-7, for the same stage of the tidal cycle.

From another series of measurements of shorter duration the turbulent intensity components $(\overline{v_x'^2})^{1/2}/v_{xt}$ and $(\overline{v_z'^2})^{1/2}/v_{xt}$ were determined from the observations of v_x' and v_z' as a function of depth and three different stages of the tide. The results are shown in Fig. 9-28, the circles representing the longitudinal intensity component and the triangles the vertical component. A favorable comparison is made with the results of Bowden and Howe (1963), discussed in Section 7-1.

Nichols and Poor (1967) investigated the circulation velocity, salinity, and suspended sediment distributions in the Rappahannock estuary. Although their report itself is fairly short, it contains some of the best and most complete measurements of these three quantities in an estuary. The measurements were taken over several tidal cycles as a function of depth at several longitudinal and transverse locations in the estuary.

Figure 9-29*a* and 9-29*b* presents the gist of the tidal averaged net circulation and salinity distribution results. It is to be noted that the surface of no net motion has an undulating form, which generally dips in a downestuary direction but is also significantly affected by the shoaling at

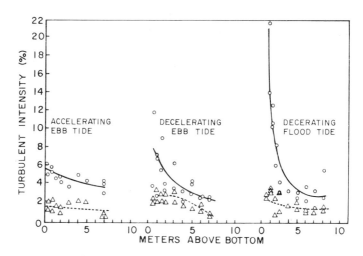

Fig. 9-28 Reproduced with permission from C. M. Gordon and C. F. Dohne, 1973, *Journal of Geophysical Research,* **78,** 1971–1978, Fig. 6.

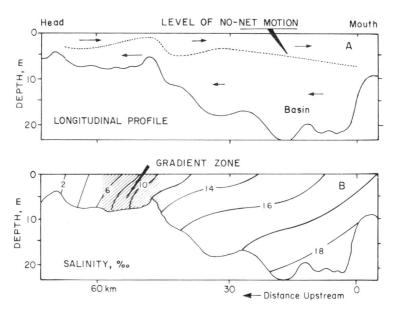

Fig. 9-29 Reproduced with permission from M. Nichols and G. Poor, 1967, *American Society of Civil Engineers, Journal of the Waterways and Harbor Division,* **93,** WW4, 83–95, Fig. 3.

the upper end of the estuary. In a lateral direction the surface of no net motion does not continue to the shoreline but intersects the bottom in areas of shoaling, so that the general effect is to have an upestuary flow near the bottom in the deeper central channel of the estuary and a downestuary flow near the bottom along the lateral shoal areas on either side of the channel.

The salinity distribution of Fig. 9-29*b* shows a particularly interesting feature. This is the region labeled the *gradient zone* and referred to as such by several investigators. It is generally associated with the mouth of a freshwater river, either as it opens out into its associated estuary or as it opens out into the ocean itself. It is not an uncommon feature of partially stratified estuaries. Upestuary of the gradient zone the waters are nearly fresh and are well mixed, representative of the river conditions. Downestuary of the gradient zone the waters develop into a partially stratified condition, and the salinity increases in both a vertical and a downestuary longitudinal direction. The gradient zone represents the confluence of the lower layer upestuary flow and the river flow, which is downestuary throughout the water column; understandably, then, it is a zone of rather

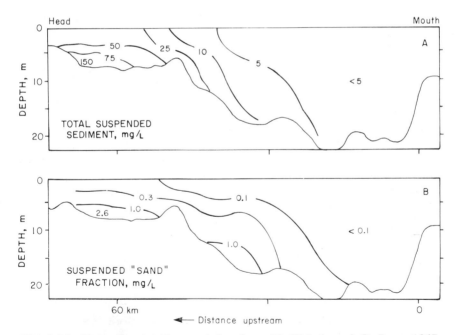

Fig. 9-30   Reproduced with permission from M. Nichols and G. Poor, 1967, *American Society of Civil Engineers, Journal of the Waterways and Harbor Division,* **93,** WW4, 83–95, Fig. 6.

intense vertical and horizontal mixing, with the resultant development of a substantial longitudinal salinity gradient.

Figure 9-30*a* and *b* shows the total suspended sediment distribution and the suspended sand fraction greater than 32 μm distribution along the same stretch of the estuary. As one might expect, the region of the highest total suspended sediment, or the turbidity maximum, corresponds in a general way with the location of the gradient zone.

From the data forming the basis of Figs. 9-29*a* and 9-30*a* the suspended sediment transport rates for a transverse section in the middle portion of the estuary were calculated and are presented in Fig. 9-31. The

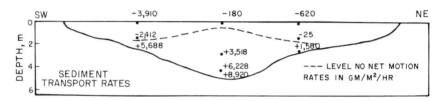

Fig. 9-31 Reproduced with permission from M. Nichols and G. Poor, 1967, *American Society of Civil Engineers, Journal of the Waterways and Harbor Division,* **93,** WW4, 83–95, Fig. 7.

rates are in grams per square meter per hour, a plus sign indicating upestuary transport and a minus sign downestuary transport. The dominant upestuary transport of the lower layer in the central channel portion of the estuary should be noted.

Kuo and Fang (1972) used the time dependent longitudinal diffusion equation in a finite difference computer model form to determine the salinity variations as a function of river flow in the York estuary. Tidal diffusion and net circulation contributions to the longitudinal dispersion coefficient were determined from empirical formulas as adjusted by the observed salinity distribution.

9-8. JAMES ESTUARY

The measurements by Pritchard (1952a) for Chesapeake bay and its tributary estuaries, discussed briefly in the previous section, were principally focused on the James estuary. One of the more interesting sequences of measurements was that of continuous observations of the longitudinal current velocity taken over several tidal cycles for three periods between 18 June and 21 July 1950. These results are summarized in Fig.

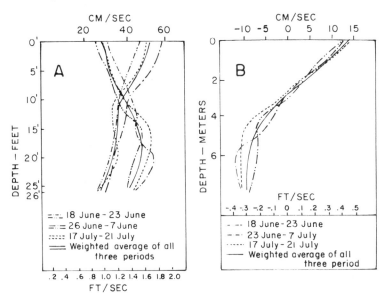

Fig. 9-32 Reproduced with permission from D. W. Pritchard, 1952, *Journal of Marine Research*, **11**, 106–123, Figs. 8 and 9.

9-32 for a deep channel station. In Fig. 9-32a the mean ebb velocity and mean flood velocity are plotted without regard to sign. The ebb velocities are relatively large at the surface and decrease with depth, while the flood velocities are relatively small at the surface, increase to a depth of about 17 ft (5 m), and then slowly decrease toward the bottom. Figure 9-32b is a plot of the net circulation velocity averaged over a tidal cycle. Above 10 ft (3 m) the current is directed downestuary; and below this depth it is directed upestuary, all of which is in accordance with what one might anticipate. The shallow water stations on either side of the estuary showed only a downestuary velocity distribution. Figure 9-33 shows the lateral variation in the level of no motion, the slope of the surface being interpreted as a Coriolis effect.

An evaluation was then made of the salt balance in the estuary. The salt flux integrals of the tidal averaged circulation velocity and the tidal averaged salinity, $\bar{v}_x$ and $\bar{s}$ of (5-90) and (5-91), were determined for the observations at several cross sections along the estuary. The resultant values were small, of the order of 1–5% of the values of the downestuary or upestuary components of the integral. It was concluded from this that the tidal diffusion salt flux, S_2 of (5-96), could be neglected. It appears that this conclusion is subject to question. The integral evaluated is that of a combination of river advection salt flux and the net circulation salt flux, S_1

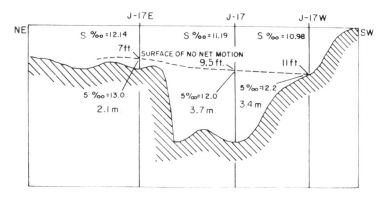

Fig. 9-33   Reproduced with permission from D. W. Pritchard, 1952, *Journal of Marine Research*, **11**, 106–123, Fig. 10.

and S_3 of (5-96); and it does not seem that a comparison of each of the components of this integral with their resultant amounts to a valid comparison between S_2 and S_3.

Pritchard (1954) continued the analysis of the same data. The tidal averaged vertical velocity $\bar{v}_z$ was determined from the mass continuity equation in finite difference form, using the data of Fig. 9-32 and a surface boundary condition that $\bar{v}_z = 0$ at $z = 0$. The resultant values are shown in Fig. 9-34 with, as expected, a maximum of $\bar{v}_z$ near the depth of the halocline. The internal consistency of the data and the calculations is indicated by the return of $\bar{v}_z$ to nearly zero at the bottom.

As has been stated, the tidal diffusion, or nonadvective longitudinal, salt flux as integrated over a vertical section was assumed to be small in comparison with the net circulation, or advective longitudinal, salt flux, which, indeed, is probably correct for the James estuary. Of more importance, however, for salt flux gradient calculations at a point, the nonadvective longitudinal salt flux gradient can be neglected in consideration of the advective longitudinal salt flux gradient if $\partial s/\partial x$ is independent of time t, which is one of the basic underlying assumptions of the derivations in Section 5-6. The vertical advective salt flux gradients were then determined from the vertical velocity values of Fig. 9-34 and the observed vertical salinity gradients. The salt continuity equation was then applied in finite difference form to determine the remaining quantity, the vertical diffusive, or nonadvective, salt flux gradient contributions. The results show that the longitudinal advective terms and the vertical nonadvective terms are most important point by point as a function of depth, in accordance with the formulation of (5-164). The vertical advective term is

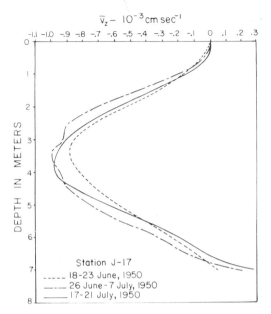

$\bar{v}_z - 10^{-3}$ cm sec^{-1}

Station J-17
----- 18-23 June, 1950
—·— 26 June-7 July, 1950
——— 17-21 July, 1950

Fig. 9-34 Reproduced with permission from D. W. Pritchard, 1954, *Journal of Marine Research*, **13**, 133–144, Fig. 2.

of importance and of a value comparable to that of the vertical nonadvective term in the small depth interval of the halocline.

Pritchard (1956) further continued this type of investigation with an examination of the relative importance of the various terms in the longitudinal and lateral equations of motion. For the longitudinal equation the variable part of the longitudinal pressure gradient term $\partial p/\partial x$ was determined from relation (4-120) from the observations of salinity and temperature as a function of depth and longitudinal distance. The inertial terms $\bar{v}_x(\partial \bar{v}_x/\partial x)$, $\bar{v}_z(\partial \bar{v}_x/\partial z)$, and $C(\partial C/\partial x)$ were determined directly from the velocity observations. The constant, or surface slope portion, of $\partial p/\partial x$ was determined by imposing the condition that $f_{zx} = 0$ at $z = 0$ or, equivalently, that there was negligible wind stress at the surface. The frictional resistance f_{zx} was then determined sequentially as a function of depth in the manner of relation (4-123), with the inclusion of the three inertial terms. The results show that the pressure gradient and internal frictional terms are the most important ones, in accordance with the formulation of (4-86). The results also show that, while the inertial terms $\bar{v}_x(\partial \bar{v}_x/\partial x)$ and $\bar{v}_z(\partial \bar{v}_x/\partial z)$ are negligible at all depths, the tidal inertial term $C(\partial C/\partial x)$ is of a value comparable to the $\partial p/\partial x$ and $\partial f_{zx}/\partial z$ terms in the depth interval of no motion near the halocline where these two terms change signs.

A similar comparison was made for the lateral equation of motion among the Coriolis, lateral pressure gradient, and lateral internal frictional terms, using the procedures mentioned above to determine f_{zy} from the calculations for $2\omega_0 \bar{v}_x \sin \varphi$ and $\partial p/\partial y$. The results show that the Coriolis term is balanced principally by the lateral pressure gradient term, the lateral frictional term being of secondary importance. Pritchard and Kent (1956) discussed interrelationships between the two determined internal frictional terms f_{zx} and f_{zy}. Stewart (1957) objected to the form of the lateral equation of motion used in this analysis because it excluded the inertial terms due to the slight lateral curvature of the essentially longitudinal motion, along the lines discussed in Section 4-9. He showed that, for the geometric configuration of the James estuary, this term was of comparable importance to those previously considered.

Kent and Pritchard (1959) discussed various mixing length concepts from the salt flux determinations of Pritchard (1954). Pritchard (1967) carried out essentially the same type of analysis in terms of K_z, calculated from the defining relation (2-64). A graph of calculated K_z as a function of depth is shown in Fig. 9-35. The results are interesting in that they show relatively strong vertical mixing in the well mixed zones above and below the halocline, and a region of relatively high stability in the vicinity of the halocline.

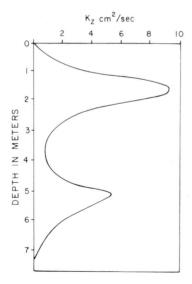

Fig. 9-35 Reproduced with permission from D. W. Pitchard, 1967, *Estuaries*, American Association for the Advancement of Science, Washington, D.C., pp. 37–44, Fig. 7.

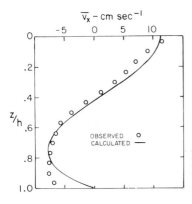

Fig. 9-36   Reproduced with permission from M. Rattray and D. V. Hansen, 1962, *Journal of Marine Research*, **20**, 131–133, Fig. 9.

Rattray and Hansen (1962) and Hansen and Rattray (1965) applied their similarity solution results, which are the same as those covered in Section 4-5 and 5-6, to Pritchard's James estuary data. By using relation (4-98) and a value of $N_z = 4$ cm^2 sec^{-1} the graphic comparison between observed and calculated values as shown in Fig. 9-36 was obtained. Undoubtedly, a better comparison between observed and calculated values near the bottom could be obtained if a bottom frictional coefficient k and relations (4-108) and (4-110) were used. A similar comparison between observed and calculated salinity values was obtained, using relation (5-173) and a value of $K_z = 2$ cm^2 sec^{-1}. In addition, calculations of the fraction ν of relations (5-193) and (5-195) were made obtaining for the James estuary $\nu = 0.1$, as compared with similar calculations for the Mersey estuary of $\nu = 0.5$ and for the Columbia estuary of $\nu = 0.8$–0.9. These calculations are of particular significance in showing the substantial variation in ν that can exist from one estuary to the next. In other words in some estuaries the circulation contribution to the total longitudinal dispersion flux is dominant, giving low values of ν; in others the tidal diffusion contribution is dominant, giving high values of ν; while in still others the two contributions are of comparable importance.

Fisher et al. (1972) derived the velocity relation given by (4-116) with the inclusion of the river runoff term and the corresponding salinity deviation relations and compared them with Pritchard's James estuary data.

O'Connor (1960, 1965) applied the procedures discussed in Sections 9-4 and 9-6 and the analytic solutions for the coupled BOD-DO equations of Section 6-7 to the James estuary with two significant sewage outfalls upestuary at Richmond and Hopewell. A comparison of the excellent agreement of the calculations with the observations of the DO concentration was given previously in Fig. 6-10.

9-9. SAVANNAH AND CHARLESTON HARBORS

From an extensive series of salinity measurements during the period
December 1937–September 1939, in the Savannah estuary, Lamar (1940)
clearly demonstrated a strong dependence of the longitudinal movement
of the salinity contours on the neap to spring tidal cycle and on river
runoff variations. The longitudinal location of the salinity contours followed the neap to spring tidal cycle, being further downestuary for the
neap tides and further upestuary for the spring tides. The salinity contours also showed a progressive downestuary movement with increasing
river runoff discharge. Both of these dependencies are indeed what one
might anticipate.

Simmons (1955, 1966) and Simmons and Brown (1969) examined the
sedimentation and shoaling characteristics of Savannah harbor following
the general procedures as discussed in Section 9-4. In the first paper an
interesting general observation is made, which was mentioned in Chapter
1 and is repeated here. This is the observation from field experience that
the physical oceanographic conditions of an estuary can normally be
ascertained from the ratio of the river runoff over a tidal cycle into an
estuary to the tidal prism of the estuary. When the ratio is of the order of
10^0, an arrested flow condition normally exists; when it is of the order of
10^{-1}, a partially stratified condition normally exists; and when it is of the
order of 10^{-2}, a well mixed condition normally exists.

Figure 9-37 shows the flow predominance for Savannah harbor. As
indicated in the figure, flow at the surface is predominantly downestuary.
That at the bottom is predominantly downestuary from the upper end of

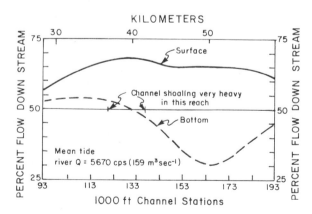

Fig. 9-37   Reprinted with permission from H. B. Simmons, 1966, *Estuary and
Coastline Hydrodynamics*, McGraw-Hill, New York, pp. 673–690, Fig. 16.9.

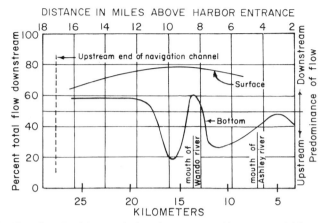

Fig. 9-38 Reprinted with permission from H. B. Simmons, 1966, *Estuary and Coastline Hydrodynamics*, McGraw-Hill, New York, pp. 673–690, Fig. 16.11.

the harbor to about station 130, and predominantly upestuary from this station to the harbor entrance. The region of shoaling is, as expected, in the vicinity of zero net bottom motion. An interesting comparison is also given of the variation in shoaling characteristics as the harbor channel was progressively deepened during 1923–1954. As the channel was deepened, salt water penetrated farther into the harbor; and the nodal point for bottom flow predominance moved farther upestuary, with consequent upestuary movement of the principal region of shoaling.

Simmons (1966) made a similar investigation for Charleston harbor. In this case there are two tributary rivers emptying into the harbor channel, with surface and bottom flow predominance results as shown in Fig. 9-38. In this case there are three major shoaling areas bracketing the three nodal points in the figure.

Harris (1965) comments on a fluff layer, presumably a flocculent layer, that occurs in the Savannah river. In this instance it was observed that the location of the fluff layer varied in response to river flow conditions, being about 45,000 ft (14 km) further downestuary during the high river discharge, spring runoff season than during the rest of the year. Figure 9-39 shows the longitudinal distribution of the flocculent layers along the Savannah river during several months and their dependence on river discharge conditions.

9-10. FLORIDA PENINSULA WATERS

Pyatt (1964) took a statistical approach in the correlation of observables from eight stations in the St. Johns estuary. Various measurements were

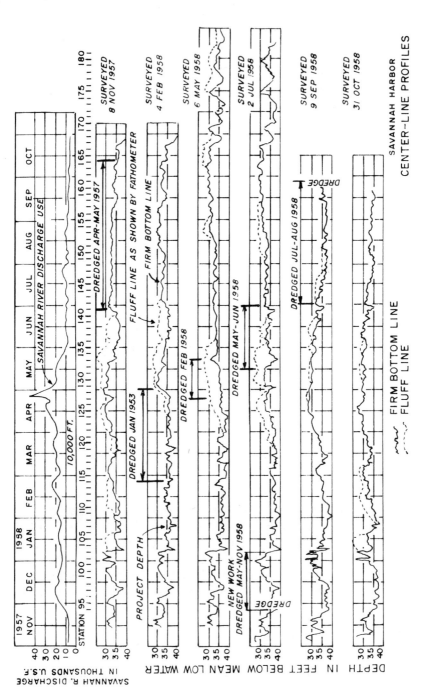

Effect of Savannah River discharge on shoaling.

Fig. 9-39 Reprinted with permission from J. W. Harris, 1965, *Proceedings of the Interagency Sedimentation Conference,* Miscellaneous Publication 970, United States Department of Argiculture, Washington, D.C., pp. 669-674, Fig. 7.

368

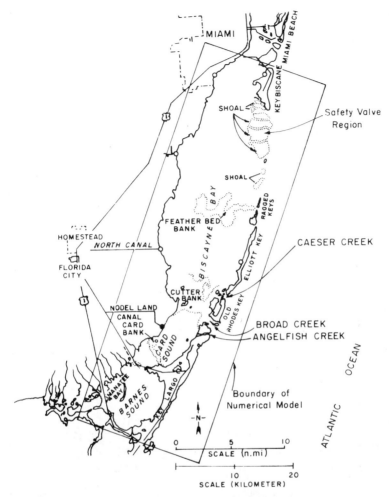

Fig. 9-40   Reproduced with permission from R. G. Dean and R. B. Taylor, 1972, *American Society of Civil Engineers, Proceedings of the Thirteenth Conference on Coastal Engineering*, pp. 2227–2249, Fig. 1.

taken five times during 1954–1955 at each station over a tidal cycle. He observed a correlation of the DO concentration with the 30 day antecedent river discharge rate.

Dean and Taylor (1972) made an interesting finite difference numerical model study of water transports in Biscayne bay and vicinity. The area of their investigation is shown in Fig. 9-40. The primary tidal flow into Biscayne bay is through the Safety Valve region. The bay system itself may be characterized as several shallow basins separated by much shallower limestone sills and mud banks, which affect the hydraulics considerably. Vertically integrated tidal equations of motion and continuity were used, and a sinusoidal tidal input function of constant phase was taken along the seaward grid of the model with an amplitude corresponding to the observed ocean tides. Verification of the model, and presumably also of the bottom frictional coefficients used, was obtained by comparison of the computed tidal variations in the bay with those of the existing tide gauge stations. The numerical model was then used to predict the flow patterns of the cooling water intake and discharge for a proposed power generating station at the southern end of the bay, both on the basis of no diffusion and on the basis of horizontal diffusion of the discharge waters with the ambient bay waters using a diffusion, or rather an exchange, coefficient obtained from a dye dispersion experiment in the bay.

Stelzenmuller (1965) has discussed the tidal, surface slope, and mixing characteristics of two estuaries in western Florida—the Waccasassa and the Fenholloway. Variations in the tidal characteristics at the mouths of the two estuaries due to wind, hurricane, and annual effects are described. Of particular interest here is the observation that the surface slopes of the two estuaries are considerably different, the Fenholloway having a substantially greater slope, as shown in Fig. 9-41. From the discussions of Sections 4-5 and 5-6 we then anticipate that under similar river runoff

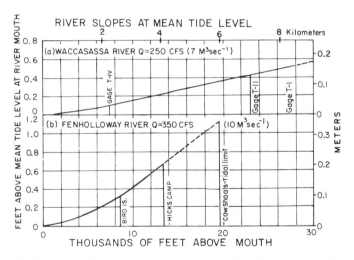

Fig. 9-41 Reproduced with permission from W. B. Stelzenmuller, 1965, *American Society of Civil Engineers, Journal of the Waterways and Harbor Division*, **91**, WW3, 25–36, Fig. 9.

conditions the Fenholloway will have a greater longitudinal salinity gradient, a greater net circulation velocity effect, and consequently a greater vertical salinity deviation effect than the Waccasassa, which is indeed what is observed.

Ichiye et al. (1961) made a very interesting and thorough investigation of two shallow bays in Florida, Alligator harbor and Ochlockonee bay, which border on the Gulf of Mexico. Alligator harbor has an essentially uniform vertical salinity distribution with a mean water depth of less than 6 ft (2 m), and no river inflow into the harbor. Ochlockonee bay has a strongly stratified vertical salinity structure with a mean water depth of less than 6 ft (2 m) in the inner bay, and depths of 6–12 ft (2–4 m) in the outer bay and with river inflow from the Sopchoppy and Ochlockonee rivers.

The temporal salinity variations in Alligator harbor show a tidal variation as well as a variation with a 2 hr period which is attributed to the seiche period of the harbor. The longitudinal salinity gradient is, as expected, small. Three examples of the tidal mean, surface salinity distributions are shown in Fig. 9-42. Normal conditions are either higher salinity toward the harbor mouth as shown in the lower portion of the figure, or a negligible salinity gradient as shown in the middle portion. The salinity distribution in the upper portion of the figure, showing a decrease in salinity toward the harbor mouth, is attributed to a fresher water contribution coming from the adjacent Ochlockonee bay.

The following instructive theoretical discussion was then given to explain some of the features of the observed salinity variations in Alligator harbor. As the salinity shows essentially no variations in a vertical section, we may take as our defining salt continuity equation from (2-65) an equation of the form

$$\frac{\partial s}{\partial t} + v_x \frac{\partial s}{\partial x} = \frac{\partial}{\partial x}\left(K_x \frac{\partial s}{\partial x}\right) \tag{9-14}$$

As the length of the harbor is small in comparison with the tidal wavelength, we may take the tidal cooscillation velocity, neglecting bottom friction, to be of the form of (5-110), or

$$v_x = v_{xt} = v_l \frac{x}{l} \cos \omega t \tag{9-15}$$

where v_l is the amplitude of the tidal velocity variations at the mouth of the harbor. If we next restrict our time scale of interest to times within a tidal cycle or over a few tidal cycles, the two terms on the left hand side of (9-14) will be dominant, and we may set the equation equal to zero. We then see that the resulting partial differential equation may be solved

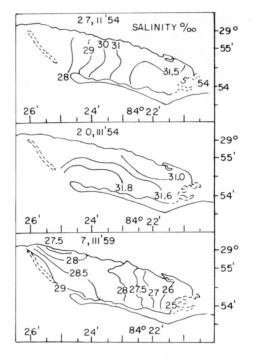

Fig. 9-42 Reproduced with permission from T. Ichiye, M. L. Jones, N. C. Hulings, and F. C. W. Olson, 1961, *Journal of the Oceanographical Society of Japan*, **17**, 1–9, Fig. 3.

by the method of separation of variables and has a solution of the form

$$s = \sum_{n=0}^{\infty} s_n(t)\left(\frac{x}{l}\right)^n \tag{9-16}$$

where we take the summation only up to $n = 2$ in accordance with our assumption that the right hand side of (9-14) can be neglected. Substituting (9-16) into the reduced form of (9-14) and equating coefficients of the same powers of x/l, we have

$$s_1 = c_1 e^{-(v_l/\omega l)\sin \omega t} = c_1\left(1 - \frac{v_l}{\omega l}\sin \omega t + \cdots\right)$$
$$s_2 = c_2 e^{-(2v_l/\omega l)\sin \omega t} = c_2\left(1 - \frac{2v_l}{\omega l}\sin \omega t + \cdots\right) \tag{9-17}$$

where c_1 and c_2 are constants of integration and where we may approximate the exponential by the first two terms in its series expansion since

$v_l/\omega l$ is small for the physical conditions of this example. Imposing next the observed physical condition that the salinity variations at the harbor mouth are small over a tidal cycle, we have

$$c_1 = -2c_2 \qquad (9\text{-}18)$$

so that the final solution is

$$s = s_0 + c_1\left(\frac{x}{l}\right)\left(1 - \frac{v_l}{\omega l}\sin \omega t\right) - \tfrac{1}{2}c_1\left(\frac{x}{l}\right)^2\left(1 - \frac{2v_l}{\omega l}\sin \omega t\right) \qquad (9\text{-}19)$$

with, from the original equation (under the assumption that K_x can be considered a constant), the approximate subsidiary relation

$$\frac{\partial s_0}{\partial t} = K_x \frac{2s_2}{l^2} = -K_x \frac{c_1}{l^2} \qquad (9\text{-}20)$$

for c_1 in terms of the longer term salinity variations $\partial s_0/\partial t$ at the harbor

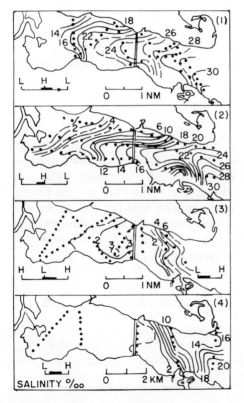

Fig. 9-43 Reproduced with permission from T. Ichiye, M. L. Jones, N. C. Hulings, and F. C. W. Olson, 1961, *Journal of the Oceanographical Society of Japan*, **17**, 1–9, Fig. 4.

mouth. We then note that the tidal mean value of salinity in excess of s_0 should vary longitudinally as $(c_1/2)(2\lambda - \lambda^2)$, where $\lambda = x/l$, from the inner end of the harbor to its mouth, and that the amplitude of the tidal variations should vary as $(c_1 v_l/\omega l)(\lambda^2 - \lambda)$. In other words the mean value of the salinity should increase or decrease longitudinally depending on the sign of c_1, and the tidal amplitude variation should be zero at both ends and a maximum at the middistance $\lambda = \frac{1}{2}$. The one set of complete observations the authors present does indeed follow this dependence.

The temporal salinity variations in Ochlockonee bay are quite different. The surface salinity variations taken on different days during March–July 1956 are shown in Fig. 9-43. The upper two portions of the figure refer to measurements taken near high tide, the third near low tide, and the fourth near midtide. The immense variations over a tidal cycle are quite evident. The vertical salinity profiles show a corresponding strong vertical gradient in the vicinity of the strong longitudinal gradient.

Ichiye and Jones (1961) made a conventional salinity, temperature, and current velocity survey in the St. Andrew bay system of western Florida, taking measurements as a function of depth over a 24 hr period at five to seven stations repeated five times in different seasons. They found that the time variations of the surface and bottom salinities followed the simple longitudinal, tidal, advective relation of (3-31) or (5-162), and that the middepth salinity variations were mainly due to the vertical motion of the halocline.

9-11. MISSISSIPPI RIVER

Stommel and Farmer (1952, 1952a) applied the two layer control relation (4-16), which they developed, to the observed conditions at the outermost jetty in Southwest pass, Mississippi river, through which most of the river flows. The total water depth and width here are 35 ft (10.7 m) and 1500 ft (460 m), respectively; and the density contrast between the upper and lower layers was such that β in (4-16) was about 0.010. A comparison of the observed and computed depths of the interface for four different river flow conditions are restated here in metric units in Table 9-2.

TABLE 9-2

	1	2	3	4
River discharge Q_1 (m^3 sec^{-1})	850	1700	2830	5660
Observed interface depth h_1 (m)	3.7	4.6	6.1	9.5
Computed interface depth h_1 (m)	3.4	5.2	7.0	10.7

The comparison is quite good, and it appears that a control condition exists at the mouth of the Mississippi river. It should also be noted that the calculations for the highest river flow condition indicate that there should be no salt wedge penetration into the river mouth.

Farmer and Morgan (1953), using a theoretical relation, the equivalent of (4-54) with $n_{2m} = \frac{3}{8}$, and the observational data of Rhodes (1950) on the longitudinal geometric configuration of the Mississippi salt wedge at Southwest pass, obtained the comparison shown in Fig. 9-44 for the

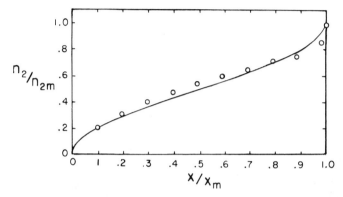

Fig. 9-44 Reproduced with permission from H. G. Farmer and G. W. Morgan, 1953, *American Society of Civil Engineers, Proceedings of the Third Conference on Coastal Engineering*, pp. 54–64, Fig. 3.

fractional thickness of the salt wedge in terms of its thickness at the mouth versus the fractional longitudinal distance from the salt wedge tip to the river mouth. The agreement is quite satisfactory but not necessarily definitive in consideration of the parameters that have been graphed.

Simmons (1955, 1966) comments that for an arrested salt wedge estuary, of which the Southwest pass of the Mississippi river is an example, the area of shoaling for bottom transported sediments is anticipated to be in the vicinity of the tip of the salt wedge where the near bottom motion changes from a downestuary direction in the freshwater upestuary of the tip to an upestuary direction in the salt wedge itself. He further observes that this is indeed the case for the Southwest pass, with the heavier particles transported along the river bed coming to rest at the tip and the deposition area showing a gradation to finer particles in a downestuary direction.

Wright (1971) examined the general hydrography and salt water intrusion characteristics for the South pass of the Mississippi. At the Head of passes the Mississippi river branches into three major distributaries, Pass

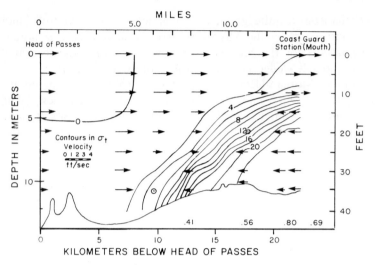

Fig. 9-45   Reproduced with permission from L. D. Wright, 1971, *American Society of Civil Engineers, Journal of the Waterways, Harbors, and Coastal Engineering Division*, **97**, WW3, 491–504, Fig. 2.

á Loutre, South pass, and Southwest pass. South pass carries approximately 15% of the total river discharge. The observations were taken with a Roberts type current meter and an induction type salinometer over a tidal cycle and as a function of depth. Figure 9-45 is a summary of observations taken during a flooding tide on 1 October 1969. The contours are for the density difference σ_t from unity in parts per thousand at atmospheric pressure, determined from standard tables with the observed salinity values. The numbers at the bottom of the figure indicate the interfacial, or densimetric, Froude number, defined in a manner similar to (2-174),

$$F_i = \Pi_i = \frac{v_1^2}{g\beta h_1} = \frac{q_1^2}{g\beta h_1^3} \tag{9-21}$$

using the notation of Section 4-1. We see from relation (4-16) that a control condition at the mouth of the estuary implies that the interfacial Froude number is equal to unity there.

Figure 9-45 is particularly instructive in regard to the general circulation and mixing condition here. We see that instead of an interface condition we have a strong halocline, indicative of substantial entrainment, or other, mixing of the more saline water into the upper fresher water portion. We also see, complementary to this, that, whereas current velocities in the fresher water portion are uniformly in a downestuary direction, current velocities in the more saline water portion are in the aggregate upestuary, rather than balancing out nearly to zero, indicative of some combination of substantial entrainment of the more saline water into the upper fresher water flow and upestuary longitudinal movement of the salt wedge. The interfacial Froude number increases toward unity at the river mouth. It should also be mentioned in this regard that, as can be seen from relation (4-13) and as given more specifically in Stommel and Farmer (1952), when there is a substantial net flow in the lower layer, the interfacial Froude number has a value less than unity for a critical flow condition, depending on the relative magnitude of the two flows and the relative depth of the interface.

Figure 9-46 shows the density and current velocity conditions at the mouth of South pass during a flooding tide and during an ebbing tide on 1 October 1969. During the flooding tide the conditions approach that of an interface, with downestuary water transport in the upper portion and upestuary transport in the lower portion. On the ebbing tide the halocline is more diffusive, and the maximum density gradient is at a lower depth. There is essentially no transport in the lower layer, and weighted means show that the freshwater discharge during ebb is approximately twice that

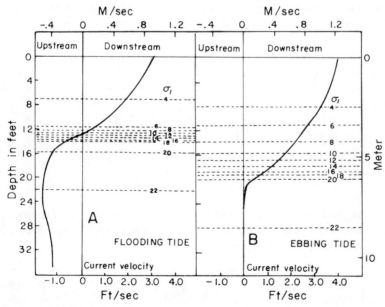

Fig. 9-46 Reproduced with permission from L. D. Wright, 1971, *American Society of Civil Engineers, Journal of the Waterways, Harbors, and Coastal Engineering Division*, **97**, WW3, 491–504, Fig. 5.

of flood. The conclusion is also made that, among the distributaries of the Mississippi river, South pass is probably the only one that shows this degree of tidal influence.

Wright and Coleman (1971) continued these investigations with an examination of the brackish water river plume that extends out from the mouth of South pass into the Gulf of Mexico. In addition to salinity and current measurements aerial photographic observations were taken. The aerial photographs show three distinct water masses: the South pass effluent, the more turbid and brackish water supplied by outlets to the northeast, and the more saline gulf water. They conclude that flow deceleration and effluent deconcentration are primarily the result of vertical rather than lateral mixing. This conclusion is made both from the observations themselves and from a rather complex and partially empirical analysis of the anticipated lateral inertial and vertical entrainment effects.

9-12. GALVESTON BAY

Reid and Bodine (1968) presented results from a two dimensional, longitudinal and lateral, numerical model calculation and verification for storm and hurricane surges in Galveston bay. Although, as stated previously, such hydrodynamic investigations are not within what is specifically intended to be covered in this book, it is of interest to go over their methodology if not the specifics of their results and conclusions. The equations of motion employed were the two dimensional vertically integrated equations of motion in the forms, as we can see from (3-54) and (8-6),

$$\frac{\partial M_x}{\partial t} + gh\frac{\partial \eta}{\partial x} = T_x - \frac{kM_x\sqrt{M_x^2+M_y^2}}{h^2}$$

$$\frac{\partial M_y}{\partial t} + gh\frac{\partial \eta}{\partial y} = T_y - \frac{kM_y\sqrt{M_x^2+M_y^2}}{h^2}$$

(9-22)

where M_x and M_y are the x and y components of volume transport per unit width, and T_x and T_y are the x and y components of wind stress. The wind stress components are in turn given by

$$T_x = k_1w^2\cos\theta$$
$$T_y = k_1w^2\sin\theta$$

(9-23)

in terms of the wind speed w at an elevation of 10 m above the water and the angle θ between the wind velocity and the x axis. The corresponding continuity equation was taken in the form

$$\frac{\partial \eta}{\partial t} + \frac{\partial M_x}{\partial x} + \frac{\partial M_y}{\partial y} = R \qquad (9\text{-}24)$$

where R is the rainfall rate. In their numerical calculations the depth h was taken as a function of t in consideration of the shallow waters of the bay bars and not as a constant for a given location.

Masch and Shankar (1969) discuss briefly a two dimensional numerical model, again longitudinal and lateral, for investigating dispersion effects in Galveston bay and other similar bays where well mixed vertical conditions can be assumed. They used motion and continuity equations similar to the above and included the Coriolis term. Similarly, the salt, or conservative pollutant, continuity equation was taken in a vertically integrated form with both K_x and K_y dispersion terms.

9-13. AMAZON RIVER

The Amazon river is unique. It is the world's largest river; having a mean discharge of $175 \times 10^3 \, \mathrm{m}^3 \, \mathrm{sec}^{-1}$, it supplies about 20% of the total river discharge into the world's oceans. The tidal influence in the river extends at least 735 km upriver. The Amazon is five times larger than the second largest river in the world, the Congo, which has a mean discharge of $38 \times 10^3 \, \mathrm{m}^3 \, \mathrm{sec}^{-1}$, and it is one to two orders of magnitude greater than the Mississippi and Columbia rivers which have mean discharges of $16 \times 10^3 \, \mathrm{m}^3 \, \mathrm{sec}^{-1}$ and $7 \times 10^3 \, \mathrm{m}^3 \, \mathrm{sec}^{-1}$, respectively, and are both normally considered rather substantial rivers.

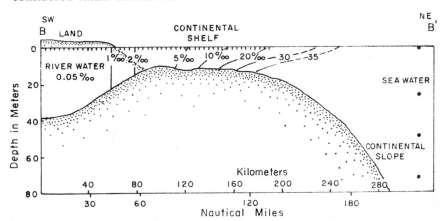

Fig. 9-47 Reproduced with permission from R. J. Gibbs, 1970, *Journal of Marine Research*, **28**, 113–123, Fig. 1.

Ryther et al. (1967) examined the hydrography and nutrient chemistry of portions of the river discharge carried by the Guiana current along the continental shelf northwest of the river. Gibbs (1970) examined the physical oceanographic conditions in the Amazon estuary and on the adjacent continental shelf. His results clearly show that, because of the large discharge of the river, there is no salt water penetration into the estuary during either low or high river discharge periods. The results for a low river discharge period, August–September 1963, are shown in Fig. 9-47. Particular note should be made of the distance scale in this figure. The region of nearly vertical isohalines near the river mouth is interpreted as being a zone of intense mixing. This region is well known locally as a zone of extremely rough, choppy seas, and the area abounds in visual evidence of great turbulence which gradually decreases seaward as a stratified condition begins to develop.

9-14. BAY OF FUNDY

The Bay of Fundy is a rather large rectangular bay which borders on the Gulf of Maine, having approximate dimensions of 250 km length, 50 km breadth, and depths varying from 50 to 200 m from its upper to its lower end. It is well known for its extreme tidal range, one of the largest in the world. Its length and average depth dimensions are such that, from the simple relation $\lambda = T\sqrt{gh}$, it is nearly at a quarter wavelength resonance for the semidiurnal, tidal cooscillating, driving force at its entrance.

Watson (1936) has described the general physical oceanographic conditions of the Bay of Fundy. Ketchum and Keen (1953) applied the flushing time and modified tidal prism concepts of Section 5-1 to these and other available data. Dividing the bay into several segments for convenience in calculations and using the observed salinity data and river flows into the bay for the period April–July, relations (5-2), (5-3), and (5-1) were sequentially applied to obtain a flushing time of 76 days for the bay as a whole.

Then, by following the procedures discussed in Section 9-5 for the Raritan estuary, the modified tidal prism concept was applied. The bay was divided into defined segments, and exchange ratios r_n were calculated for each from (5-17). The exchange ratios varied from 0.05 to 0.15 from the entrance of the bay to its upper portion, where it branches into two arms, Chignecto bay and Minas basin, and thence increased from 0.20 to 0.95 to the upper reaches of each, indicating the decreasing water depths and the increasing tidal amplitudes. A comparison was made of the calculated values of r_n with those determined from the observed river runoff volumes per tidal cycle V_R and the volume fractions of fresh water

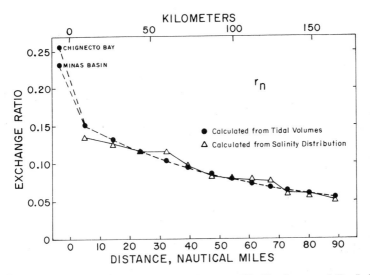

Fig. 9-48   Reproduced with permission from B. H. Ketchum and D. J. Keen, 1953, *Journal of the Fisheries Research Board of Canada*, **10**, 97–124, Fig. 9.

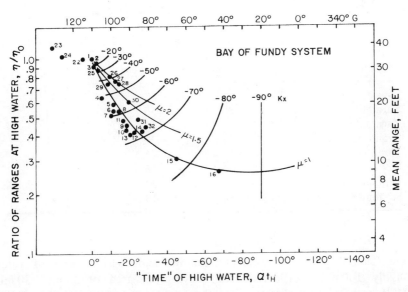

Fig. 9-49   Reproduced with permission from A. C. Redfield, 1950, *Papers in Physical Oceanography and Meteorology* (Massachusetts Institute of Technology and Woods Hole Oceanographic Institution), **11**(4) 1–36 Fig. 9.

in each modified tidal prism segment V_{fn}, using the observed salinity values and relation (5-19). The results are shown in Fig. 9-48.

Redfield (1950) applied the concepts of Section 3-4, which he developed, to tidal cooscillation in the Bay of Fundy, using standard tide table and current table data for the stations there. By combining the calculations of relations (3-79) and (3-81) or, equivalently, Figs. 3-9 and 3-10, a comparison can be made between the theoretical and observed amplitude ratio η/η_0 and the local time of high water ωt_H, as shown in Fig. 9-49. The agreement is quite good for a frictional coefficient of $\mu = 1$, particularly considering the geometric and boundary condition simplifications of the theory. The higher amplitudes for stations 25 to 32 on the eastern, or Nova Scotia, side of the bay are attributed to the anticipated geostrophic effect. Similarly, the calculations of relations (3-79) and (3-83) or, equivalently, Figs. 3-9 and 3-11, can be combined to give a comparison of the observed local time angle between high water and slack water $\omega t_H - \omega t_s$ and the phase difference relative to the upper end of the bay, κx, taken from Fig. 9-49. A comparison for the more limited number of current stations in the bay is as given in Table 9-3; the agreement is again quite good. Ippen and Harleman (1966) made an analysis similar to the above and obtained similar results.

TABLE 9-3

Reference station	Phase difference, κx (deg)	Time angle between high water and slack water, $\omega t_H - \omega t_s$ (deg)	
		Theory	Observation
Lower East Pubnico	−84	68	63
Yarmouth	−80	55	56
Yarmouth	−76	43	46
Grand passage	−68	27	31
Three islands	−66	23	24
Moose cove	−66	26	21
Lepreau bay	−61	22	25
St. John	−56	19	29

9-15. LONG ISLAND SOUND

Long Island sound is a relatively shallow, semienclosed body of water roughly 100 n mi (190 km) long, having an area of about 930 mi^2 (2400 km^2). It has a maximum depth of 100 m near its eastern end, but elsewhere there is little water more than 30 m deep. At its eastern end there is free exchange between Long Island sound and the adjoining

Block Island sound through a series of passes. At its western end there is a more limited exchange with the waters of New York harbor. One of the most thorough investigations of the physical, chemical, and biological characteristics of a coastal body of water were the investigations made by personnel at the Bingham Oceanographic Laboratory of Long Island and the adjoining, and more open, Block Island sounds. These investigations were carried out during 1940–1955 and have been outlined in brief by Riley (1955). We are concerned here only with a discussion of the physical oceanographic characteristics of Long Island sound.

Riley (1952) discusses the general hydrography of Long Island sound as determined from nine cruises in 1946 on which measurements of temperature and salinity as a function of depth were made, and from other available data on tides, currents, and freshwater drainage. The net circulation is similar to that for an estuary with a net flow out, or to the east, in the upper and fresher portion of the water column, and a net flow in, or to the west, in the lower and more saline portion of the water column. In addition, based on more limited data, there appear to be three large eddies within the sound, counterclockwise at the eastern and western ends and clockwise in between.

The simple volume and salt continuity relations of (2-37) and (2-38), the Knudsen relations, were applied to the flow conditions at the eastern end of the sound with the inclusion of a time dependent term for the net increase or decrease in the volume-averaged salinity in Long Island sound

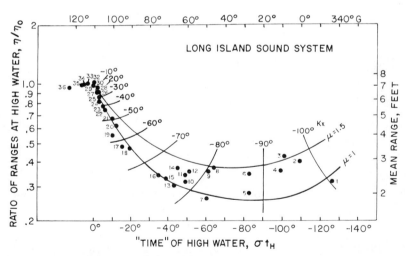

Fig. 9-50 Reproduced with permission from A. C. Redfield, 1950, *Papers in Physical Oceanography and Meteorology* (Massachusetts Institute of Technology and Woods Hole Oceanographic Institution), **11**(4) 1–36 Fig. 6.

to estimate volume transports in and out of the sound. In the calculations it was assumed that the principal freshwater contribution was from rivers draining into the sound, that the rainfall and evaporation contributions essentially canceled out one another, and that there was negligible exchange at the western end of the sound. The resultant calculations show near bottom volume transfer rates of 3000–12,000 m^3 sec^{-1} and near surface volume transfer rates of 4000–13,000 m^3 sec^{-1} for river discharge rates of 300–1200 m^3 sec^{-1} for the nine cruises.

Riley (1955a, 1955b) continued with a general description of the hydrography of Long Island sound from measurements taken during 1952–1954. Also included in these measurements are those from an anchor current meter station, which show ebb tide velocities stronger than the flood tide velocities near the surface but weaker than the flood near the bottom, as one might have anticipated.

Redfield (1950) used the procedures discussed in the previous section to examine tidal cooscillation in Long Island and Block Island sounds. The results are summarized in Fig. 9-50, which is similar to Fig. 9-49. The agreement is good for a frictional coefficient of $\mu = 1$.

The station locations and the tidal cophase lines, taken from Fig. 9-50, are shown in Fig. 9-51. The higher amplitude stations in Fig. 9-50 are in

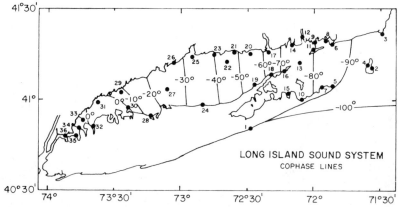

Fig. 9-51 Reproduced with permission from A. C. Redfield, 1950, *Papers in Physical Oceanography and Meteorology* (Massachusetts Institute of Technology and Woods Hole Oceanographic Institution), **11**(4) 1–36 Fig. 7.

general on the northern side of the sound, and again their increased amplitudes are attributed to the anticipated geostrophic effect. A comparison of the calculated and observed local time angles for high water and slack water, corresponding to Table 9-3, also shows good agreement, although some stations show larger differences than those of Table 9-3.

CHAPTER 10

Americas, West Coast

10-1. ALASKA INLETS

The west coast of North America between about 49°N and 59°N is indented by many long, narrow, fjord type inlets. Pickard (1967) has described the general oceanographic conditions of several of the larger inlets of southeastern Alaska. The locations of these inlets are shown in Fig. 10-1. Their average length and width is 117 and 10 km, respectively; their mean and maximum depths are 350 and 840 m, respectively; and their average outer sill depth is 150 m.

The annual mean freshwater discharge into the inlets varies considerably from values as high as $500 \, m^3 \, sec^{-1}$ to values of $10 \, m^3 \, sec^{-1}$. In addition, the distribution of the discharge can be quite different from one inlet to the next. Three categories are distinguished. For type A the stored runoff from glacier and snowfield melt in summer is important, the maximum runoff occurring in June or July. Type C is associated with a low island watershed where there is little influence from stored runoff; the maximum runoff is associated with the precipitation maximum, usually occurring in October. Type B is simply a combination of types A and C. The larger mean annual discharges are associated with type A runoff. The relative annual differences among the three are shown in Fig. 10-2.

Observations of salinity, temperature, and dissolved oxygen as a function of depth were made in 15 inlets at the locations shown in Fig. 10-1 during June 1964, August 1965, and May 1966. Most of the inlets showed a marked decrease to low salinity values at the head, the exceptions being those inlets having icebergs. Salinity increased with depth, reaching 90% of the deep water values by 20 m or less. Dissolved oxygen decreased with depth in most cases, a conspicuous exception being the high deep water oxygen values in the iceberg inlets.

McAlister et al. (1959), as also reported in Dyer (1973), made a rather thorough study of Silver bay, a fjord type estuary. The fjord is about 8 km in length and 0.8–1.2 km in width. There is no entrance sill, but a small sill at a depth of 70 m occurs about 1.5 km inside the entrance. The main basin of the fjord has a depth of about 80 m, and the bay opens at its

385

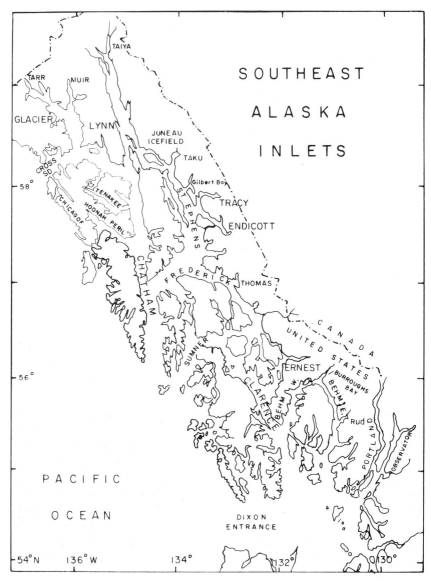

Fig. 10-1 Reproduced with permission from G. L. Pickard, 1967, *Journal of the Fisheries Research Board of Canada*, **24,** 1475–1506, Fig. 1.

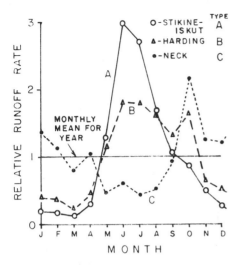

Fig. 10-2 Reproduced with permission from G. L. Pickard, 1967, *Journal of the Fisheries Research Board of Canada*, **24**, 1475–1506, Fig. 4.

entrance into Sitka sound with depths of 120–140 m. High runoff with discharges of 86 m^3 sec^{-1} occur in late spring and summer and low runoff with discharges of less than 3 m^3 sec^{-1} in winter.

Measurements were completed on three sections across the fjord near its mouth over a number of tidal cycles at both high and low river discharge. Salinity and temperature measurements were completed at stations on all three sections, but current measurements were made only for the central station on the middle section. Values for the mean salinity, temperature, and current velocity at this station are shown in Fig. 10-3. The circle values are for high runoff conditions in July 1956, and the cross values for low runoff in March 1957. During high discharge conditions the outflowing surface water layer was about 5 m thick, and its salinity ranged from zero at the head to 28‰ at the mouth. A compensating upestuary flow occurred in the depth interval 5–30 m. During the low discharge period the conditions were more homogeneous. The amplitude of the tidal velocity was about 2 cm sec^{-1}, considerably smaller than the mean flow in the surface layer.

By following procedures similar to Pritchard's analysis of the James river data for a partially stratified estuary, an analysis of the relative importance of the various terms in the longitudinal equation of motion for this case of a fjord type estuary was also made. Here, following the discussion of Section 4-2 for fjord type entrainment flow we anticipate that the inertial terms are of importance. Accordingly, the analysis was made with respect to a reduced equation of motion from (2-134) of the

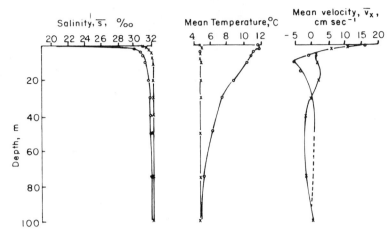

Fig. 10-3   Reproduced with permission from W. B. McAlister, M. Rattray, and C. A. Barnes, 1959, Technical Report 62, Department of Oceanography, University of Washington, Seattle, Figs. 9 and 10.

form

$$\bar{v}_x \frac{\partial \bar{v}_x}{\partial x} + \bar{v}_z \frac{\partial \bar{v}_x}{\partial z} = -\frac{1}{\rho}\frac{\partial p}{\partial x} - \frac{\partial}{\partial z}\overline{(v'_x v'_z)} \qquad (10\text{-}1)$$

In order to obtain the longitudinal gradient of the mean longitudinal velocities it was assumed that all the salt added to the upper surface layer between sections came from below and was added entirely by upward vertical advection. Thus a knowledge of the surface layer salinities at each section and the salinity at depth permitted the calculation of a vertical advection velocity and the longitudinal velocity gradient. Further, as the form of the salinity profiles was similar at all sections, it was assumed that the velocity profiles were also similar and that the longitudinal velocities at any depth had the same ratio to the surface velocity. The term $\bar{v}_z(\partial \bar{v}_x/\partial z)$ was calculated by applying continuity principles to the calculated horizontal velocities. The pressure gradient term was calculated using the dynamic height methodology of (4-177) and assuming a depth of no motion at 100 m, a depth at which the mean longitudinal velocity was zero. The turbulent term was then obtained from the balance specified by (10-1). The pressure gradient and the longitudinal inertial terms were dominant in the upper 5 m of the surface layer and decreased rapidly with depth. The other two terms were of comparable magnitude and of opposite sign; it is difficult to ascertain the relative importance of their calculated magnitudes, because of the various assumptions made in the calculations for $\bar{v}_z$.

10-2. BRITISH COLUMBIA INLETS

Pickard (1956, 1961) discusses the general physical oceanographic characteristics of several inlets in British Columbia. The various inlets of the northern portion of this coast are shown in Fig. 10-4, and those of the southern portion in Fig. 10-5. As is the case for southeastern Alaska, the inlets of British Columbia are morphologically fjords but few possess sills of depths less than 15 m. In this regard they are quite different from the Norwegian fjords, which leads to a different circulation pattern at depth and a lack of stagnation, or oxygen depletion, in the deeper water layers.

The most significant influence in these inlets is freshwater runoff, chiefly from rivers. It is large in many of the inlets fed by rivers from glaciers, is seasonal in flow, and determines the estuarine character and circulation of the inlets and thereby the distribution of their water characteristics. A minority of the inlets, mostly short ones, are not fed from stored runoff and show an annual cycle of the type C category of Fig. 10-2.

The salinity versus depth profiles observed in the various inlets have been separated into two types, as shown in Fig. 10-6. In type 1, associated with high runoff conditions, the salinity profile varies from (*a*) to (*b*) to (*c*) from the inlet head to its entrance, showing a relatively homogeneous surface layer underlain by a strong halocline in the inner to middle reaches of the inlet. For type 2, associated with low runoff conditions, there is no surface layer; and the salinity increases with depth from the surface at a rate that decreases as the depth increases.

Generally the large runoff inlets show less variable dissolved oxygen values along the inlet at any depth than the small runoff inlets. Supersaturation of the upper layers is common, and there is often an oxygen maximum just below the halocline of the larger runoff inlets. A few small runoff inlets have a middepth oxygen minimum, and the lowest values are at the inlet head. DO values of less than $2 \, \text{ml} \, \text{l}^{-1}$ are uncommon, and no zero values have been obtained.

Cameron (1951) made an interesting study of the transverse effects in Portland inlet. Portland inlet is long, narrow, and deep; and its side walls plunge quickly to depths approaching 400 fm (700 m). We might then expect that the geostrophic force is the dominant transverse force and that the simplified concepts of (4-179) for the surface velocity and (4-180) for the flow volume apply. These relations were applied to the observed salinity conditions at the mouth of Portland inlet, an example of which is shown in Fig. 10-7 where station 30 is to the right looking in a downestuary direction and station 31 is to the left. The freshwater discharge was then determined from the calculations of (4-180), using a relation similar

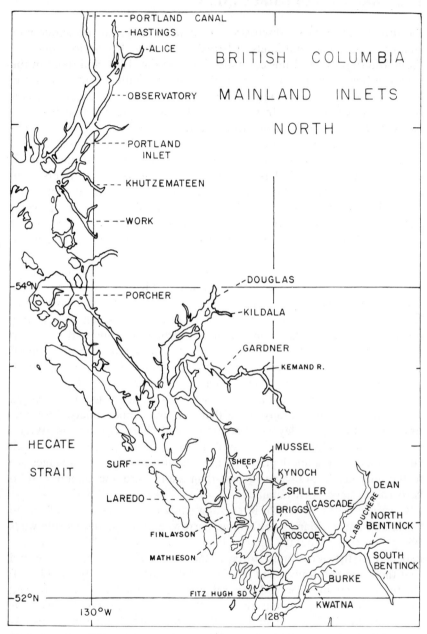

Fig. 10-4   Reproduced with permission from G. L. Pickard, 1961, *Journal of the Fisheries Research Board of Canada*, **18**, 907–999, Fig. 1.

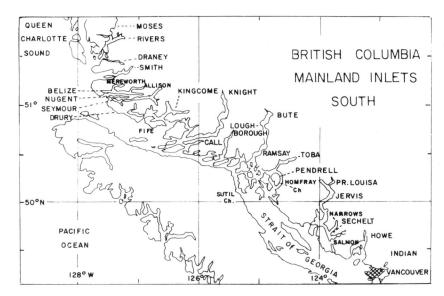

Fig. 10-5 Reproduced with permission from G. L. Pickard, 1961, *Journal of the Fisheries Research Board of Canada*, **18**, 907–999, Fig. 2.

to (5-3) under the assumption that the salinity increase in the upper layer was due entirely to entrainment effects. A comparison of these computed freshwater flows with those observed at gauge stations on the tributary Nass river and a comparison of the computed surface velocities from (4-179) with those observed at a nearby anchor station were satisfactory.

Pickard and Trites (1957) applied the concepts of mass, salt, and heat conservation to a determination of the freshwater transport in estuaries. The assumption is made, along the lines of Section 5-3, that the addition of salt and heat from below to the surface layer is by entrainment only. The assumption is also made that heat flux addition through the surface can be considered a constant and can be determined from meteorological observations and known empirical formulas. From (5-45) we then have

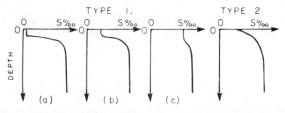

Fig. 10-6 Reproduced with permission from G. L. Pickard, 1961, *Journal of the Fisheries Research Board of Canada*, **18**, 907–999, Fig. 15.

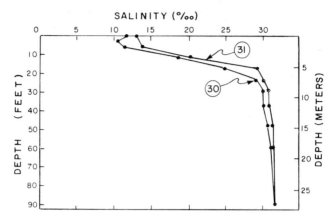

Fig. 10-7 Reproduced with permission from W. M. Cameron, 1951, *Proceedings of the Royal Society of Canada*, **45**(5) 1–8 Fig. 2.

for the volume continuity equation

$$\frac{dq_1}{dx} = w \tag{10-2}$$

where q_1 is the upper layer flow per unit breadth, and w is the entrainment velocity, and where we have considered the thickness of the upper layer h_1 to be a variable also. From (5-46) we have correspondingly for the salt continuity equation

$$\frac{d}{dx}(q_1 s_1) = w s_2 \tag{10-3}$$

and correspondingly for the heat continuity equation

$$\frac{d}{dx}(q_1 T_1) = w T_2 + \frac{F}{c} \tag{10-4}$$

where F is the net heat flux through the surface, c is the specific heat of sea water, T_1 is the temperature of the upper layer, and T_2 is that of the lower layer, considered constant. Substituting (10-2) into (10-3) and (10-4) we obtain

$$q_1 \frac{ds_1}{dx} = (s_2 - s_1)w$$

and

$$q_1 \frac{dT_1}{dx} = (T_2 - T_1)w + \frac{F}{c}$$

(10-5)

Eliminating w from (10-5) we then have finally

$$\frac{dT_1}{dx} + \frac{T_1 - T_2}{s_2 - s_1}\frac{ds_1}{dx} = \frac{F}{cq_1} \qquad (10\text{-}6)$$

From observations of the quantities on the left hand side of (10-6) and a knowledge of F, a value for q_1 can be determined. Equation (10-6) was applied to the upper reaches of several estuaries and the determined values of the total surface layer flow, $Q_1 = bq_1$, compared with the gauged values of the river discharge rate R with satisfactory agreement.

Alternatively we might consider (10-6) a relation between T_2 and s_2, presuming that q_1 is known. From (10-6) we see that this is a linear relation of the form

$$T_2 = c_1 - c_2 s_2 \qquad (10\text{-}7)$$

where c_1 and c_2 are given by

$$c_1 = \frac{(F/cq_1) - (dT_1/dx)}{ds_1/dx}s_1 + T_1$$

and $\qquad\qquad\qquad\qquad\qquad\qquad\qquad\qquad\qquad (10\text{-}8)$

$$c_2 = \frac{(F/cq_1) - (dT_1/dx)}{ds_1/dx}$$

An example of the application of this alternative method is illustrated in Figs. 10-8 and 10-9. Figure 10-8 shows plots of the salinity and temperature versus depth profiles for three locations in Knight inlet during the

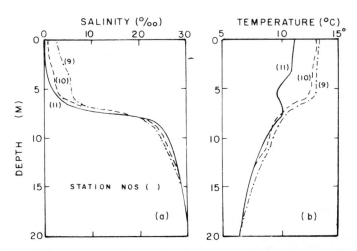

Fig. 10-8   Reproduced with permission from G. L. Pickard and R. W. Trites, 1957, *Journal of the Fisheries Research Board of Canada*, **14**, 605–616, Fig. 5.

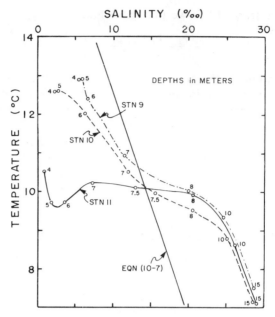

Fig. 10-9   Reproduced with permission from G. L. Pickard and R. W. Trites, 1957, *Journal of the Fisheries Research Board of Canada,* **14,** 605–616, Fig. 6.

summer. Figure 10-9 shows the resultant temperature versus salinity plots determined from the values of Fig. 10-8. The circle numbers are the corresponding depths. The straight line is from the determinations of (10-7) and (10-8). We conclude then that the entrained water is drawn from a depth of about 7 m. The application of this method is probably limited, except for illustration purposes, in that Q_1 varies as a function of x and cannot be taken as equal to its determined value at the upper end of the inlet, or equal to R, as was done in the above example.

Tabata and Pickard (1957) made an extensive examination of the physical oceanographic conditions of Bute inlet from 12 surveys conducted during the period August 1950 to July 1953. The results show much the same general characteristics as found in other inlets. During the heavy runoff period, summer, there is a strong salinity stratification with an increase in surface salinity from 1 to 15‰ from the inlet head to its mouth. During the reduced runoff period, winter, there is no surface layer; and the salinity increases from 20 to 29‰ from the inlet head to its mouth. The surface water on the right hand side looking downestuary is almost always observed to be less saline than that on the left hand side, and this is attributed to the Coriolis force. Examples of the salinity

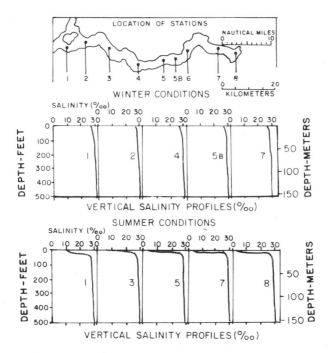

Fig. 10-10 Reproduced with permission from S. Tabata and G. L. Pickard, 1957, *Journal of the Fisheries Research Board of Canada*, **14**, 487–520, Figs. 3 and 4.

profiles as a function of longitudinal distance under these two conditions are shown in Fig. 10-10.

Pickard and Rodgers (1959) made a series of current measurements in Knight inlet during July 1956 at two locations, using a current drag for the measurements in the upper 20 m and an Ekman current meter for the measurements below 20 m. Knight inlet consists of an outer basin, an inner sill, and an inner basin. The first station was located over the inner sill with a water depth of 75 m, and the second station was located just inside the inner basin with a water depth of 350 m. The salinity profiles at these two stations were not typical of what one usually associates with high runoff summer conditions for inlets. The outer basin is essentially well mixed from the surface to the bottom. The inner basin shows a progressive stratification development in an upestuary direction. At the first station the salinity profile showed only a slight change in salinity from surface to bottom of a few parts per thousand. At the second station there was a slight stratification with a surface salinity of 20‰.

At the first station the current measurements showed a tidal variation essentially uniform as a function of depth with an amplitude of about 50 cm sec^{-1}. During the first 25 hr period of the measurements there was no wind, and during the second 25 hr period there was a wind in an upestuary direction of 12 knots (620 cm sec^{-1}). The mean, or net, circulation velocity distribution for each measurement period was quite different, as shown in Fig. 10-11. During the first period there was a net

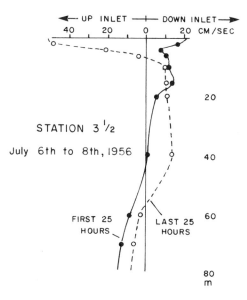

Fig. 10-11 Reproduced with permission from G. L. Pickard and K. Rodgers, 1959, *Journal of the Fisheries Research Board of Canada*, **16,** 635–678, Fig. 13.

downestuary flow at all depths of measurement down to 40 m, while the flow at 60 and 70 m was upestuary. During the second period the surface flow down to 6 m was reversed, and the downestuary flow in the mid-depth range was increased.

At the second station the current measurement at the deeper depths from 50 to 300 m showed a considerably reduced tidal variation with amplitudes of 10–15 cm sec^{-1}. In the upper 20 m the tidal variations were much larger, ranging from 120 cm sec^{-1} in a downestuary direction to 45 cm sec^{-1} in an upestuary direction. Again at this station the 50 hr measurement period could be divided into two parts, the first with a downestuary wind of 10 knots (510 cm sec^{-1}) and the second with little wind. The mean, or net, circulation velocity distributions were somewhat different for the two periods, as shown in Fig. 10-12. The near surface

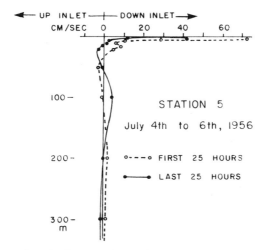

Fig. 10-12 Reproduced with permission from G. L. Pickard and K. Rodgers, 1959, *Journal of the Fisheries Research Board of Canada*, **16**, 635–678, Fig. 16.

currents down to 20 m depth were substantially reduced from the first to the second measurement period.

Waldichuk (1965, 1966) described the general oceanographic conditions and some of the pollution effects in Port Moody, a comparatively small appendage to Burrard inlet, and in Porpoise harbor, Wainwright basin, and Morse basin.

10-3. VANCOUVER ISLAND INLETS

Pickard (1963) made a study of the inlets on the west coast of Vancouver island, similar to those discussed in the previous two sections for Alaska and British Columbia. Generally the Vancouver island inlets are shorter and shallower than those on the British Columbia coast and have shallower sills. The river runoff into the inlets is considerably less than that into the ones on the coast; and the runoff has a winter maximum, related to the precipitation maximum, in contrast to the summer maximum generally found in the inlets on the coast. During the summer the low salinity surface layer generally has a thickness of less than 2 m or is nonexistent. The dissolved oxygen values vary considerably along the inlets. At depths greater than 100 m the DO content is usually less than 4 ml l^{-1} and in many cases less than 1 ml l^{-1}, and these low oxygen values are attributed to the effect of the shallow sills in limiting deep water circulation.

Herlinveaux (1962) studied the oceanographic properties of Saanich inlet, located on the southeast coast of Vancouver island. Saanich inlet is 24 km in length with a basin depth of 234 m and a sill depth at its entrance of 75 m. It is unusual in the sense that there is negligible river runoff into the inlet. Above the sill depth the properties of the water are normal and continuous with those in the approaches that connect with the Strait of Georgia. The waters below the sill depth are isolated, oxygen deficient, and usually contain hydrogen sulfide. In general, it appears that the lower structure displays reasonable constancy with time, whereas the upper structure varies markedly on a seasonal basis. In particular the dissolved oxygen profiles show two oxyclines, one associated with the surface and the other with the sill depth. Examples of DO profiles along the inlet for June 1959 are shown in Fig. 10-13, where station H is over the sill, and station B is in the approaches to the inlet. It is conjectured that from time to time there is an intrusion of water from the approaches over the sill to flush or flush partially the deeper waters of the basin when these waters become more dense than the waters below the sill depth.

Anderson and Devol (1973) continued the study of the intermittent flushing of the deep waters of Saanich inlet. They ascertained that this

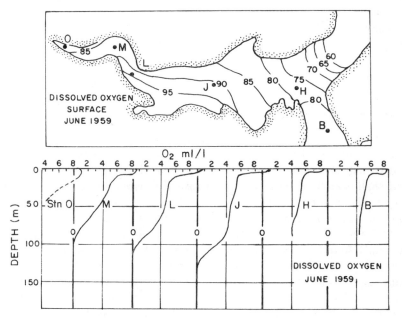

Fig. 10-13   Reproduced with permission from R. H. Herlinveaux, 1962, *Journal of the Fisheries Research Board of Canada*, **19**, 1–37, Fig. 18.

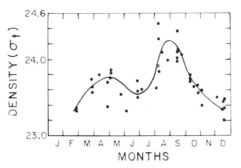

Fig. 10-14   Reproduced with permission from J. J. Anderson and A. H. Devol, 1973, *Estuarine and Coastal Marine Science*, **1,** 1–10, Fig. 6.

intrusion of water over the sill, to flow down into the basin and mix with the resident deep waters there, occurs on an annual basis in late summer or early fall when dense oxygenated water reaches the sill of Saanich inlet from nearby Haro strait. A summary of the annual density cycle in units of σ_t for the lower 10 m on the inlet sill as determined from measurements between 1961 and 1970 is shown in Fig. 10-14. A characteristic density of the resident deep water before flushing may be taken as $\sigma_t = 24.20$.

Prior to the commencement of flushing, the deep waters of the basin are devoid of oxygen and nitrate. By considering nitrate a quasiconservative quantity over the flushing period of a few months the volume of water intruded can be determined simply by dividing the total mass of nitrate introduced, as determined from observations of the deeper waters in the basin during the flushing period, by the concentration of nitrate in the flushing water before it enters. An example of the nitrate distribution at a station in the inlet prior to flushing on 6 August 1969 and after water was intruded on 23 October 1969 is shown in Fig. 10-15. It is to be noted that the intruded water has descended and mixed with the deeper water of the basin, displacing a residual portion of the deoxygenated layer upward. Using the above method for intruded volume calculations, Anderson and Devol (1973) obtained intruded volume estimates for 2 years, 1962 and 1969. Up to 12 September 1962 they determined a volume equivalent to 46% of the basin water below a depth of 150 m was intruded, and that from 12 September to 26 October 1962 an additional amount of 18% was intruded. For 1969 they determined that up to 23 September 1969 an amount of 33% was intruded; and their measurements on 18 October 1969 indicated that there had been no further intrusion in the intervening time interval.

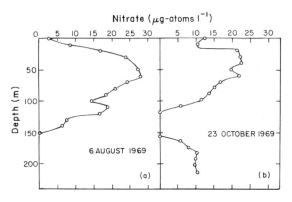

Nitrate (μg-atoms l^{-1})

Fig. 10-15   Reproduced with permission from J. J. Anderson and A. H. Devol, 1973, *Estuarine and Coastal Marine Science*, **1,** 1–10, Fig. 3.

Waldichuk (1958) investigated some of the oceanographic features of Neroutses inlet, which is heavily polluted by a sulfide process pulp mill near its head. There is little river runoff into this inlet, and it consequently does not possess the freshwater characteristics that might provide for adequate flushing of the surface waters. The mill effluent is discharged on the surface at the shore and is largely confined to a thin surface layer about 5 m thick extending for a considerable distance from the outfall in both a downestuary and an upestuary direction.

In connection with a study whose purpose was to give rough estimates of the dispersion of a surface-discharged, sulfate paper mill effluent Waldichuk (1964) described in some detail the rather unusual tidal conditions in Northumberland channel. Northumberland channel is a tidal passage between Gabriola and Vancouver Islands. It is 6.5 km in length and has an average width of 1.5 km and a mean depth of 60 m. At its northwestern end it is open to the Strait of Georgia. At its southeastern end the flow is limited to two restricted waterways, Dodd and False narrows, which connect it with Stuart and Pylades channels. The major portion of the surface water funnels out through Dodd and False narrows; the deep water exchange is limited to the northwest. From the salinity and temperature profiles the waters of the channel can be described as a partially stratified tidal system. There is no river runoff, and the surface salinity is relatively uniform along the entire channel length. The tide is mixed, being a combination of both semidiurnal and diurnal components. An illustration of the variation in the tidal current is given in Fig. 10-16, taken from a midchannel station. It is to be noted that at the surface the tidal current is always in an ebb, easterly, direction. The tidal currents in Dodd narrows, which is the dominant discharge passage, are generally

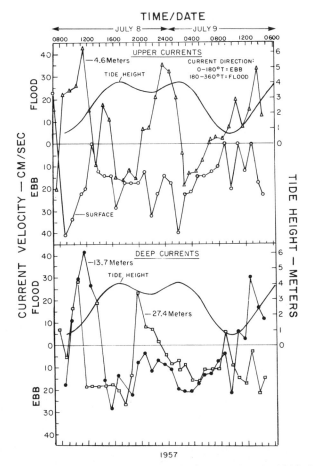

Fig. 10-16 Reproduced with permission from M. Waldichuk, 1964, *International Conference on Water Pollution Research*, **3**, 133–166, Fig. 4.

sinusoidal in character, are of large amplitude of about 200 cm sec^{-1} related to the tide height difference at the two ends of the passage, and have an easterly net flow of 32 cm sec^{-1}.

Grant et al. (1962) made a detailed study of the turbulence energy spectrum of the tidal currents in Discovery passage. Discovery passage is between Vancouver and Quadra islands, and the tidal currents are relatively large, having amplitudes of the order of 100 cm sec^{-1}. The measurements were concentrated in the high frequency range. The turbulent velocities were measured with a towed hot wire flowmeter. The turbulent spectral amplitudes were determined by conventional spectral analysis techniques; and the wave number k was given by the usual relation, $k = 2\pi f/v_{xt}$, where f is the observed frequency, and v_{xt} is the tidal velocity of flow past the meter. The observational times for the record analyses were of the order of 30 min. In the wave number range $k = 0.02$–1.0 cm^{-1} the energy spectrum decreased as $k^{-5/3}$ in accordance with relation (2-100) of the inertial subrange, or wave number range in which there is no dissipation of the turbulent energy. At wave numbers greater than $k = 1.0$ cm^{-1} the spectrum decreased more rapidly than $k^{-5/3}$, and a discussion is given of the dissipation processes involved at these wave numbers.

10-4. ALBERNI INLET

One of the more thorough studies of an inlet was that by Tully (1949) for Alberni inlet on the western coast of Vancouver Island. The measurements, essentially of temperature and salinity as a function of depth, longitudinal distance along the inlet, and phase of tidal cycle were taken during the summer of 1941 in connection with a proposed pulp mill to be located at the upper end of the inlet. The geography and bathymetry of the inlet and the station locations are shown in Fig. 10-17. The inlet has

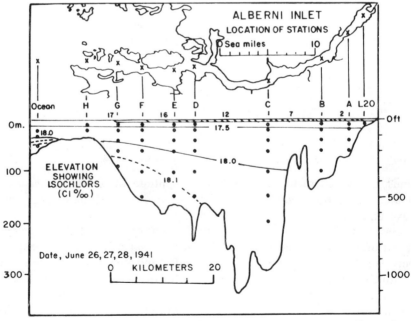

Fig. 10-17   Reproduced with permission from J. P. Tully, 1949, *Bulletin of the Fisheries Research Board of Canada*, No. 83, pp. 1–169, Fig. 3.

an outer sill at its entrance at a depth of 120 ft (37 m) and an inner sill near its head at a depth of 138 ft (42 m), and its sides are precipitous and rocky.

Figure 10-17 also shows a characteristic chlorinity distribution for the measurement period, illustrating clearly the vertical stratification and the longitudinal gradient for the near surface flow. Three zones have been delimited from such observations: a surface zone whose lower boundary is marked by the halocline, a middle zone between this boundary and the sill depth where the water is continuous with the adjacent sea, and a deep zone below the sill depth where the chlorinity is different from that in the adjacent sea at corresponding depths.

Figure 10-18 is an interesting illustration of the difference in the water structure during an ebb tide and during a flood tide at the upper end of

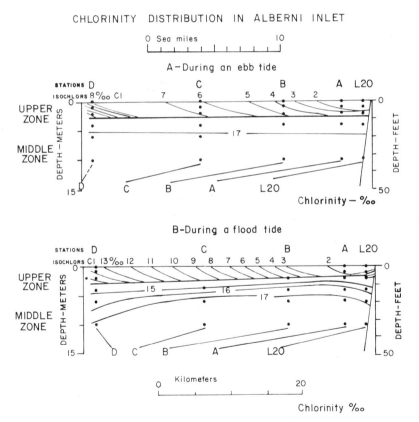

Fig. 10-18 Reproduced with permission from J. P. Tully, 1949, *Bulletin of the Fisheries Research Board of Canada*, No. 83, pp. 1–169, Fig. 4.

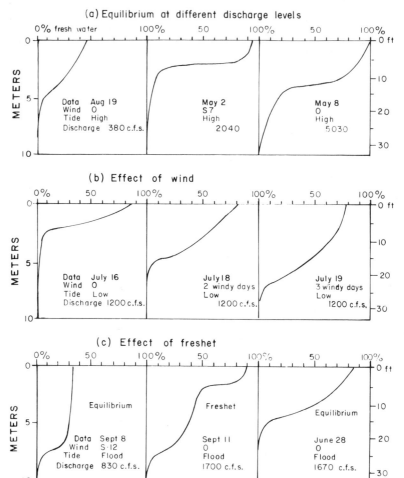

Fig. 10-19   Reproduced with permission from J. P. Tully, 1949, *Bulletin of the Fisheries Research Board of Canada*, No. 83, pp. 1–119, Fig. 15.

the inlet. During the ebb tide there is a strong and distinct separation between the two water masses across the stratification boundary. During the flood tide, although the stratification boundary still exists, there appears to be greater vertical mixing between the two water masses.

Figure 10-19 is a further illustration of the variations in the water structure as a function of other parameters. The measurements are all for the station at the head of the inlet. The upper portion of the figure shows the variation in the water structure as a function of the river runoff rate. For a relatively low runoff rate there is a rather gradual decrease in the freshwater fraction as a function of depth. As the runoff rate increases, a strong halocline is established at a shallow depth; and as the runoff rate increases still further, the strong halocline remains but is depressed in depth. The middle portion of the figure shows the effect of a wind in a direction opposite the surface flow. For constant runoff conditions the effect of the wind is to increase the depth of mixing. The lower portion of the figure shows the effect of a freshet in establishing a temporary secondary halocline and freshwater zone at the water surface.

In his analysis of the mixing, Tully (1949) introduces some of the concepts discussed in more detail in Section 5-1. In particular he introduces the concept of a displacement, which in our notation is the same as the exchange ratio r_n of (5-19). Ketchum (1951) continued this analysis using the modified tidal prism concept of Section 5-1, as applied to the upper surface layer only, and obtained agreement for the predicted river water accumulated along the inlet with that observed.

Stommel (1951) applied relation (5-52), derived by him, to the observed salinity data, using a value of $C = 3.5 \times 10^{-4}$ from Section 5-3 and an averaged depth for the upper layer. The calculated values for the freshwater fraction f and the observed values are shown in Fig. 10-20 with quite good agreement.

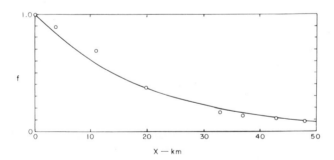

Fig. 10-20   Reproduced with permission from H. Stommel, 1951, Reference 51–33, Woods Hole Oceanographic Institution, Woods Hole, Mass., Fig. 3.

Stommel and Farmer (1952b) applied relation (4-26) to the observed data in the following manner. From one set of longitudinal salinity observations for the upper layer a value for the density difference ratio β of (4-12) of $\beta = 0.001$ at the inlet mouth was chosen, with a corresponding value of $\beta_0 = 0.025$ at the head of the inlet. For this observation period the river runoff rate was approximately $60 \text{ m}^3 \text{ sec}^{-1}$, giving a value of $r = q_0 = 400 \text{ cm}^2 \text{ sec}^{-1}$ for an average inlet width of 1500 m. From the simple relation (4-29) the upper layer flow per unit breadth at the inlet mouth was calculated to be $q_1 = 10^4 \text{ cm}^2 \text{ sec}^{-1}$. On the assumption that a critical flow condition existed at the inlet mouth, a value for the upper layer thickness was then calculated from relation (4-16), giving $h_1 = 4.65$ m. The basic relation (4-26) states that the quantity in the brackets remains constant as a function of longitudinal distance. From the determined values of q_1, h_1, and β at the estuary mouth we then have the value of this constant. We can then apply relation (4-26) at various points upestuary for various values of β, using also relation (4-29) for q_1 in (4-26), to determine the corresponding values of h_1. A comparison of the computed values of h_1 as a function of β and the upper layer salinity s_1 is shown in Table 10-1, along with the observed values of h_1 and their associated values of s_1.

TABLE 10-1

Computed layer depths			Actual layer depths		
β	h_1(m)	s_1(‰)	s_1(‰)	h_1(m)	Station
0.001	4.65	31.0	31	4.5	H
0.0015	5.6	30.0	29.5	2.0	G
0.002	5.2	29.5	29	2.0	F
0.004	3.9	27.0	28	1.5	E*
0.012	2.3	16.6	26	4.0	D*
0.020	1.8	6.4	19	2.0	C
0.025	0.5	0.0	11	2.2	B
			5	2.0	A

* Station is in widening of the channel.

10-5. DUWAMISH AND SNOHOMISH RIVERS

The Duwamish and Snohomish rivers are two small tidal rivers that empty into Puget sound. They traverse the industrial and residential areas of Seattle and Everett, Washington. Both are partially polluted. During the period of high river runoff from November to May the Duwamish river shows a salt wedge characteristic, and during the period of low river runoff from July to September it is partially stratified. Presumably similar conditions apply to the Snohomish river.

Santos and Stoner (1972) describe some of the physical and chemical characteristics of the Duwamish river. The section of the lower Duwamish river investigated is approximately 20 km in length, 130 m in average width, and has depths varying from 1.5 m at its upper end to 12 m at its entrance into Puget sound. During the high runoff period the mean river flows are about 50 m^3 sec^{-1}, and during the low runoff period about 10 m^3 sec^{-1}. The observed vertical and longitudinal salinity distributions are much as one might expect under the varying river runoff conditions. Flushing times were calculated using the modified tidal prism concept of Section 5-1, rather than the simpler relation (5-1), and taking a lower boundary at the salt wedge interface or at the halocline. For the 21 km section of the river investigated the calculated flushing times varied from about 20 hr at a river flow of 150 m^3 sec^{-1} to 60 hr at a river flow of 10 m^3 sec^{-1}. DO and BOD measurements were taken in the upper fresher water layer and in the lower more saline water layer, and related in a general way to the river runoff and tidal conditions and with respect to the principal secondary sewage treatment plant at the upper end of the section of the river investigated.

Stoner (1972) looked at the mass and salt continuity flow relations across a section about 8 km from the mouth of the river. Measurements of the longitudinal current velocity and the salinity as a function of depth over a tidal cycle were taken on three occasions, two during low river runoff conditions and one during a high river runoff condition. The salinity was established to be uniform laterally, and the salinity measurements were taken at midchannel. Velocity measurements were taken along two verticals in the main channel. By using averaged current and cross section areas for the upper and lower layers, rather than integrated product values, over the observed tidal cycles, an essential balance was obtained for the difference between the upper layer, downestuary flow and the lower layer, upestuary flow as compared to the net river runoff flow. For the salt balance, by ignoring longitudinal diffusion effects which should be permissible for this type of analysis, an essential balance was also obtained between the salt fluxes in the two layers.

Dawson and Tilley (1972) developed an empirical technique for estimating the longitudinal excursion of the salt wedge during a tidal cycle by observing the longitudinal shift in the lower layer, longitudinal DO profile during the tidal cycle. For a tidal amplitude range of 1.3 m typical salt wedge excursion values of 1 km were obtained, and for a 3 m tide range excursion values of 3 km.

Eldridge and Orlob (1951) made a study of the water conditions in the lower Snohomish river. Their concern was related to the oxygen demand of the spent sulfite effluent from two pulp mills at the mouth of the river and the effect of this additional oxygen demand in creating a possible pollution barrier to young migratory fish at the surface. During the period of investigation from September to November 1949, the flooding water from Port Gardner bay where the two pulp mills are located is carried into the Snohomish river, and at the maximum period of flood the current is completely reversed from top to bottom, the polluted bay water occupying the entire river channel. At such times the surface waters in the river channel had DO values of less than 1 ppm, as compared with normal upriver values of about 8 ppm. The conclusion was reached that indeed a pollution barrier did exist, and an example of a fish kill during the observation period was also discussed.

10-6. COLUMBIA RIVER ESTUARY

The Columbia river is the second largest river in the United States in terms of discharge, and the largest on the west coast of the Americas. Its mean annual discharge is about 7300 m^3 sec^{-1}, ranging annually between approximately 3000 and 20,000 m^3 sec^{-1}. At its lower end the river broadens into an estuary, still maintaining nevertheless a rather well defined riverlike channel, as shown in Fig. 10-21.

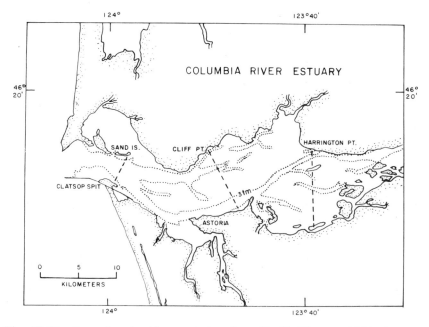

Fig. 10-21 Reproduced with permission from D. V. Hansen, 1965, *Marine Technology Society and American Society of Limnology and Oceanography, Joint Conference on Ocean Science and Ocean Engineering*, pp. 943–955, Fig. 1.

The estuary is also characterized by strong tidal currents, which can propagate nearly 300 km upriver during low river stages. The tides at the mouth of the estuary are of the mixed type, having a diurnal range of 2.3 m and an extreme range of 4 m. The tides can reverse the river current in the lower 40–80 km, depending on the tide range and river stage. The mean tidal prism is 0.75 km^3. Thus, although the tides are moderately strong, the flow ratio, being defined as the ratio of the river discharge during a 12.4 hr tidal cycle to the tidal prism, is of the order of 1 to $\frac{1}{5}$, making the mean river current $\frac{1}{2}$ to $\frac{1}{10}$ the mean tidal current at the mouth.

As stated above, the Columbia river estuary is characterized by an unusual combination of large river discharge and strong tidal currents. One might then anticipate that, instead of a salt wedge condition, which we associated with some of the lower portions of the Mississippi river with a large river discharge and weak tidal currents, there would be a strongly stratified condition due to the strong tidal conditions, and that there would be a substantial variation in the observed salinity condition over a cross section in the lower portions of the estuary during a tidal cycle. We might also anticipate that for the lower portions of the estuary, where the tidal currents are strong, that the tidal mixing would be substantial and the longitudinal dispersion coefficient large.

O'Brien (1952) investigated the current structure at the mouth of the estuary. He used data obtained by the Corps of Engineers in 20–22 June 1932 at five boat stations on a section from Sand island to Clatsop spit; measurements were taken continuously every 20 min at five depths at each of the stations over the 2 day period. The river gauged freshwater flow during the period was 580,000 ft^3 sec^{-1} (16,000 m^3 sec^{-1}); the freshwater flow obtained from the averaged values at the five stations was 530,000 ft^3 sec^{-1} (15,000 m^3 sec^{-1}). In order to ascertain the density gradient contribution to the observed tidal averaged velocities the depth and tidal averaged velocity $\langle \bar{v}_x \rangle$ at each station, as corrected by an observed velocity depth relation for the river flow taken from a station in the upper reaches of the estuary, was subtracted from the observed tidal averaged values at each depth for each station. The velocity deviations v_{x1} determined in this manner at the surface and the bottom are summarized in Table 10-2. For the bottom velocities v_{x1b} a comparison was also made with values computed from an equation including the bottom friction and horizontal density gradient effects but omitting the surface slope contribution, that is, using (4-108) for $z = h$ without the second term in the parentheses under the radical. In the calculations observed values of the horizontal density gradient λ and an assumed value for the bottom friction coefficient k were used.

TABLE 10-2

Station	Depth, total (m)	Depth, $v_{x1} = 0$ (m)	$\langle \bar{v}_x \rangle$ (cm sec^{-1})	v_{x1s} (cm sec^{-1})	v_{x1b} (cm sec^{-1})
A	7.6	3.7	28	21	-18
B	12.5	4.3	57	55	-43
C	9.5	3.4	42	24	-21
D	11.9	4.6	31	24	-18
E	11.3	4.9	58	21	-37

Hansen (1965) made a thorough study of the circulation and mixing effects in the Columbia river estuary. A measurable salinity intrusion extends only about 10 km into the estuary at a high river discharge, rarely extends greater than 20 km for any river flow, and is almost of the same order as the tidal excursion. Because the salinity intrusion is so short and its variation in time so great, a sampling program adequate to describe the salinity distribution would have to be very dense in time and space. The measurements show that the surface to bottom salinity differences can exceed 20‰, as might be expected for the large river discharge rates. For measurements on a cross section at the estuary mouth the salinity can vary from zero at low tide and high discharge across the entire section to oceanic salinity at high tide and low discharge. The currents show similar variability. The net upestuary near bottom flow, characteristic of estuarine circulations with generally lower river discharge rates, is found at the cross section at the mouth of the estuary only during low river discharge conditions. Even then it is relatively weak in conjunction with the river currents, although it is comparable to density currents found in other estuaries.

A salt flux analysis, following the general procedures given in Sections 5-4 and 5-7, was applied. In this case averaged values for the cross sectional area were used; and the salinity and velocity deviation terms were taken with respect to their area averaged values without any further averaging with respect to the longitudinal or lateral dimensions, so that the resultant circulation, or shear, flux term represented a combination of longitudinal and lateral circulation effects. The resulting calculations showed that about 40% of the net river advection flux was balanced by the tidal diffusion term and 45% by the circulation term. Additional terms involving cross sectional area variations were also included and found to be small.

Hansen and Rattray (1965), using the procedures discussed in Section

9-8, obtained a value for the vertical eddy diffusion coefficient of $K_z =$ $10 \text{ cm}^2 \text{ sec}^{-1}$ and for the longitudinal dispersion coefficient of $K_x =$ $50 \times 10^6 \text{ cm}^2 \text{ sec}^{-1}$. They also obtained values for the fraction ν of relations (5-193) and (5-195) of $\nu = 0.8$–0.9. In the theoretical derivations used to obtain this latter relation the lateral effects and the bottom frictional effects were not considered as separate items but were indirectly included in the tidal diffusion portion of ν. Then, if this analysis and that of the previous paragraph are both taken to be correct, which is probably the case, it must be concluded that some combination of lateral circulation and bottom frictional effects are important in a total salt flux analysis.

Neal (1966) applied the modified tidal prism concept of Section 5-1 and the fraction of freshwater method of (5-1) to determine the flushing time for the length of the salinity intrusion under various tidal and river discharge conditions for the Columbia river estuary. The first method gave flushing times of 1–5 days depending on the river flow, and the second method gave about half these times. Relations (5-23) and (5-24) were also used to determine pollutant distributions from outfalls at various locations along the estuary. In addition values for the longitudinal dispersion coefficient K_x were determined from the familiar relation (5-100); and the simple, one dimensional longitudinal dispersion equation was used in finite difference form to determine pollutant distributions. The results of the two methods were similar, as indeed we would expect in consideration of the fact that relations (5-105) and (5–106), derived from the one dimensional dispersion equation, are the same as relations (5-23) and (5-24).

10-7. GRAYS HARBOR AND YAQUINA ESTUARIES

Beverage and Swecker (1969) have described the physical oceanographic features of Grays harbor, Washington. Grays harbor is the estuary of the Chehalis river. The low river flow period of less than $30 \text{ m}^3 \text{ sec}^{-1}$ is from July through September, and the high river flow period of 300-400 m^3 sec^{-1} or more is from November through March. During the low flow period the estuary is well mixed, and during the high flow period it is stratified. They also discuss the relations between the observed longitudinal salinity distributions and the results from dye diffusion experiments.

Burt and Marriage (1957) investigated the longitudinal dispersion properties of the Yaquina estuary, Oregon. They followed the procedures of Stommel (1953), which were discussed in Section 7-3, for determination of the longitudinal dispersion coefficient K_x, and from that the distribution of a conservative and nonconservative pollutant at various locations along the estuary. The determined values of K_x are summarized

TABLE 10-3

Distance (km)	$K_x(\times 10^6 \, cm^2 \, sec^{-1})$		Distance (km)	$K_x(\times 10^6 \, cm^2 \, sec^{-1})$	
	High flow	Low flow		High flow	Low flow
0	—	—	17.7	0.66	0.15
1.6	—	0.21	19.3	0.68	0.14
3.2	—	0.20	20.9	0.68	0.13
4.8	—	0.25	22.5	0.78	0.14
6.4	—	0.27	24.1	0.76	0.13
8.0	—	0.26	25.7	1.14	0.18
9.7	—	0.27	27.4	1.65	0.16
11.3	—	0.22	29.0	2.96	0.22
12.9	—	0.19	30.6	5.03	0.59
14.5	0.60	0.23	32.2	8.53	0.99
16.1	0.79	0.18			

in Table 10-3 for a high river discharge condition of $17 \, m^3 \, sec^{-1}$ in February 1956, and for a low river discharge condition in August 1955. Distances were measured downestuary from Elk City to the estuary mouth. The longitudinal dispersion coefficient values are substantially greater for the high river flow condition than for the low river flow condition; and both show a significant increase as the estuary mouth is approached, much the same as that found in comparable determinations elsewhere. The conservative and nonconservative pollutant distributions calculated for various outfall locations using these values of K_x show much the same distribution as that illustrated in Fig. 6-2, again with a large reduction in the nonconservative values, using a half life of 4 days, over the corresponding conservative values and a large reduction in the upestuary distributions for the high river flow condition.

10-8. SAN FRANCISCO BAY AND ASSOCIATED WATERS

The San Francisco bay area is a rather complex estuarine system. As shown in Fig. 10-22, the principal river flow is from the Sacramento and San Joaquin rivers into the northern arm of the system. The freshwater flow is first into Suisun bay, then through the constriction of Carquinez strait into San Pablo bay and the northern portion of San Francisco bay, and finally out through the Golden Gate to the Pacific ocean. The flow into the southern arm of the system, or the southern portion of San Francisco bay, is quite modest. The waters of Suisun bay are well mixed and of low salinity; the waters of San Pablo bay are partially stratified,

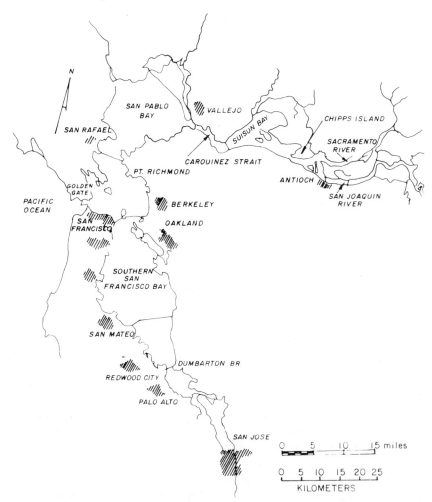

Fig. 10-22 Reproduced with permission from B. Glenne and R. E. Selleck, 1969, *Water Research*, **3**, 1–20, Fig. 1.

and those of the southern portion of San Francisco bay are well mixed. The mean annual flow of the Sacramento and San Joaquin rivers is about 20,000 ft^3 sec^{-1} (560 m^3 sec^{-1}), is highly variable, and can be as low as 2000 ft^3 sec^{-1} (56 m^3 sec^{-1}) during the summer. As is the case for the New York harbor and Delaware estuary areas, San Francisco bay is located in a heavily populated and industrialized area and has understandably been the subject of numerous engineering studies related to possible pollution and water supply problems, much the same as discussed for the former two areas in Sections 9-4 and 9-6.

Simmons (1955) discusses the flow predominance and bottom sediment deposition characteristics in the vicinity of Carquinez strait. At the eastern end of the strait in Suisun bay the net flow predominance as averaged over a tidal cycle is in a downestuary direction at all depths, which we may characterize as a river flow condition. At the western end of the strait in San Pablo bay the net flow predominance is in a downestuary direction in the upper third of the water column and in an upestuary direction in the lower two thirds of the water column, which we may characterize as an estuarine flow condition. The strait itself is subject to great turbulence as a result of the strong tidal currents there, so that well mixed conditions exist in the vertical of the essentially freshwater flow from Suisun bay and the more saline, near bottom flow from San Pablo bay. At the western end of the strait, as it opens into San Pablo bay, partially stratified conditions appear. Further, the flow characteristics indicate that the western end of Carquinez strait should be a sediment accumulation area. The tidal currents in the strait itself are much too great to permit significant deposition; however, Mare Island strait, which joins Carquinez strait at its western end, is the location of the most serious shoaling problem in the San Francisco bay area, requiring regular maintenance dredging.

Einstein and Krone (1961, 1962) made an interesting laboratory study of the flocculation process and the conditions determining the formation of flocculent layers, using sediments from San Francisco bay, where flocculent layers occur. They also conducted a companion field study. In this study they ascertained that the flocculated sediments deposited on the tidal flats during the winter were at times resuspended by the summer winds, formed a flocculent layer or a dense near bottom suspension, and were then transported by the bottom drift currents into quiet areas of the artificially deepened channels, where they again settled out to form part of the normal sediment deposit.

Bailey (1966) discusses instantaneous dye release experiments in the Sacramento and San Joaquin river system and in Suisun bay. Assuming that the simple relation (2-68) with the replacement of x^2 by $(x - \bar{v}_x t)^2$ in the exponential to account for the net advection flow applies to the observations, determinations of the longitudinal dispersion coefficient K_x were made from the observed longitudinal dye distributions with measurements at slack water over one to a few days. In the rivers the determined values of K_x varied from 100 to 1000 $\text{ft}^2 \text{ sec}^{-1}$ (0.09×10^6–$0.9 \times 10^6 \text{ cm}^2 \text{ sec}^{-1}$). In Suisun bay the determined values varied from 600 to 15,000 $\text{ft}^2 \text{ sec}^{-1}$ (0.6×10^6–$14 \times 10^6 \text{ cm}^2 \text{ sec}^{-1}$).

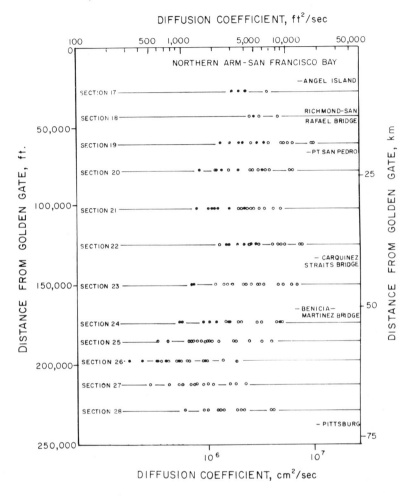

Fig. 10-23 Reproduced with permission from B. Glenne and R. E. Selleck, 1969, *Water Research*, **3**, 1–20, Fig. 5.

Glenne and Selleck (1969) used the familiar fraction of freshwater method of relation (5-100) to determine longitudinal dispersion coefficients in the northern and in the southern arms of the San Francisco bay system. The measurement program was conducted from 1959 to 1964, and the measurements in each case and at each location were taken over three stages of the tidal cycle and at two or more depths. Their determined values are summarized in Figs. 10-23 and 10-24 for the northern

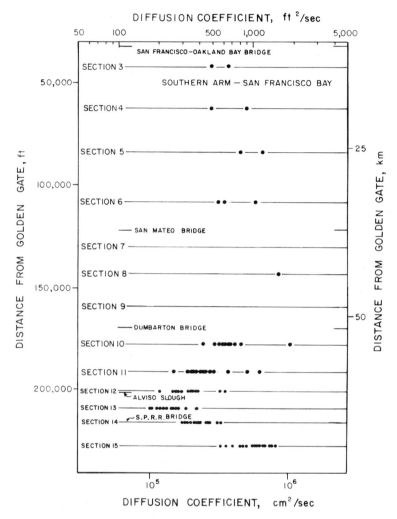

Fig. 10-24 Reproduced with permission from B. Glenne and R. E. Selleck, 1969, *Water Research*, **3**, 1–20, Fig. 3.

and southern arms, respectively. The open circles indicate measurements made when the net advection flow was greater than the mean for the measurement period, and the closed circles refer to those made when the net advection flow was less than the mean. Determinations of K_x were also made using silica as a tracer and assuming that all the silica was introduced by the river flows and that it was a conservative quantity during its passage through the system. These values show much the same distribution as that given in Figs. 10-23 and 10-24, with a somewhat larger amount of scatter.

Inspection of Figs. 10-23 and 10-24 shows that several deductions can be made regarding the behavior of the longitudinal dispersion coefficient in the bay system. In the southern arm, which is well mixed and has a relatively small advective flow, the values of K_x are smaller than those observed in the northern arm. The values of K_x increase in Carquinez strait and are larger in San Pablo bay than in Suisun bay. Further, the values increase with the net advective flow in the northern arm and in general increase in a downestuary direction toward the Golden Gate in both arms.

Orlob et al. (1968, 1969) discuss in a general way the application, with a computer program, of the longitudinal dispersion equation to the determination of nonconservative pollutant distributions in the San Francisco bay system. Thomann et al. (1970) discuss the application of their nitrogen cycle equations and model, which were discussed in more detail in Section 9-6 and 9-7 for the Delaware and Potomac estuaries, to the San Joaquin river.

The most recent investigation in San Francisco bay, which is also one of the more interesting and thorough estuarine investigations conducted anywhere, is one reported by Conomos and Peterson (1974) and by Peterson et al. (1975, 1975a). Among the items investigated was the dissolved silica distribution in the northern arm of the bay. Plots of dissolved silica versus salinity, similar to those of Figs. 9-11 and 9-12, were used to ascertain the conservative, or nonconservative, properties of the dissolved silica distribution. When the rate at which the river inflow supplied silica to the bay was large compared with the rate at which silica was used within the estuary, the silica versus salinity plot was essentially linear, or conservative, as one might have expected. When the silica utilization rate increased significantly, in relation to the uptake of silica by the diatoms in the phytoplankton population, the dissolved silica concentration in the estuary was considerably less than that predicted by simple mixing, and was often less than the dissolved silica in the adjacent, near surface ocean water.

The authors further discuss in detail, the significance of the *null zone* which they define as the confluence of the essentially riverine type flow located upestuary with the estuarine, or density gradient, flow located downestuary from the null zone. The null zone is also the region of the greatest *residence time* for a water particle, where residence time is defined by them in terms of the reciprocal of the integral with respect to depth of the absolute value of the longitudinal circulation velocity. The null zone is, then, the location of the phytoplankton maximum as it is, also, the location of the suspended sediment turbidity maximum.

10-9. INLETS OF CHILE

Pickard (1971) made a general oceanographic study of the inlets of Chile, similar to previous studies made by him of the inlets of Alaska, British Columbia, and Vancouver island, which were discussed in the first three sections of this chapter. The study was based on a survey made during March 1970. In mean and extreme values the dimensions of the inlets in Chile are similar to those in British Columbia and Alaska. The salinity and DO profiles were also similar. Stagnant basins are uncommon, and the absence of shallow sills is considered the main reason for this.

10-10. STRAITS OF JUAN DE FUCA AND GEORGIA

The Straits of Juan de Fuca and Georgia are the bodies of water between Vancouver island and the mainland of the State of Washington and the Province of British Columbia, respectively. They are open to the Pacific ocean at the mouth of Juan de Fuca strait and have restricted access through Discovery passage at the head of Georgia strait.

Redfield (1950) used the same procedures as discussed in Sections 9-14 and 9-15 to examine the tidal cooscillation in the Juan de Fuca and Georgia straits. This investigation is of particular interest in that the tides are of the mixed type, consisting of both a semidiurnal and a diurnal component, with differing tidal cophase relations. A striking result of this is the change along the straits of the relative magnitudes of the diurnal and semidiurnal components. This leads to near suppression of the semidiurnal component in the neighborhood of Victoria, where the tides are principally diurnal for most of the lunar month. A comparison of the amplitudes of the M_2 semidiurnal and K_1 diurnal components and their

TABLE 10-4

Number	Station	M_2 (m)	K_1 (m)	M_2/K_1
2	Neah bay	0.81	0.49	1.67
3	Port Angeles	0.52	0.66	0.78
4	Victoria	0.37	0.63	0.58
7	Friday harbor	0.56	0.76	0.74
11	Point Atkinson	0.91	0.86	1.05
12	Whaleton	1.01	0.93	1.09

ratio at several locations along the straits are given in Table 10-4. The
station locations are shown in Fig. 10-25.

By following the same analytical procedures as before, a comparison
can be made between the theoretical and observed relations between the
amplitude ratio η/η_0 and the local time of high water wt_H, as given before
for the Bay of Fundy and for Long Island and Block Island sounds in Figs.
9-49 and 9-50. This is shown in Fig. 10-26 for the K_1 diurnal component,
and in Fig. 10-27 for the M_2 semidiurnal component. There is good
agreement in both cases for a friction coefficient μ of between 1.5 and
2.0. The substantial difference in the tidal cophase relations between the

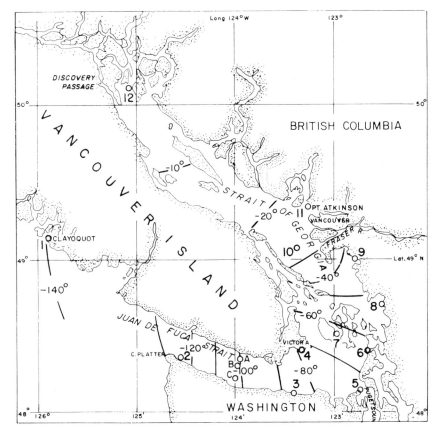

Fig. 10-25   Reproduced with permission from R. H. Herlinveaux, 1954, *Journal
of the Fisheries Research Board of Canada*, **11**, 14–31, Fig. 1; A. C. Redfield,
1950, *Papers in Physical Oceanography and Meteorology*, (Massachusetts Institute
of Technology and Woods Hole Oceanographic Institution), **11**(4) 1–36 Fig. 11.

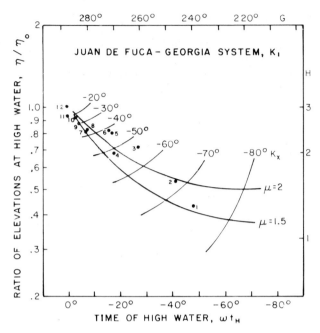

Fig. 10-26 Reproduced with permission from A. C. Redfield, 1950, *Papers in Physical Oceanography and Meteorology* (Massachusetts Institute of Technology and Woods Hole Oceanographic Institution), **11**(4), 1–36 Fig. 12.

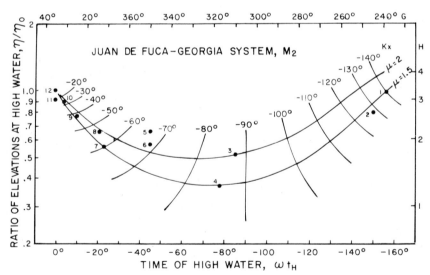

Fig. 10-27 Reproduced with permission from A. C. Redfield, 1950, *Papers in Physical Oceanography and Meteorology* (Massachusetts Institute of Technology and Woods Hole Oceanographic Institution), **11**(4), 1–36 Fig. 13.

two should be noted; in particular the M_2 semidiurnal component passes through a node in the vicinity of Port Angeles and Victoria. The tidal cophase lines for the M_2 component, taken from Fig. 10-27, are included in Fig. 10-25. In the case of the K_1 component the cophase line distribution is quite similar, except that the lines separate phase differences of 5° rather than 10° as drawn.

Herlinveaux (1954, 1954a) measured the tidal current as a function of depth at the three locations A, B, and C, shown in Fig. 10-25. An empirical linear relation for the current velocity in terms of the tide height difference between Clayoquot and Port Atkinson was developed. In general, the ebb current was stronger at the surface and the flood current stronger at the bottom, as expected.

Tully and Dodimead (1957) and Waldichuk (1957) give a thorough description of the general oceanographic features of Georgia strait, and Herlinveaux and Tully (1961) present a similar description of the oceanographic features of Juan de Fuca strait.

CHAPTER 11

Asia, Australia, and Japan

11-1. VELLAR ESTUARY

The Vellar estuary is situated at Porto Novo, India, and opens into the Bay of Bengal. Dyer and Ramamoorthy (1969) studied the physical oceanographic characteristics of the Vellar in some detail; their findings are also reported and analyzed in Dyer (1973, 1974). The Vellar is a rather unusual estuary, both as regards its geographic description and its seasonal variation in river runoff. The section of the estuary investigated was at the lowermost 6 km, which is straight and about 300 m wide; along this section the estuary is quite shallow, averaging 1.5 m in depth. The river runoff shows extreme seasonal variations. During the drought period, May–July, coastal waters occupy almost the entire estuary. During the monsoon period, October–November, the estuary is almost completely filled with fresh water.

Numerous physical oceanographic measurements of temperature, salinity, and current speed and direction were taken as a function of depth over a tidal cycle with *in situ* instrumentation at various of the locations shown in Fig. 11-1 during January and February 1967. Three surveys were completed. On 20 January a survey of only the lower 10 stations was made. On 27–28 January and 9–10 February surveys were completed over all the stations. On 15 February a detailed survey to analyze the mixing processes at stations 5, 6, and 7 and 8, 9, and 10 was carried out.

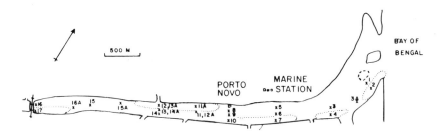

Fig. 11-1   Reproduced with permission from K. R. Dyer and K. Ramamoorthy, 1969, *Limnology and Oceanography,* **14,** 4–15, Fig. 1.

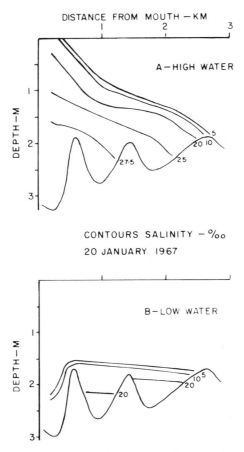

Fig. 11-2   Reproduced with permission from K. R. Dyer and K. Ramamoorthy, 1969, *Limnology and Oceanography*, **14**, 4–15, Fig. 2.

The river runoff flow varied substantially during this period, being 375 m³ sec⁻¹ on 20 January, 90 m³ sec⁻¹ on 27 January, 86 m³ sec⁻¹ on 9 February, and 3 m³ sec⁻¹ on 15 February. The observed conditions of the estuary are described as being that of a salt wedge at the high river discharge period and well stratified at the low river discharge period. The measurements provide one of the most striking examples of the variations in circulation and mixing that can occur when river runoff flow is the prime determining quantity.

The longitudinal distributions of salinity at high and low tides are shown in Figs. 11-2 to 11-4. The difference in characteristics from those of a salt wedge during the high river discharge period of Fig. 11-2 to a

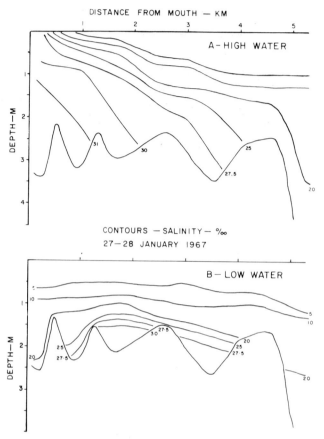

Fig. 11-3   Reproduced with permission from K. R. Dyer and K. Ramamoorthy, 1969, *Limnology and Oceanography*, **14**, 4–15, Fig. 3.

well stratified estuary during the lower discharge periods of Figs. 11-3 and 11-4 should be noted. In each case during the flood tide there is intrusion of a saline wedge along the bottom. The distance of penetration of the wedge increases with time as the river runoff discharge decreases. At about high tide the impounded river water is released and quickly establishes a homogeneous surface layer with a virtually horizontal halocline throughout the length of the estuary.

   The relationship between the salinity and longitudinal velocity variations during a tidal cycle is shown in Figs. 11-5 and 11-6 for the measurements from stations 2 and 7 on 17 January. At station 2 for the flood tide the current was maximum at middepth. About an hour before high tide the surface current reversed, and the salinity started to decrease.

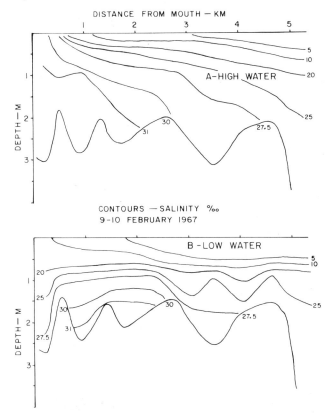

DISTANCE FROM MOUTH — KM

DEPTH — M

A–HIGH WATER

CONTOURS — SALINITY ‰
9–10 FEBRUARY 1967

B –LOW WATER

DEPTH — M

Fig. 11-4   Reproduced with permission from K. R. Dyer and K. Ramamoorthy, 1969, *Limnology and Oceanography*, **14**, 4–15, Fig. 4.

The bottom current, however, did not reverse until high tide. During the ebb tide the maximum velocity was at the surface, with appreciable velocities extending down to the bottom. About $\frac{3}{4}$ hr before low tide the saline water started to flow in near the bottom, but the surface current continued to flow seaward until about $\frac{1}{2}$ hr after low tide. At station 7 the pattern was similar, except that the current reversed throughout the water column before high tide and after low tide. During the ebb tide the bottom velocities were immeasurably low, and this coincided with the period of high bottom salinities.

Figure 11-7 shows the mean salinities and longitudinal velocities over a tidal cycle for station 7 during the four measurement periods. The mean salinity for the 20 January period showed a freshwater surface layer overlying a uniform salinity gradient to the bottom. This was associated with a net seaward velocity throughout the water column. As the river

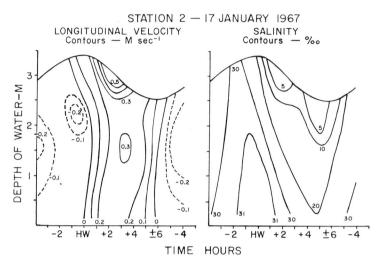

Fig. 11-5 Reproduced with permission from K. R. Dyer and K. Ramamoorthy, 1969, *Limnology and Oceanography*, **14,** 4–15, Fig. 6.

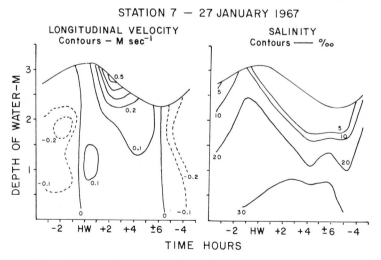

Fig. 11-6 Reproduced with permission from K. R. Dyer and K. Ramamoorthy, 1969, *Limnology and Oceanography*, **14,** 4–15, Fig. 7.

426

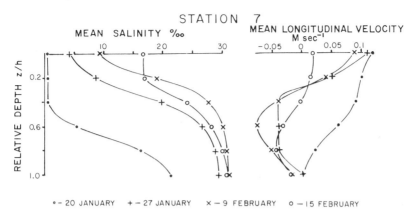

Fig. 11-7 Reproduced with permission from K. R. Dyer and K. Ramamoorthy, 1969, *Limnology and Oceanography*, **14**, 4–15, Fig. 9.

discharge decreased and the mixing increased the salinity of the upper layer, the saline bottom water became thicker and more homogeneous; and a net landward velocity developed from the middepths to the bottom, the net seaward flow being confined to the surface layer. With the still further reduction in river discharge for the 15 February measurements there was an abrupt decrease in the net velocity near the surface, and the surface water of intermediate salinity became thicker and more uniform. For the latter measurements at station 7 the net seaward surface flow was then considerably less than the net landward bottom flow, the necessary mean seaward flow due to the river discharge being maintained by an increased seaward surface flow and a decreased landward bottom flow at the shallower water stations 5 and 6.

By using the same methodology as discussed previously in Section 7-4 for the Southampton estuary, the continuity equation (7-18) was applied to the observed values of the longitudinal and lateral velocity components to obtain $\partial \bar{v}_z / \partial z$. Then, applying the boundary condition of zero vertical velocity at the surface, the tidal mean vertical velocities were obtained by vertical integration. As in the previous calculations, the determined vertical velocity components did not close back to zero at the bottom. Values of the order of 2×10^{-2} cm sec^{-1} were obtained, sometimes positive on one side of the estuary and negative on the other, and sometimes reversed in sequence. In addition the salt flux terms in the form $\partial(\bar{v}_x \bar{s})/\partial x$, $\partial(\bar{v}_y \bar{s})/\partial y$, and $\partial(\bar{v}_z \bar{s})/\partial z$ were calculated as before, the lateral and vertical terms being dominant.

11-2. HOOGHLY ESTUARY

Gole and Thakar (1969) investigated progressive salinity intrusion during the dry season in the Hooghly estuary. The Hooghly estuary, much the same as the Vellar estuary but on a larger scale, shows immense changes in its physical characteristics from the wet season when the estuary is almost entirely fresh to the dry season when there is progressive intrusion of saline water upestuary. Besides being of general scientific interest the dry season salinity intrusion seriously affects the supply of drinking water to the city of Calcutta.

From measurements at various gauging stations along the estuary the simple relation (5-162) was used to describe the salinity variations at a given location within a tidal cycle, and relation (5-163) to describe the steady state spatial relations as averaged over a tidal cycle. The time dependent form of (5-163), namely, relation (8-7) in terms of salinity, was then used to describe the time dependent seasonal variations of salinity. In the latter analysis an exponentially dependent analytic solution form for s in terms of $(x - at + b)^2$, where a and b are constants, was used, the solution being dependent on an assumed, empirical form for the longitudinal dispersion coefficient K_x, which is itself a function of x and t.

11-3. CHAO PHYA ESTUARY

Allersma et al. (1966) investigated sediment transport in the Chao Phya estuary, Thailand. Here the yearly cycle of river flow shows a dry period between January and July with discharges of 25–250 m^3 sec^{-1} and with a minimum in May, and a wet period from July to December with a maximum discharge of about 4000 m^3 sec^{-1} at the end of October or in the first half of November.

They observed flocculent layers with concentrations of $100{,}000$–$300{,}000$ ppm and densities of 1.1–1.25 gm cm^{-3} and inferred that they were an important part of the sediment transport process. From echo sounder records their thicknesses were 0.5–2.5 m. They occurred at places and times of heavy silting of flocculated sediments from a dense suspension in relatively quiet water, which was in the northern part of the estuary channel during the first half of the wet season and in the southern part of the channel during the latter half of the wet season and during April and May. Resuspension of flocculated sediments took place when the tidal velocities reached 20–100 cm sec^{-1}. From laboratory experiments these investigators ascertained that consolidation of the flocculent layers was very slow, layers of 0.5–2.5 m thickness remaining fluid over periods of several weeks in a settling tube.

11-4. AUSTRALIAN ESTUARIES

Rochford (1951) gives a rather complete description of the physical oceanographic features of several of the estuaries in Australia. In general the estuaries of eastern Australia have good tidal mixing and show partially stratified to well mixed estuarine conditions. In southwestern Australia the estuaries have little tidal mixing. Those estuaries that have no bar at their entrance show strongly stratified to arrested salt wedge conditions, and those estuaries in southwestern Australia that have a bar at their entrance show fjord like conditions.

The annual rainfall and consequent river discharge cycles show considerable variation around the coast of the continent of Australia, as one might expect. Figure 11-8 shows the mean rainfall in inches for January, a

MEAN MONTHLY RAINFALL

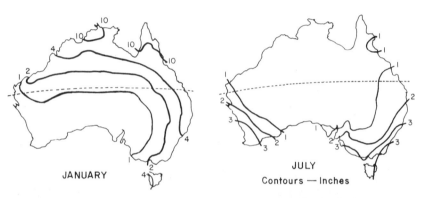

Fig. 11-8 Reproduced with permission from D. J. Rochford, 1951, *Australian Journal of Marine and Freshwater Research*, **2,** 1–116, Fig. 5.

midsummer month, and for July, a midwinter month. The Australian estuaries can be divided into three principal groups on the basis of their rainfall characteristics. In the first group, covering the sector from northwestern to eastern Australia, river discharge is dominated by a summer flood and a winter drought condition. In the second group in southwestern Australia the discharge is dominated by a smaller winter flood and a summer drought condition. In the third group in southeastern Australia the discharge has characteristics of both the first and the second groups, producing a more evenly distributed annual pattern.

An interesting and useful longitudinal division is also given, based on the longitudinal salinity profile. As shown in Fig. 11-9, the estuary is divided into four zones. The freshwater and marine zones are self

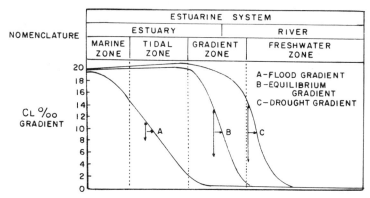

Fig. 11-9   Reproduced with permission from D. J. Rochford, 1951, *Australian Journal of Marine and Freshwater Research*, **2,** 1–116, Fig. 6.

explanatory. The tidal zone is the region in which the full effect of tidal mixing can be seen. The longitudinal gradient zone is the region in which there is a fairly rapid change in the salinity from nearly freshwater values to nearly oceanic values, presumably related to the change in circulation conditions from that dominated by the river flow to that dominated by the estuarine density gradient flow. For any given estuary any one or more of these zones may be curtailed or nonexistent. This division scheme has been applied to Australian estuaries and presumably is of general applicability elsewhere.

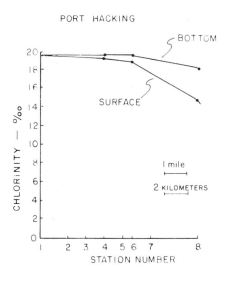

Fig. 11-10 Reproduced with permission from D. J. Rochford, 1951, *Australian Journal of Marine and Freshwater Research*, **2,** 1–116, Figs. 14 and 16.

Examples of estuary description in terms of this zonal classification are given below in terms of the mean annual longitudinal chlorinity profiles for the surface and bottom waters. The chlorinity profiles for Port Hacking are shown in Fig. 11-10, where the longitudinal distance is plotted in an upestuary direction from the estuary mouth. The right hand end of the figure corresponds to the entrance of the Audley river. The river water enters the port over a weir at a fairly constant, small discharge rate. This estuarine section is classified as being entirely in the marine zone.

The corresponding profiles for the George's river and Botany bay estuarine system are shown in Fig. 11-11. In this case the longitudinal salinity variation is slight between stations 1 and 9, and this region is classified as within the marine-tidal zone. The longitudinal gradient zone occupies the region between stations 9 and 21.

The profiles for the Hawkesbury river system are shown in Fig. 11-12. Here there is no marine or tidal zone. The longitudinal salinity gradient zone, which is about 50 mi (80 km) long, is the chief feature of the salinity distribution. The freshwater zone occupies the upper portion of the river section.

A surface chlorinity profile for the Macleay river taken on 27 January 1948 is shown in Fig. 11-13. The absence of any marine or tidal zone is evident from the salinity distribution, and the domination of the freshwater zone is evident from the steepness of the longitudinal salinity gradient zone even during this period of low river discharge.

The mean annual chlorinity profiles for the surface and bottom waters of the Swan river are shown in Fig. 11-14. Here there is no marine or

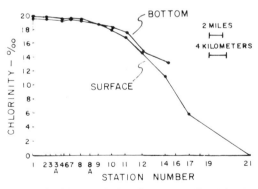

Fig. 11-11   Reproduced with permission from D. J. Rochford, 1951, *Australian Journal of Marine and Freshwater Research*, **2,** 1–116, Figs. 18 and 20.

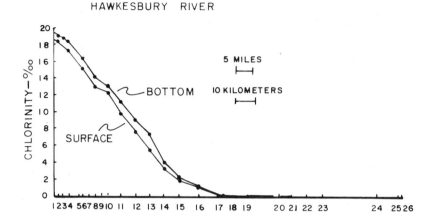

Fig. 11-12   Reproduced with permission from D. J. Rochford, 1951, *Australian Journal of Marine and Freshwater Research*, **2**, 1–116, Figs. 23 and 25.

tidal zone associated with the surface waters, and the longitudinal salinity gradient zone extends over the entire length of the section, gradually changing into what could be referred to as a freshwater zone. The first four examples are located in eastern or southeastern Australia; this last example of the Swan river is located in southwestern Australia.

Fischer (1972a) applied a two dimensional, longitudinal and lateral, time dependent computer model to an examination of possible pollutant distributions in Botany bay. An influx of fresh water from the George river was included, and an efflux of mixed water out the entrance of Botany bay to the Pacific ocean. The vertically integrated motion and

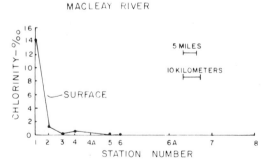

Fig. 11-13   Reproduced with permission from D. J. Rochford, 1951, *Australian Journal of Marine and Freshwater Research*, **2**, 1–116, Fig. 27.

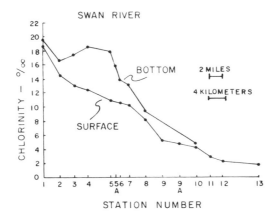

Fig. 11-14   Reproduced with permission from D. J. Rochford, 1951, *Australian Journal of Marine and Freshwater Research*, **2,** 1–116, Figs. 31 and 33.

pollutant concentration equations were used in the numerical calculations. The calculations did not predict the observed salinity distribution in the bay, and this may be related to the values of the longitudinal and lateral diffusion coefficient values used in the calculations of $K_x = K_y = 10^4$ cm^2 sec^{-1}, which are considerably smaller than the values usually associated with tidal mixing and exchange phenomena.

11-5. RIVERS OF JAPAN

Many of the rivers of Japan come down to the coast with a relatively small change in cross sectional area, thus maintaining a rather substantial runoff flow per unit breadth to the river mouth. In addition the tides are often small, particularly for rivers coming down to the Sea of Japan. It is not uncommon then to have a salt wedge condition in the river near its mouth, with a plume of brackish water extending offshore.

Takano (1954) applied the theory he developed for the characteristics of a brackish water plume extending out from a river mouth, which was discussed in Section 4-10, to the river entering Port Ajiro. The measurements were of the track of a surface buoy in its passage along the center line of the plume. The measurements of the buoy position were taken every 2 minutes, and the successive locations are shown in Fig. 11-15. The longitudinal velocity v_x was calculated from the velocity potential relation (4-197) and is plotted in Fig. 11-16 in terms of the scaled distance ξ of (4-200). The comparison of the theoretical curve with the observed velocities is good over the range interval of the observations.

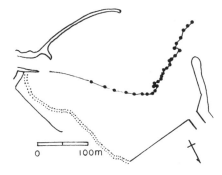

Fig. 11-15 Reproduced with permission from K. Takano, 1954, *Journal of the Oceanographical Society of Japan*, **10**, 60–64, Fig. 4.

Unfortunately the measurements do not extend back toward the river mouth into the range of the steep velocity gradient.

Fukushima et al. (1964, 1966, 1969) made one of the most interesting investigations of the characteristics of a salt wedge. They obtained high frequency reflections in the form of a continuous profiling record of the salt wedge interface in much the same manner as one obtains a continuous record of the bottom with a conventional echo sounder. The profiling frequency was $100-200$ kc sec^{-1}. The reader is encouraged to consult these references for illustrations of these excellent, interesting, and informative records. An example is shown in Fig. 11-17.

All the observations were taken in the Ishikari river, the location and physical characteristics of which are shown in Fig. 11-18. The Ishikari river has a normal discharge of $300-500$ m^3 sec^{-1}. The tidal range at the river mouth in the Sea of Japan is quite small, with a yearly maximum of 30 cm.

Examples of the salinity and longitudinal current velocity structure at the time of the measurements of Fig. 11-17 are shown in Fig. 11-19. The river discharge rate at this time was at a seasonal low value of

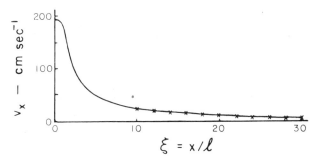

Fig. 11-16 Reproduced with permission from K. Takano, 1954, *Journal of the Oceanographical Society of Japan*, **10**, 60–64, Fig. 6.

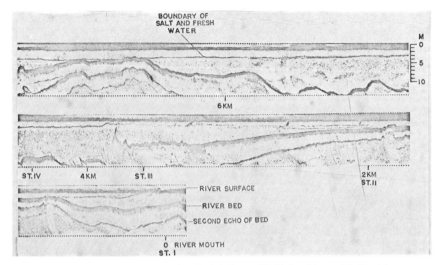

Fig. 11-17 Reproduced with permission from H. Fukushima, M. Kashiwamura, and I. Yakuwa, 1966, *American Society of Civil Engineers, Proceedings of the Tenth Conference on Coastal Engineering*, pp. 1435–1447, Fig. 10.

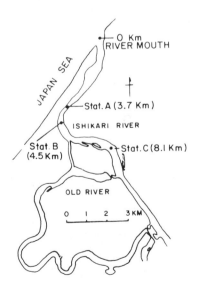

Fig. 11-18 Reproduced with permission from H. Fukushima, I. Yakuwa, and S. Takahashi, 1969, *International Association for Hydraulic Research, Proceedings of the Thirteenth Congress*, **3**, 191–197, Fig. 1B.

435

STATION I — AT THE RIVER MOUTH
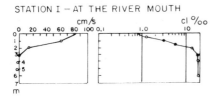

STATION Ⅱ – 2.0 KM UPSTREAM FROM THE MOUTH
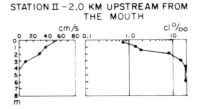

STATION Ⅲ – 3.7 KM UPSTREAM FROM THE MOUTH
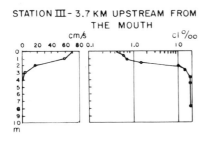

STATION Ⅳ – 4.5 KM UPSTREAM FROM THE MOUTH
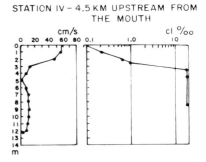

Fig. 11-19 Reproduced with permission from H. Fukushima, M. Kashiwamura, and I. Yakuwa, 1966, *American Society of Civil Engineers, Proceedings of the Tenth Conference on Coastal Engineering*, pp. 1435–1447, Fig. 12.

$190 \, m^3 \, sec^{-1}$, and the salt wedge extends for a considerable distance upriver. It is to be noted from this figure that the salt interface is quite sharp, particularly in an upriver direction, with a salinity change of several parts per thousand over a depth interval of a fraction of a meter. It is also to be noted in Fig. 11-17 that, whereas the interface, as shown on the echo sounder records, is quite sharp upriver from the bottom projection at 3.7 km, it shows a jump and becomes more diffuse immediately downriver from the projection, gradually regaining its distinctive characteristics as the river mouth is approached. The comment is also made that this diffuse characteristic on the record is indicative of entrainment mixing effects across the interface associated with flow over the bottom projection, which seems to be a reasonable conclusion.

For the measurements taken on 4 June 1964, when the river discharge rate was $720 \, m^3 \, sec^{-1}$, the salt wedge did not penetrate into the river mouth. On 5 June 1965, when the discharge rate was $550 \, m^3 \, sec^{-1}$, the salt wedge formed a tongue penetrating 3 km upriver; and on 3 August 1967, when the discharge rate was $190 \, m^3 \, sec^{-1}$, the salt wedge penetrated a distance of 13 km upriver. The profiles of the interface for the latter two cases are shown in Fig. 11-20. The comment is also made that, when the salt wedge begins to decay from an upriver extension as the

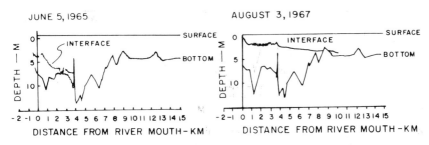

Fig. 11-20   Reproduced with permission from H. Fukushima, I. Yakuwa, and S. Takahashi, 1969, *International Association for Hydraulic Research, Proceedings of the Thirteenth Congress,* **3,** 191–197, Fig. 3.

result of an increase in river discharge, the interface as shown on the profiling records becomes diffusive.

11-6. OSAKA AND ARIAKE BAYS

Wada (1966, 1969) investigated analytically the cooling water circulation for a 2000 MW electric power generating station in Osaka bay. The cooling water flow in this instance is substantial and amounts to 100 m^3 sec^{-1}. The investigation provides an interesting and careful application of physical oceanographic principles and procedures to an engineering problem.

The problem is relatively straightforward, involving a surface outfall of known dimensions and a flow of known amount and initial temperature rise into an open bay. The solution to the problem is neither simple nor straightforward. The motion equations used included the viscosity terms and the horizontal pressure gradient terms resulting from the horizontal temperature gradients from the outfall. In this regard the motion equations used are quite different from those often considered for the near field effect from a high velocity jet flow from an outfall, in that the inertial terms were neglected. The heat continuity equation included the usual advection and diffusion terms and a surface heat loss term, and the mass continuity equation was taken in its usual form. The surface heat loss term was taken to be proportional to the temperature rise above ambient temperature; and the density was taken to be given in the form $\rho = \rho_0(1 - \alpha T)$, or the density gradient to be proportional to the temperature gradient, both approximations being the ones usually taken for such calculations.

The method of solution is in a sense one of successive approximations of either the motion or heat continuity equation under certain simplifying assumptions and then the substitution of the solution into the defining terms for the other equation. The waters of Osaka bay have a natural thermal stratification. In the summer the thermocline is located at a depth of 3–4 m, with a temperature difference from surface to bottom of 4–5°C. In consideration of the buoyant surface plume of warm water from the outfall the assumption is made that the thermocline represents a surface across which there is neither mixing nor friction. The motion equation is then solved in two dimensional, horizontal form with a computer program using much the same formulation procedure as discussed in Section 4-10, with given source and boundary conditions and neglecting the horizontal pressure gradient terms.

The heat continuity equation is then solved, also with a computer program, using these determined values for v_x and v_y. The eddy diffusion coefficients K_x and K_y used in these calculations were taken from values determined from observations of the outfall at another electric power generating station in Mizushima bay. The values determined there were $K_x = K_y = 0.5 \times 10^4$ cm^2 sec^{-1} and $K_z = 100$ cm^2 sec^{-1}. These values for K_x and K_y are considerably smaller than the values usually associated with tidal mixing phenomena but are in order with the scale of the temperature distribution observed, as shown in Fig. 11-21. Again this points out the importance of the scale of the process being observed in consideration of the effective diffusion coefficients that are appropriate. Finally, the temperature gradients from these calculations were used in the motion equation, including the vertical viscosity term and assuming a ratio of $N_z/N_x = 0.1$ to redetermine the velocity distribution.

Higuchi (1967) investigated the mixing characteristics of Ariake bay. Using the formulation of (5-116), (5-118), and (5-115) he determined a longitudinal dispersion coefficient of $K_x = 6.7 \times 10^6$ cm^2 sec^{-1} from the observed salinity distribution in the bay created by the freshwater flow from the adjoining Chikugo river. From dye patch releases in the bay he

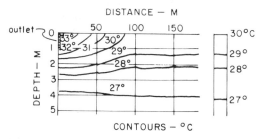

Fig. 11-21   Reproduced with permission from A. Wada, 1966, *American Society of Civil Engineers, Proceedings of the Tenth Conference on Coastal Engineering,* pp. 1388–1411, Fig. 3.

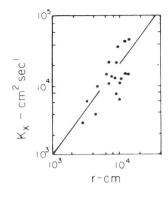

Fig. 11-22 Reproduced with permission from H. Higuchi, 1967, *International Association for Hydraulic Research, Proceedings of the Twelfth Congress,* **4,** 79–88, Fig. 6.

determined effective diffusion coefficients for this smaller scale process, assuming that a relation of the form (2-68) applied to the observed distribution. From observations of the spread of the rim of the dye patch, values of K_x were determined from this form at successive time intervals as the dye patch spread. The calculated results for this form of K_x as a function of the radius of the dye patch are shown in Fig. 11-22. The line plotted in this figure has a slope of $\frac{4}{3}$. Although this formulation and methodology are not exactly the same as those given by the neighbor separation theory of Section 2-6, it is similar and shows the same $\frac{4}{3}$ power dependence on the separation distance.

Iwai et al. (1969) continued the discussion of mixing and possible pollution effects in Ariake bay. Solutions are given for various pollutant source distributions, assuming the applicability of the Fickian diffusion equation, and the results from additional dye experiments with formulations similar to that given above and shown in Fig. 11-22 are described.

Bibliography

Abbott, M. R., 1960, Boundary layer effects in estuaries, *Journal of Marine Research*, **18,** 83–100.

Abbott, M. R., 1960a, Salinity effects in estuaries, *Journal of Marine Research*, **18,** 101–111.

Allersma, E., A. J. Hoekstra, and E. W. Byker, 1966, Transport patterns in the Chao Phya estuary, *American Society of Civil Engineers, Proceedings of the Tenth Conference on Coastal Engineering*, **1,** 632–650.

Anderson, J. J. and A. H. Devol, 1973, Deep water renewal in Saanich inlet, an intermittently anoxic basin, *Estuarine and Coastal Marine Science*, **1,** 1–10.

Aris, R., 1956, On the dispersion of a solute in a fluid flowing through a tube, *Proceedings of the Royal Society of London*, **A235,** 67–77.

Arons, A. B. and H. Stommel, 1951, A mixing length theory of tidal flushing, *Transactions of the American Geophysical Union*, **32,** 419–421.

Bailey, T. E., 1966, Fluorescent tracer studies of an estuary, *Journal of the Water Pollution Control Federation*, **38,** 1986–2001.

Bassindale, R., 1943, Studies on the biology of the Bristol channel, *Journal of Ecology*, **31,** 1–29.

Batchelor, G. K., 1949, Diffusion in a field of homogeneous turbulence, *Australian Journal of Scientific Research*, **A2,** 437–450.

Beverage, J. P. and M. N. Swecker, 1969, Estuarine studies in Upper Grays harbor, Washington, Water Supply Paper 1873-B, United States Geological Survey, Washington, D.C., pp. 1–90.

Bousfield, E. L., 1955, Some physical features of the Miramichi estuary, *Journal of the Fisheries Research Board of Canada*, **12,** 342–361.

Bowden, K. F., 1947, Some observations of waves and other fluctuations in a tidal current, *Proceedings of the Royal Society of London*, **A192,** 403–425.

Bowden, K. F., 1948, The processes of heating and cooling in a section of the Irish sea, *Monthly Notices of the Royal Astronomical Society, Geophysics Supplement*, **5,** 270–281.

Bowden, K. F., 1950, Processes affecting the salinity of the Irish sea, *Monthly Notices of the Royal Astronomical Society, Geophysics Supplement*, **6,** 63–87.

Bowden, K. F., 1960, Circulation and mixing in the Mersey estuary, *Proceedings of the International Association of Scientific Hydrology*, **51,** 352–360.

Bowden, K. F., 1962, Turbulence, in *The Sea*, Vol. I, John Wiley, New York, pp. 802–825.

Bowden, K. F., 1962a, The mixing processes in a tidal estuary, Proceedings of the First International Conference on Water Pollution Research, in *Advances in Water Pollution Research*, Vol. 3, Pergamon Press, London, pp. 329–346.

Bowden, K. F., 1962b, Measurements of turbulence near the sea bed in a tidal current, *Journal of Geophysical Research*, **67,** 3181–3186.

Bowden, K. F., 1963, The mixing processes in a tidal estuary, *International Journal of Air and Water Pollution*, **7,** 343–356.

Bowden, K. F., 1965, Horizontal mixing in the sea due to a shearing current, *Journal of Fluid Mechanics*, **21,** 83–95.

Bowden, K. F., 1967, Circulation and diffusion, in *Estuaries*, Publication No. 83, American Association for the Advancement of Science, Washington, D.C., pp. 15–36.

Bowden, K. F. and L. A. Fairbairn, 1952, Further observations of the turbulent fluctuations in a tidal current, *Philosophical Transactions of the Royal Society of London*, **A244**, 335–356.

Bowden, K. F. and L. A. Fairbairn, 1952a, A determination of the frictional forces in a tidal current, *Proceedings of the Royal Society of London*, **A214**, 371–392.

Bowden, K. F. and L. A. Fairbairn, 1956, Measurements of turbulent fluctuations and Reynolds stresses in a tidal current, *Proceedings of the Royal Society of London*, **A237**, 422–438.

Bowden, K. F. and R. M. Gilligan, 1971, Characteristic features of estuarine circulation as represented in the Mersey estuary, *Limnology and Oceanography*, **16**, 490–502.

Bowden, K. F. and M. R. Howe, 1963, Observations of turbulence in a tidal current, *Journal of Fluid Mechanics*, **17**, 271–284.

Bowden, K. F. and J. Proudman, 1949, Observations on the turbulent fluctuations of a tidal current, *Proceedings of the Royal Society of London*, **A199**, 311–327.

Bowden, K. F. and S. H. Sharaf El Din, 1966, Circulation, salinity and river discharge in the Mersey estuary, *Geophysical Journal of the Royal Astronomical Society*, **10**, 383–399.

Bowden, K. F. and S. H. Sharaf El Din, 1966a, Circulation and mixing processes in the Liverpool bay area of the Irish sea, *Geophysical Journal of the Royal Astronomical Society*, **11**, 279–292.

Bowden, K. F., L. A. Fairbairn, and P. Hughes, 1959, The distribution of shearing stresses in a tidal current, *Geophysical Journal of the Royal Astronomical Society*, **2**, 288–305.

Boyle, E., R. Collier, A. T. Dengler, J. M. Edmond, A. C. Ng, and R. F. Stallard, 1974, On the chemical mass balance in estuaries, *Geochimica et Cosmochimica Acta*, **38**, 1719–1728.

Burt, W. V. and L. D. Marriage, 1957, Computation of pollution in the Yaquina river estuary, *Sewage and Industrial Wastes*, **29**, 1385–1389.

Cameron, W. M., 1951, On the transverse forces in a British Columbia inlet, *Proceedings of the Royal Society of Canada*, **45**(5), 1–8.

Cameron, W. M., 1951a, On the dynamics of inlet circulation, Doctoral dissertation, University of California, Los Angeles.

Cameron, W. M. and D. W. Pritchard, 1965, Estuaries, in *The Sea*, Vol. II, John Wiley, New York, pp. 306–324.

Cannon, G. A., 1971, Statistical characteristics of velocity fluctuations at intermediate scales in a coastal plain estuary, *Journal of Geophysical Research*, **76**, 5852–5858.

Carslaw, H. S. and J. C. Jaeger, 1959, *Conduction of Heat in Solids*, Oxford University Press, London.

Carstens, T., 1970, Turbulent diffusion and entrainment in two layer flow, *American Society of Civil Engineers, Journal of the Waterways and Harbor Division*, **96**, WW1, 97–104.

Conomos, T. J. and D. H. Peterson, 1974, Biological and chemical aspects of the San Francisco bay turbidity maximum, *Mémoires de l'Institut de Geologie du Bassin d'Aquitaine*, **7**, 45–52.

Cromwell, T. and J. L. Reid, 1956, A study of oceanic fronts, *Tellus*, **8**, 94–101.

Csanady, G. T., 1973, *Turbulent Diffusion in the Environment*, D. Reidel, Dordrecht, The Netherlands.

Darbyshire, J. and A. Edwards, 1972, Seasonal formation and movement of the thermocline in lakes, *Pure and Applied Geophysics*, **93**, 141–150.

Darbyshire, J. and G. L. Jones, 1974, A further note on the seasonal formation and movement of the thermocline in lakes, *Pure and Applied Geophysics*, **112**, 955–966.

Dawson, W. A. and L. J. Tilley, 1972, Measurement of salt wedge excursion distance in the

Duwamish river estuary, Seattle, Washington, by means of the dissolved oxygen gradient, Water Supply Paper 1873-D, United States Geological Survey, Washington, D.C., pp. 1–27.

Dean, R. G. and R. B. Taylor, 1972, Numerical modeling of constituent transport in bay systems, *American Society of Civil Engineers, Proceedings of the Thirteenth Conference on Coastal Engineering*, pp. 2227–2249.

Defant, A., 1930, Die bewegungen und der thermo-haline aufbau der wassermassen im meeresstrassen, *Sitzungsberichte der Preussischen Akademie der Wissenschaften, Physikalisch-Mathematische Klasse* (Berlin), pp. 191–208.

Defant, A., 1960, *Physical Oceanography*, Vol. II, Pergamon Press, New York.

Defant, A., 1961, *Physical Oceanography*, Vol. I, Pergamon Press, New York.

Dietrich, G., 1954, Einfluss der gezeitenstromturbulenz auf die hydrographische schichtung der Nordsee, *Archiv für Meteorologie, Geophysik, und Bioklimatologie*, **7A**, 391–405.

Dietrich, G., 1957, *General Oceanography*, John Wiley, New York.

Dorrestein, R., 1960, A method of computing the spreading of matter in the water of an estuary, *Disposal of Radioactive Wastes (International Atomic Energy Agency)*, **2**, 163–166.

Dorrestein, R. and L. Otto, 1960, On the mixing and flushing of the water in the Ems estuary, *Verhandelingen Nederlandsch Geologisch Mijbouwkundig Genootschap, Geological Series*, D1, **19**, 83–102.

Dronkers, J. J., 1964, *Tidal Computations in Rivers and Coastal Waters*, North-Holland, Amsterdam.

Duke, C. M., 1961, Shoaling of the lower Hudson river, *American Society of Civil Engineers, Journal of the Waterways and Harbor Division*, **87**, WW1, 29–45.

Dyer, K. R., 1972, Sedimentation in estuaries, in *The Estuarine Environment*, Applied Science Publishers, London, pp. 10–32.

Dyer, K. R., 1973, *Estuaries: A Physical Introduction*, John Wiley, London.

Dyer, K. R., 1974, The salt balance in stratified estuaries, *Estuarine and Coastal Marine Science*, **2**, 273–281.

Dyer, K. R. and H. L. King, 1975, The residual water flow through the Solent, South England, *Geophysical Journal of the Royal Astronomical Society*, **42**, 97–106.

Dyer, K. R. and K. Ramamoorthy, 1969, Salinity and water circulation in the Vellar estuary, *Limnology and Oceanography*, **14**, 4–15.

Dyer, K. R. and P. A. Taylor, 1973, A simple segmented prism model of tidal mixing in well-mixed estuaries, *Estuarine and Coastal Marine Science*, **1**, 411–418.

Eggink, H. J., 1966, Predicted effects of future discharges of industrial wastes into the Eems estuary, *Journal of the Water Pollution Control Federation*, **38**, 373–374.

Einstein, H. A. and R. B. Krone, 1961, Estuarial sediment transport processes, *American Society of Civil Engineers, Journal of the Hydraulics Division*, **87**, HY2, 51–59.

Einstein, H. A. and R. B. Krone, 1962, Experiments to determine modes of cohesive sediment transport in salt water, *Journal of Geophysical Research*, **67**, 1451–1461.

Ekman, V. W., 1905, On the influence of the earth's rotation on ocean currents, Arkiv foer Matematik, Astronomi och Fysik (Stockholm), **211**, 1–53.

Eldridge, C. F. and G. T. Orlob, 1951, Investigation of pollution of Port Gardner bay and Snohomish river estuary, *Sewage and Industrial Wastes*, **23**, 782–795.

Farmer, H. G. and G. W. Morgan, 1953, The salt wedge, *American Society of Civil Engineers, Proceedings of the Third Conference on Coastal Engineering*, pp. 54–64.

Fischer, H. B., 1972, Mass transport mechanisms in partially stratified estuaries, *Journal of Fluid Mechanics*, **53**, 671–687.

Fischer, H. B., 1972a, A numerical model of estuarine pollutant transport, *American Society of Civil Engineers, Proceedings of the Thirteenth Conference on Coastal Engineering*, pp. 2265–2274.

Fisher, J. S., J. D. Ditmars, and D. R. F. Harleman, 1972, Analytical modeling of estuarine circulation, *American Society of Civil Engineers, Proceedings of the Thirteenth Conference on Coastal Engineering*, pp. 2307–2326.

Fjeldstat, J. E., 1933, Wärmleitung im meere, *Geofysiske Publikasjoner*, **10**(7), 1–20.

Francis, J. R. D. and H. Stommel, 1953, How much does a gale mix the surface layers of the ocean, *Quarterly Journal of the Royal Meteorological Society*, **79**, 534–536.

Francis, J. R. D., H. Stommel, H. G. Farmer, and D. Parson, 1953, Observations of turbulent mixing processes in a tidal estuary, Reference No. 53-22, Woods Hole Oceanographic Institution, Woods Hole, Mass.

Frenkiel, F. N. and P. A. Sheppard, Eds., 1959, *Atmospheric Diffusion and Air Pollution*, Academic Press, New York.

Fukushima, H., M. Kashiwamura, I. Yakuwa, and S. Takahashi, 1964, A study on salt water wedge at river mouth by ultrasonic method, *Coastal Engineering in Japan*, **7**, 101–107.

Fukushima, H., M. Kashiwamura, and I. Yakuwa, 1966, Studies on salt wedge by ultrasonic method, *American Society of Civil Engineers, Proceedings of the Tenth Conference on Coastal Engineering*, pp. 1435–1447.

Fukushima, H., I. Yakuwa, and S. Takahashi, 1969, Salinity diffusion at the interface of stratified flow in an estuary, *International Association for Hydraulic Research, Proceedings of the Thirteenth Congress*, **3**, 191–197.

Gade, H. G., 1963, Some hydrographic observations of the inner Oslofjord during 1959, *Det Norske Videnskaps-Akademi I Oslo, Hvalradets Skrifter*, **46**, 1–61.

Gameson, A. L. H., H. Hall, and W. S. Preddy, 1957, Effects of heated discharges on the temperature of the Thames estuary, *The Engineer*, **204**, 816–819, 850–852, 893–896.

Gameson, A. L. H., M. J. Barrett, and W. S. Preddy, 1964, Predicting the condition of a polluted estuary, Proceedings of the Second International Conference on Water Pollution, in *Advances in Water Pollution Research*, Vol. 3, Pergamon Press, London, pp. 167–192.

Garvine, R. W., 1974, Physical features of the Connecticut river outflow during high discharge. *Journal of Geophysical Research*, **79**, 831–846.

Garvine, R. W., 1974a, Dynamics of small scale oceanic fronts, *Journal of Physical Oceanography*, **4**, 557–569.

Garvine, R. W. and J. D. Monk, 1974, Frontal structure of a river plume, *Journal of Geophysical Research*, **79**, 2251–2259.

Gibbs, R. J., 1970, Circulation in the Amazon river estuary and adjacent Atlantic ocean, *Journal of Marine Research*, **28**, 113–123.

Glenne, B. and R. E. Selleck, 1969, Longitudinal estuarine diffusion in San Francisco bay, California, *Water Research*, **3**, 1–20.

Goldstein, S., Ed., 1938, Turbulence, in *Modern Developments in Fluid Dynamics*, Vol. I, Oxford University Press, London, pp. 191–233.

Gole, G. V. and V. S. Thakar, 1969, Progressive salinity intrusion during the dry season in the Hooghly estuary, *International Association for Hydraulic Research, Proceedings of the Thirteenth Congress*, **3**, 275–282.

Gordon, C. M. and C. F. Dohne, 1973, Some observations of turbulent flow in a tidal estuary, *Journal of Geophysical Research*, **78**, 1971–1978.

Grace, S. F., 1931, The influence of friction on the tidal motion of the Gulf of Suez, *Monthly Notices of the Royal Astronomical Society, Geophysics Supplement*, **2**, 309–318.

Grace, S. F., 1936, Friction in the tidal currents of the Bristol channel, *Monthly Notices of the Royal Astronomical Society, Geophysics Supplement*, **3**, 388–395.

Grant, H. L., R. W. Stewart, and A. Moilliet, 1962, Turbulence spectra from a tidal channel, *Journal of Fluid Mechanics,* **12,** 241–268.

Hachey, H. B., 1935, Tidal mixing in an estuary, *Journal of the Biology Board of Canada,* **1,** 171–178.

Hachey, H. B., 1939, Hydrographic features of the waters of Saint John harbour, *Journal of the Fisheries Research Board of Canada,* **4,** 424–440.

Haefner, P. A., 1967, Hydrography of the Penobscot river (Maine) estuary, *Journal of the Fisheries Research Board of Canada,* **24,** 1553–1571.

Hansen, D. V., 1965, Currents and mixing in the Columbia river estuary, *Marine Technology Society and American Society of Limnology and Oceanography, Joint Conference on Ocean Science and Ocean Engineering,* pp. 943–955.

Hansen, D. V., 1967, Salt balance and circulation in partially mixed estuaries, in *Estuaries,* Publication No. 83, American Association for the Advancement of Science, Washington D.C., pp. 45–51.

Hansen, D. V. and J. F. Festa, 1974, Inlet circulation induced by mixing of stratified water masses, *International Council for the Exploration of the Sea, Rapports,* **167,** 163–170.

Hansen, D. V. and M. Rattray, 1965, Gravitational circulation in straits and estuaries, *Journal of Marine Research,* **23,** 104–122.

Hansen, D. V. and M. Rattray, 1966, New dimensions in estuary classification, *Limnology and Oceanography,* **11,** 319–326.

Hansen, D. V. and M. Rattray, 1972, Estuarine circulation induced by diffusion, *Journal of Marine Research,* **30,** 281–294.

Harleman, D. R. F., 1961, Stratified flow, in *Handbook of fluid dynamics,* McGraw-Hill, New York, 26:1–26:21.

Harleman, D. R. F., 1966, Diffusion processes in stratified flows, in *Estuary and Coastline Hydrodynamics,* McGraw-Hill, New York, pp. 575–597.

Harleman, D. R. F., 1966a, Pollution in estuaries, in *Estuary and Coastline Hydrodynamics,* McGraw-Hill, New York, pp. 630–647.

Harleman, D. R. F., 1969, Mechanics of condenser-water discharge from thermal-power plants, in *Engineering Aspects of Thermal Pollution,* Vanderbilt University Press, Nashville, Tenn., pp. 144–176.

Harris, J. W., 1965, Means and methods of inducing sediment deposition and removal, *Proceedings of the Federal Interagency Sedimentation Conference,* Miscellaneous Publication 970, United States Department of Agriculture, Washington, D.C., pp. 669–674.

Heisenberg, W., 1948, Zur statischen theorie der turbulenz, *Zeitschrift für Physik,* **124,** 628–657.

Herlinveaux, R. H., 1954, Surface tidal currents in Juan de Fuca strait, *Journal of the Fisheries Research Board of Canada,* **11,** 14–31.

Herlinveaux, R. H., 1954a, Tidal currents in Juan de Fuca strait, *Journal of the Fisheries Research Board of Canada,* **11,** 799–815.

Herlinveaux, R. H., 1962, Oceanography of Saanich inlet in Vancouver island, British Columbia, *Journal of the Fisheries Research Board of Canada,* **19,** 1–37.

Herlinveaux, R. H. and J. P. Tully, 1961, Some oceanographic features of Juan de Fuca strait, *Journal of the Fisheries Research Board of Canada,* **18,** 1027–1071.

Hetling, L. J. and R. L. O'Connell, 1965, Estimating diffusion characteristics of tidal waters, *Water and Sewage Works,* **110,** 378–380.

Hetling, L. J. and R. L. O'Connell, 1966, A study of tidal dispersion in the Potomac river, *Water Resources Research,* **2,** 825–841.

Higuchi, H., 1967, Hydraulic model experiment on the diffusion due to the tidal current,

International Association for Hydraulic Research, Proceedings of the Twelfth Congress, **4,** 79–88.

Hinze, J. O., 1959, *Turbulence,* McGraw-Hill, New York.

Howard, C. S., 1940, Salt water intrusion in the Connecticut river, *Transactions of the American Geophysical Union,* **21,** 455–457.

✓Hughes, P., 1958, Tidal mixing in the narrows of the Mersey estuary, *Geophysical Journal of the Royal Astronomical Society,* **1,** 271–283.

Hunt, J. N., 1964, Tidal oscillations in estuaries, *Geophysical Journal of the Royal Astronomical Society,* **8,** 440–455.

Ichiye, T., 1951, Theory of oceanic turbulence, *Oceanographical Magazine,* **3,** 83–87.

Ichiye, T. and M. L. Jones, 1961, On the hydrography of the St. Andrew bay system, Florida, *Limnology and Oceanography,* **6,** 302–311.

Ichiye, T. and F. C. W. Olson, 1960, Über die neighbour diffusivity im ozean, *Deutsche Hydrographische Zeitschrift,* **13,** 13–23.

Ichiye, T., M. L. Jones, N. C. Hulings, and F. C. W. Olson, 1961, Salinity changes in Alligator harbor and Ochlockonee bay, Florida, *Journal of the Oceanographical Society of Japan,* **17,** 1–9.

Inglis, C. C. and F. H. Allen, 1957, The regimen of the Thames estuary as affected by currents, salinities, and river flows, *Proceedings of the Institution of Civil Engineers,* **7,** 827–868.

Ippen, A. T., Ed., 1966, *Estuary and Coastline Hydrodynamics,* McGraw-Hill, New York.

Ippen, A. T. and D. R. F. Harleman, 1966, Tidal dynamics in estuaries, in *Estuary and Coastline Hydrodynamics,* McGraw-Hill, New York, pp. 493–545.

Iselin, C. O'D., 1946, The application of oceanography to subsurface warfare, Summary Technical Report 6A, National Defense Research Committee, Washington, D.C.

Iwai, S., Y. Inoue, and H. Higuchi, 1969, Survey and prediction of pollution in the Omuta industrial harbor, Proceedings of the Fourth International Conference on Water Pollution Research, in *Advances in Water Pollution Research,* Pergamon Press, London, pp. 883–899.

Jacobsen, J. P., 1913, Beitrag zur hydrographie der dänischen gewasser, *Meddelelser Kommission Havundersogelser, Series Hydrographie* (Copenhagen), **2**(2), 1–94.

Jacobsen, J. P., 1918, Hydrographische intersuchungen in Randersfjord, *Meddelelser Kommission Havundersogelser, Series Hydrographie* (Copenhagen), **2**(7), 1–46.

Jacobsen, J. P., 1930, Remarks on the determination of the movement of the water and the intermixing of the watersheets in a vertical direction, *International Council for the Exploration of the Sea, Rapports,* **64,** 59–68.

Jakhellyn, A., 1936, The water transport of gradient currents, *Geofysiske Publikasjoner,* **11**(11), 1–14.

Joseph, von J. and H. Sendner, 1958, Über die horizontale diffusion im meere, *Deutsche Hydrographische Zeitschrift,* **11,** 49–77.

Kent, R., 1960, Diffusion in a sectionally homogeneous estuary. *American Society of Civil Engineers, Journal of the Sanitary Engineering Division,* **86,** SA2, 15–47.

Kent, R. E. and D. W. Pritchard, 1959, A test of mixing length theories in a coastal plain estuary, *Journal of Marine Research,* **18,** 62–72.

Ketchum, B. H., 1950, Hydrographic factors involved in the dispersion of pollutants introduced into tidal waters, *Journal of the Boston Society of Civil Engineers,* **37,** 296–314.

Ketchum, B. H., 1951, The exchange of fresh and salt waters in tidal estuaries, *Journal of Marine Research,* **10,** 18–37.

Ketchum, B. H., 1951a, The flushing of tidal estuaries, *Sewage and Industrial Wastes*, **23**, 198–209.

Ketchum, B. H., 1953, Circulation in estuaries, *American Society of Civil Engineers, Proceedings of the Third Conference on Coastal Engineering*, pp. 65–73.

Ketchum, B. H., 1955, Distribution of coliform bacteria and other pollutants in tidal estuaries, *Sewage and Industrial Wastes*, **27**, 1288–1296.

Ketchum, B. H. and D. J. Keen, 1953, The exchange of fresh and salt waters in the Bay of Fundy and in Passamaquoddy bay, *Journal of the Fisheries Research Board of Canada*, **10**, 97–124.

Ketchum, B. H., J. C. Ayers, and R. F. Vaccaro, 1952, Processes contributing to the decrease of coliform bacteria in a tidal estuary, *Ecology*, **33**, 247–258.

Keulegan, G. H., 1949, Interfacial instability and mixing in stratified flows, *Journal of Research, National Bureau of Standards*, **43**, 487–500.

Kirby, R. and W. R. Parker, 1973, Fluid mud in the Severn estuary and Bristol channel and its relevance to pollution studies, Paper presented at a Symposium on Estuarine and Coastal Pollution sponsored by the Institution of Chemical Engineers.

Kirby, R. and W. R. Parker, 1974, Seabed measurements related to echo sounder records, *Dock and Harbor Authority*, **54**, 423–424.

Kraus, E. B. and J. S. Turner, 1967, A one dimensional model of the seasonal thermocline, II, *Tellus*, **19**, 98–105.

Kullenberg, G., 1971, Vertical diffusion in shallow waters, *Tellus*, **23**, 129–135.

Kullenberg, G., 1972, Apparent horizontal diffusion in stratified vertical shear flow, *Tellus*, **24**, 17–28.

Kullenberg, G., 1974, Investigations on dispersion in stratified vertical shear flow, *International Council for the Exploration of the Sea, Rapports*, **167**, 86–92.

Kullenberg, G. and J. W. Talbot, Ed., 1974a, Physical processes responsible for dispersal of pollutants in the sea, *International Council for the Exploration of the Sea, Rapports*, **167**, 1–259.

Kuo, A. Y. and C. S. Fang, 1972, A mathematical model for salinity intrusion, *American Society of Civil Engineers, Proceedings of the Thirteenth Conference on Coastal Engineering*, pp. 2275–2289.

Lamar, W. L., 1940, Salinity of the lower Savannah river in relation to stream flow and tidal action, *Transactions of the American Geophysical Union*, **21**, 463–470.

Lamb, H., 1932, *Hydrodynamics*, Cambridge University Press, Cambridge.

Lauff, G. H., Ed., 1967, *Estuaries*, Publication No. 83, American Association for the Advancement of Science, Washington, D.C.

Lindsay, R. B., 1941, *Introduction to Physical Statistics*, John Wiley, New York.

Lofquist, K., 1960, Flow and stress near an interface between stratified liquids, *Physics of Fluids*, **3**, 138–175.

Macagno, E. O. and H. Rouse, 1962, Interfacial mixing in stratified flow, *Transactions of the American Society of Civil Engineers*, **127**, 102–128.

Masch, F. D. and N. J. Shankar, 1969, Mathematical simulation of two dimensional horizontal convective dispersion in well mixed estuaries, *International Association for Hydraulic Research, Proceedings of the Thirteenth Congress*, **3**, 293–301.

Mason, W. D. and W. H. Pietsch, 1940, Salinity movement and its causes in the Delaware river estuary, *Transactions of the American Geophysical Union*, **21**, 457–463.

McAlister, W. B., M. Rattray, and C. A. Barnes, 1959, The dynamics of a fjord estuary, Silver bay, Alaska, Technical Report 62, Department of Oceanography, University of Washington, Seattle.

McPherson, M. B., 1960, Diffusion in a sectionally homogeneous estuary, *American Society of Civil Engineers, Journal of the Sanitary Engineering Division,* **86,** SA5, 69–79.

Meade, R. H., 1966, Salinity variations in the Connecticut river, *Water Resources Research,* **2,** 567–579.

Melchoir, P. J., 1966, *The Earth Tides,* Pergamon Press, New York.

Munk, W. H. and E. R. Anderson, 1948, Notes on a theory of the thermocline, *Journal of Marine Research,* **7,** 276–295.

Neal, V. T., 1966, Predicted flushing times and pollution distribution in the Columbia river estuary, *American Society of Civil Engineers, Proceedings of the Tenth Conference on Coastal Engineering,* pp. 1463–1480.

Nelson, B. W., Ed., 1973, Environmental framework of coastal plain estuaries, *Geological Society of America, Memoir* 133.

Neu, H. A., 1969, Salinity variations, density currents, and silt transport in the Saint John estuary, *International Association for Hydraulic Research, Proceedings of the Thirteenth Congress,* **3,** 241–248.

Neumann, G. and W. J. Pierson, 1966, *Principles of Physical Oceanography,* Prentice-Hall, Englewood Cliffs.

Newcombe, C. L., W. A. Horne, and B. B. Shepherd, 1939, Studies on the physics and chemistry of estuarine waters in Chesapeake bay, *Journal of Marine Research,* **3,** 87–116.

Nichols, M. and G. Poor, 1967, Sediment transport in a coastal plain estuary, *American Society of Civil Engineers, Journal of the Waterways and Harbor Division,* **93,** WW4, 83–95.

O'Brien, M. P., 1952, Salinity currents in estuaries, *Transactions of the American Geophysical Union,* **33,** 520–522.

O'Connor, D. J., 1960, Oxygen balance of an estuary, *American Society of Civil Engineers, Journal of the Sanitary Engineering Division,* **86,** SA3, 35–55.

O'Connor, D. J., 1962, Organic pollution of New York harbor—Theoretical considerations, *Journal of the Water Pollution Control Federation,* **34,** 905–919.

O'Connor, D. J., 1965, Estuarine distribution of nonconservative substances, *American Society of Civil Engineers, Journal of the Sanitary Engineering Division,* **91,** SA1, 23–42.

O'Connor, D. J., 1966, An analysis of the dissolved oxygen distribution in the East river, *Journal of the Water Pollution Control Federation,* **38,** 1813–1830.

O'Connor, D. J., J. P. St. John, and D. M. Di Toro, 1968, Water quality analysis of the Delaware river estuary, *American Society of Civil Engineers, Journal of the Sanitary Engineering Division,* **94,** SA6, 1225–1252.

Officer, C. B., 1974. *Introduction to Theoretical Geophysics,* Springer-Verlag, New York.

Okubo, A., 1962, A review of theoretical models for turbulent diffusion in the sea, *Journal of the Oceanographical Society of Japan,* 20th Anniversary Volume, pp. 286–320.

Okubo, A., 1964, Equations describing the diffusion of an introduced pollutant in a one-dimensional estuary, in *Studies on oceanography,* University of Washington Press, Seattle, pp. 216–226.

Okubo, A., 1967, The effect of shear in an oscillatory current on horizontal diffusion from an instantaneous source, *International Journal of Oceanology and Limnology,* **1,** 194–204.

Okubo, A., 1968, Some remarks on the importance of the shear effect on horizontal diffusion, *Journal of the Oceanographical Society of Japan,* **24,** 60–69.

Okubo, A., 1973, Effect of shoreline irregularities on streamwise dispersion in estuaries and other embayments, *Netherlands Journal of Sea Research,* **6,** 213–224.

Olson, F. C. W., 1952, An empirical expression for horizontal turbulence, *Florida State University Studies*, No. 7.

Olson, F. C. W. and T. Ichiye, 1959, Horizontal diffusion, *Science*, **130**, 1255.

Orlob, G. T., 1961, Eddy diffusion in homogeneous turbulence, *Transactions of the American Society of Civil Engineers*, **126**, 397–438.

Orlob, G. T., R. Selleck, A. Yeiser, R. Walsh, and E. Stann, 1968, Modeling of water quality in an estuarial environment, *Water Research*, **2**, 122–124.

Orlob, G. T., R. Selleck, R. P. Shubinski, F. Walsh, and E. Stann, 1969 Modeling of water quality in an estuarial environment, Proceedings of the Fourth International Conference on Water Pollution Research, in *Advances in Water Pollution Research*, Pergamon Press, London, pp. 845–862.

Page, L., 1935, *Introduction to Theoretical Physics*, D. Van Nostrand, New York.

Palmer, M. D. and J. B. Izatt, 1971, Lake hourly dispersion estimates from a recording current meter, *Journal of Geophysical Research*, **76**, 688–693.

Pence, G. D., J. M. Jeglic, and R. V. Thomann, 1968, Time-varying dissolved-oxygen model, *American Society of Civil Engineers, Journal of the Sanitary Engineering Division*, **94**, SA2, 381–402.

Peterson, D. H., T. J. Conomos, W. W. Broenkow, and E. P. Scrivani, 1975, Processes controlling the dissolved silica distribution in San Francisco bay, in *Estuarine Research*, Vol. I, Academic Press, New York, pp. 153–187.

Peterson, D. H., T. J. Conomos, W. W. Broenkow, and P. C. Doherty, 1975a, Location of the nontidal current null zone in Northern San Francisco bay, *Estuarine and Coastal Marine Science*, **3**, 1–11.

Phelps, E. B., 1944, *Stream Sanitation*, John Wiley, New York.

Pickard, G. L., 1956, Physical features of British Columbia inlets, *Transactions of the Royal Society of Canada*, **50**, 47–58.

Pickard, G. L., 1961, Oceanographic features of inlets in the British Columbia mainland coast, *Journal of the Fisheries Research Board of Canada*, **18**, 907–999.

Pickard, G. L., 1963, Oceanographic characteristics of inlets of Vancouver island, British Columbia, *Journal of the Fisheries Research Board of Canada*, **20**, 1109–1144.

Pickard, G. L., 1967, Some oceanographic characteristics of the larger inlets of southeast Alaska, *Journal of the Fisheries Research Board of Canada*, **24**, 1475–1506.

Pickard, G. L., 1971, Some physical oceanographic features of inlets of Chile, *Journal of the Fisheries Research Board of Canada*, **28**, 1077–1106.

Pickard, G. L. and K. Rodgers, 1959, Current measurements in Knight inlet, British Columbia, *Journal of the Fisheries Research Board of Canada*, **16**, 635–678.

Pickard, G. L. and R. W. Trites, 1957, Fresh water transport determination from the heat budget with applications to British Columbia inlets, *Journal of the Fisheries Research Board of Canada*, **14**, 605–616.

Poincaré, H,, 1910, *Leçons de méchanique céleste*, Vol. 3, *Theorie des marées*, Gauthier-Villars, Paris.

Preddy, W. S., 1954, The mixing and movement of water in the estuary of the Thames, *Journal of the Marine Biological Association of the United Kingdom*, **33**, 645–662.

Preddy, W. S. and B. Webber, 1963, The calculation of pollution of the Thames estuary by a theory of quantized mixing, *International Journal of Air and Water Pollution*, **7**, 829–843.

Price, W. A. and M. P. Kendrick, 1963, Field and model investigation into the reasons for siltation in the Mersey estuary, *Proceedings of the Institution of Civil Engineers*, **24**, 473–518.

Pritchard, D. W., 1952, Estuarine hydrography, *Advances in Geophysics,* **1,** 243–280.

Pritchard, D. W., 1952a, Salinity distribution and circulation in the Chesapeake bay estuarine system, *Journal of Marine Research,* **11,** 106–123.

Pritchard, D. W., 1954, A study of the salt balance in a coastal plain estuary, *Journal of Marine Research,* **13,** 133–144.

Pritchard, D. W., 1956, The dynamic structure of a coastal plain estuary, *Journal of Marine Research,* **15,** 33-42.

Pritchard, D. W., 1957, Discussion on estimating streamflow into a tidal estuary, *Transactions of the American Geophysical Union,* **38,** 581–584.

Pritchard, D. W., 1967, Observations of circulation in coastal plain estuaries, in *Estuaries,* Publication No. 83, American Association for the Advancement of Science, Washington, D.C., pp. 37–44.

Pritchard, D. W., 1969, Dispersion and flushing of pollutants in estuaries, *American Society of Civil Engineers, Journal of the Hydraulics Division,* **95,** HY1, 115–124.

Pritchard, D. W. and J. H. Carpenter, 1960, Measurements of turbulent diffusion in estuarine and inshore waters, *Bulletin of the International Association of Scientific Hydrology,* **20,** 37–50.

Pritchard, D. W. and R. E. Kent, 1956, A method for determining mean longitudinal velocities in a coastal plain estuary, *Journal of Marine Research,* **15,** 81–91.

Proudman, J., 1925, Tides in a channel, *Philosophical Magazine,* **49,** 465.

Proudman, J., 1946, On the distribution of tides over a channel, *Proceedings of the London Mathematical Society,* **49,** 211.

Proudman, J., 1948, On the mixing of sea water by turbulence, *Proceedings of the Royal Society of London,* **A195,** 300–309.

Proudman, J., 1953, *Dynamical Oceanography,* Methuen, London.

Pyatt, E. E., 1964, On determining pollutant distribution in tidal estuaries, Water Supply Paper 1586-F, United States Geological Survey, Washington, D.C., pp. 1–56.

Ramming, H. G., 1972, Reproduction of physical processes in coastal areas, *American Society of Civil Engineers, Proceedings of the Thirteenth Conference on Coastal Engineering,* pp. 2207–2225.

Rao, G. V. and T. S. Murty, 1973, Some case studies of vertical circulations associated with oceanic fronts, *Journal of Geophysical Research,* **78,** 549–557.

Rattray, M., 1967, Some aspects of the dynamics of circulation in fjords, in *Estuaries,* Publication No. 83, American Association for the Advancement of Science, Washington, D.C., pp. 52–62.

Rattray, M. and D. V. Hansen, 1962, A similarity solution for circulation in an estuary, *Journal of Marine Research,* **20,** 121–133.

Redfield, A. C., 1950, The analysis of tidal phenomena in narrow embayments, *Papers in Physical Oceanography and Meteorology* (Massachusetts Institute of Technology and Woods Hole Oceanographic Institution), **11**(4), 1–36.

Reid, R. O. and B. R. Bodine, 1968, Numerical model for storm surges in Galveston bay, *American Society of Civil Engineers, Journal of the Waterways and Harbors Division,* **94,** WW1, 33–57.

Reynolds, O., 1894, On the dynamical theory of incompressible viscous fluids and the determination of the criterion, *Philosophical Transactions of the Royal Society of London,* **A186,** 123.

Rhodes, R. F., 1950, Effect of salinity on current velocities, Report No. 1, United States Army Corps of Engineers, Committee on Tidal Hydraulics, Washington, D.C.

Richardson, L. F., 1920, The supply of energy from and to atmospheric eddies, *Proceedings of the Royal Society of London,* **A97,** 354–373.

Richardson, L. F., 1926, Atmospheric diffusion shown on a distance-neighbor graph, *Proceedings of the Royal Society of London*, **A110,** 709–737.

Richardson, L. F. and H. Stommel, 1948, Note on eddy diffusion in the sea, *Journal of Meteorology*, **5,** 238–240.

Riley, G. A., 1952, Hydrography of the Long island and Block island sounds, *Bulletin of the Bingham Oceanographic Collection*, **13,** 5–39.

Riley, G. A., 1955, Oceanography of Long island sound, I, Introduction, *Bulletin of the Bingham Oceanographic Collection*, **15,** 9–14.

Riley, G. A., 1955a, Oceanography of Long island sound, II, Physical oceanography, *Bulletin of the Bingham Oceanographic Collection*, **15,** 15–46.

Riley, G. A., 1955b, Review of the oceanography of Long island sound, *Deep Sea Research*, **3**(Suppl.), 224–238.

Rochford, D. J., 1951, Studies in Australian estuarine hydrology, I, Introductory and comparative features, *Australian Journal of Marine and Freshwater Research*, **2,** 1–116.

Rossby, C. G. and R. B. Montgomery, 1935, The layer of frictional influence in wind and ocean currents, *Papers in Physical Oceanography and Meteorology* (Massachusetts Institute of Technology and Woods Hole Oceanographic Institution), **3**(3), 1–101.

Rouse, H., Ed., 1950, *Engineering Hydraulics*, John Wiley, New York.

Ryther, J. H., D. W. Menzel, and N. Corwin, 1967, Influence of the Amazon river outflow on the ecology of the western tropical Atlantic, I, Hydrography and nutrient chemistry, *Journal of Marine Research*, **25,** 69–83.

Saelen, O. H., 1967, Some features of the hydrography of Norwegian fjords, in *Estuaries*, Publication No. 83, American Association for the Advancement of Science, Washington, D.C., pp. 63–70.

Saffman, P. G., 1962, The effect of wind shear on horizontal spread from an instantaneous source, *Quarterly Journal of the Royal Meteorological Society*, **88,** 382–393.

Santos, J. F. and J. D. Stoner, 1972, Physical, chemical and biological aspects of the Duwamish river estuary, King county, Washington, 1963–67, Water Supply Paper 1873-C, United States Geological Survey, Washington, D.C., pp. 1–74.

Schijf, J. B. and J. C. Schonfeld, 1953, Theoretical considerations on the motion of salt and fresh water, *American Society of Civil Engineers and International Association for Hydraulic Research, Proceedings of the Minnesota International Hydraulics Convention*, pp. 321–333.

Schlichting, H., 1955, *Boundary Layer Theory*, McGraw-Hill, New York.

Schubel, J. R., 1969, Size distribution of the suspended particles of the Chesapeake bay turbidity maximum, *Netherlands Journal of Sea Research*, **4,** 283–309.

Schubel, J. R., 1971, Tidal variation of the size distribution of suspended sediment at a station in the Chesapeake bay turbidity maximum, *Netherlands Journal of Sea Research*, **5,** 252–266.

Simmons, H. B., 1955, Some effects of upland discharge on estuarine hydraulics, *Proceedings of the American Society of Civil Engineers*, **81,** 792, 1–20.

Simmons, H. B., 1966, Field experience in estuaries, in *Estuary and Coastline Hydrodynamics*, McGraw-Hill, New York, pp. 673–690.

Simmons, H. B. and F. R. Brown, 1969, Salinity effects on estuarine hydraulics and sedimentation, *International Association for Hydraulic Research, Proceedings of the Thirteenth Congress*, **3,** 311–325.

Stelzenmuller, W. B., 1965, Tidal characteristics of two estuaries in Florida, *American Society of Civil Engineers, Journal of the Waterways and Harbor Division*, **91,** WW3, 25–36.

Stewart, R. W., 1957, A note on the dynamic balance in estuarine circulation, *Journal of Marine Research*, **16**, 34–39.

Stoker, J. J., 1957, *Water Waves*, Interscience, New York.

Stommel, H., 1949, Horizontal diffusion due to oceanic turbulence, *Journal of Marine Research*, **8**, 199–225.

Stommel, H., 1951, Recent developments in the study of tidal estuaries, Reference No. 51-33, Woods Hole Oceanographic Institution, Woods Hole, Mass.

Stommel, H., 1953, Computation of pollution in a vertically mixed estuary, *Sewage and Industrial Wastes*, **25**, 1065–1071.

Stommel, H. and H. G. Farmer, 1952, Abrupt change in width in two layer open channel flow, *Journal of Marine Research*, **11**, 205–214.

Stommel, H. and H. G. Farmer, 1952a, On the nature of estuarine circulation, Part I, Reference 52-51, Woods Hole Oceanographic Institution, Woods Hole, Mass.

Stommel, H. and H. G. Farmer, 1952b, On the nature of estuarine circulation, Part I, Reference 52-88, Woods Hole Oceanographic Institution, Woods Hole, Mass.

Stommel, H. and H. G. Farmer, 1952c, On the nature of estuarine circulation, Part III, Reference No. 52-63, Woods Hole Oceanographic Institution, Woods Hole, Mass.

Stommel H. and H. G. Farmer, 1953, Control of salinity in an estuary by a transition, *Journal of Marine Research*, **12**, 13–20.

Stommel, H. and A. H. Woodcock, 1951, Diurnal heating of the surface of the Gulf of Mexico in the spring of 1942, *Transactions of the American Geophysical Union*, **32**, 565–571.

Stoner, J. D., 1972, Determination of mass balance and entrainment in the stratified Duwamish river estuary, King county, Washington, Water Supply Paper 1873-F, United States Geological Survey Washington, D.C., pp. 1–17.

Streeter, V. L., 1948, *Fluid Dynamics*, McGraw-Hill, New York.

Streeter, V. L., 1958, *Fluid Mechanics*, McGraw-Hill, New York.

Sutton, O. G., 1949, *Atmospheric Turbulence*, Methuen, London.

Sverdrup, H. V., M. W. Johnson, and R. H. Fleming, 1942, *The Oceans*, Prentice-Hall, New York.

Szekielda, K. H., S. L. Kupferman, V. Klemas, and D. F. Polis, 1972, Element enrichment in organic films and foam associated with aquatic frontal systems, *Journal of Geophysical Research*, **77**, 5278–5282.

Tabata, S. and G. L. Pickard, 1957, The physical oceanography of Bute inlet, British Columbia, *Journal of the Fisheries Research Board of Canada*, **14**, 487–520.

Takano, K., 1954, On the velocity distribution off the mouth of a river, *Journal of the Oceanographical Society of Japan*, **10**, 60–64.

Takano, K., 1954a, On the salinity and velocity distribution off the mouth of a river, *Journal of the Oceanographical Society of Japan*, **10**, 92–98.

Takano, K., 1955, A complementary note on the diffusion of the seaward river flow off the mouth, *Journal of the Oceanographical Society of Japan*, **11**, 147–149.

Talbot, J. W. and G. A. Talbot, 1974, Diffusion in shallow seas and in English coastal and estuarine waters, *International Council for the Exploration of the Sea, Rapports*, **167**, 93–110.

Taylor, G. I., 1919, Tidal friction in the Irish sea, *Philosophical Transactions of the Royal Society of London*, **A220**, 1–33.

Taylor, G. I., 1920, Tidal oscillations in gulfs and rectangular basins, *Proceedings of the London Mathematical Society*, **20**, 148–181.

Taylor, G. I., 1921, Diffusion by continuous movements, *Proceedings of the London Mathematical Society*, **20**, 196–212.

Taylor, G. I., 1921a, Tides in the Bristol channel, *Proceedings of the Cambridge Philosophical Society,* **20,** 320–325.

Taylor, G. I., 1931, Internal waves and turbulence in a fluid of variable density, *International Council for the Exploration of the Sea, Rapports,* **76,** 35–43.

Taylor, G. I., 1938, The spectrum of turbulence, *Proceedings of the Royal Society of London,* **A164,** 476–490.

Taylor, G. I., 1953, Dispersion of soluble matter in solvent flowing slowly through a tube, *Proceedings of the Royal Society of London,* **A219,** 186–203.

Taylor, G. I., 1954, The dispersion of matter in turbulent flow through a pipe, *Proceedings of the Royal Society of London,* **A223,** 446–468.

Thomann, R. V., 1963, Mathematical model for dissolved oxygen, *American Society of Civil Engineers, Journal of the Sanitary Engineering Division,* **89,** SA5, 1–30.

Thomann, R. V., 1965, Recent results from a mathematical model of water pollution control in the Delaware estuary, *Water Resources Research,* **1,** 349–359.

Thomann, R. V., 1967, Tine-series analyses of water-quality data, *American Society of Civil Engineers, Journal of the Sanitary Engineering Division,* **93,** SA1, 1–23.

Thomann, R. V., 1972, *Systems analysis and water quality management,* Environmental Research and Applications, New York.

Thomann, R. V. and M. J. Sobel, 1964, Estuarine water quality management and forecasting, *American Society of Civil Engineers, Journal of the Sanitary Engineering Division,* **90,** SA5, 9–36.

Thomann, R. V., D. J. O'Connor, and D. M. Di Toro, 1970, Modeling of the nitrogen and algal cycles in estuaries, Proceedings of the Fifth International Conference on Water Pollution Research, in *Advances in Water Pollution Research,* Vol. 3, Pergamon Press, London, pp. 9/1–9/14.

Trites, R. W., 1960, An oceanographical and biological reconnaissance of Kennebecasis bay and the Saint John river estuary, *Journal of the Fisheries Research Board of Canada,* **17,** 377–408.

Tully, J. P., 1949, Oceanography and prediction of pulp mill pollution in Alberni inlet, *Bulletin of the Fisheries Research Board of Canada,* No. 83, pp. 1–169.

Tully, J. P., 1958, On structure, entrainment, and transport in estuarine embayments, *Journal of Marine Research,* **17,** 523–535.

Tully, J. P., 1964, Oceanographic regions and processes in the seasonal zone of the North Pacific Ocean, in *Studies on Oceanography,* University of Washington Press, Seattle, pp. 68–84.

Tully, J. P. and A. J. Dodimead, 1957, Properties of the water in the Strait of Georgia, British Columbia, and influencing factors, *Journal of the Fisheries Research Board of Canada,* **14,** 241–319.

Turner, J. S. and E. B. Kraus, 1967, A one dimensional model of the seasonal thermocline, I, *Tellus,* **19,** 88–97.

de Turville, C. M. and R. T. Jarman, 1965, The mixing of warm water from the Uskmonth power station in the estuary of the river Usk, *International Journal of Air and Water Pollution,* **9,** 239–251.

Vreugdenhill, C. B., 1969, Numerical computation of fully stratified flow, *International Association for Hydraulic Research, Proceedings of the Thirteenth Congress,* **3,** 37–44.

Wada, A., 1966, A study on phenomena of flow and thermal diffusion caused by outfall of cooling water, *American Society of Civil Engineers, Proceedings of the Tenth Conference on Coastal Engineering,* pp. 1388–1411.

Wada, A., 1969, Numerical analysis of distribution of flow and thermal diffusion caused by outfall of cooling water, *International Association for Hydraulic Research, Proceedings of the Thirteenth Congress,* **3,** 335–342.

Waldichuk, M., 1957, Physical oceanography of the Strait of Georgia, British Columbia, *Journal of the Fisheries Research Board of Canada,* **14,** 321–486.

Waldichuk, M., 1958, Some oceanographic characteristics of a polluted inlet in British Columbia, *Journal of Marine Research,* **17,** 536–551.

Waldichuk, M., 1964, Estimation of flushing rates from tide heights and current data in an inshore marine channel of the Canadian Pacific coast, Proceedings of the Second International Conference on Water Pollution Research, in *Advances in Water Pollution Research,* Vol. 3, Pergamon Press, London, pp. 133–166.

Waldichuk, M., 1965, Water exchange in Port Moody, British Columbia, and its effect on waste disposal, *Journal of the Fisheries Research Board of Canada,* **22,** 801–822.

Waldichuk, M., 1966, Effects on sulfite wastes in a partially enclosed marine system in British Columbia, *Journal of the Water Pollution Control Federation,* **38,** 1484–1505.

Waldichuk, M., 1974, Application of oceanographic information to the design of sewer and industrial waste outfalls, *International Council for the Exploration of the Sea, Rapports,* **167,** 236–259.

Ward, G. H. and W. H. Espey, Ed., 1971, *Estuarine modeling, An assessment,* No. 16070 DZV 02/71, Environmental Protection Agency, Washington, D.C.

Watson, E. E., 1936, Mixing and residual currents in tidal waters as illustrated in the Bay of Fundy, *Journal of the Biology Board of Canada,* **2,** 141–208.

von Weizsäcker, C. F., 1948, Das spektrum der turbulenz bei grossen Reynoldsschen zahlen, *Zeitschrift für Physik,* **124,** 614–627.

West, J. R. and D. J. A. Williams, 1972, An evalution of mixing in the Tay estuary, *American Society of Civil Engineers, Proceedings of the Thirteenth Conference on Coastal Engineering,* pp. 2153–2169.

Winter, D. F., 1973, A similarity solution for steady state gravitational circulation in fjords, *Estuary and Coastal Marine Science,* **1,** 387–400.

Wright, L. D., 1971, Hydrography of South pass, Mississippi river, *American Society of Civil Engineers, Journal of the Waterways, Harbors and Coastal Engineering Division,* **97,** WW3, 491–504.

Wright, L. D. and J. M. Coleman, 1971, Effluent expansion and interfacial mixing in the presence of a salt wedge, Mississippi river delta, *Journal of Geophysical Research,* **76,** 8649–8661.

Author Index

Abbott, M. R., 132, 284, 285
Allen, F. H., 283
Allersma, E., 428
Anderson, E. R., 210, 213, 218, 225
Anderson, J. J., 398, 399, 400
Aris, R., 245
Arons, A. B., vi, 182, 191
Ayers, J. C., 343

Bailey, T. E., 414
Barnes, C. A., 388
Bassindale, R., 266
Batchelor, G. K., 30
Beverage, J. P., 411
Bodine, B. R., 378
Bousfield, E. L., 324
Bowden, K. F., vi, 53, 124, 178, 191, 201,
  214, 218, 240, 245, 254, 255, 256, 257,
  260, 262, 264, 265, 267, 282, 287, 288,
  289, 291, 292, 293, 295, 296, 309, 357
Boyle, E., 332, 334
Brown, F. R., 366
Burt, W. V., 411

Cameron, W. M., vi, 1, 172, 203, 350, 351,
  389, 392
Cannon, G. A., 355
Carpenter, J. H., 349, 350, 351
Carslaw, H. S., 26, 221, 223
Carstens, T., 303, 304, 306
Coleman, J. M., 378
Collier, R., 334
Conomos, T. J., 417
Cromwell, T., 139
Csanady, G. T., vi, 30, 245

Darbyshire, J., 222, 223, 225
Dawson, W. A., 407
Dean, R. G., 369
Defant, A., vi, 84, 112, 116, 144, 150, 151,
  154, 310, 311, 312, 313

Dengler, A. T., 334
Devol, A. H., 398, 399, 400
Dietrich, G., 225
Dodimead, A. J., 421
Dohne, C. F., 356, 357, 358
Dorrenstein, R., 310
Dronkers, J. J., 93, 96, 309
Duke, C. M., 335, 336
Dyer, K. R., vi, 162, 178, 203, 230, 235,
  271, 274, 275, 276, 277, 282, 351, 385,
  422, 423, 424, 425, 426, 427

Edmond, J. M., 334
Edwards, A., 222, 223
Eggink, H. J., 310
Einstein, H. A., 271, 272, 414
Ekman, V. W., 151
Eldridge, C. F., 407
Espey, W. H., vi

Fairbairn, L. A., 256, 257, 287, 288, 289,
  292, 293
Fang, C. S., 360
Farmer, H. G., vi, 97, 102, 106, 112, 162,
  163, 166, 171, 185, 191, 322, 324, 374,
  375, 377, 405
Festa, J. F., 127, 203, 207
Fischer, H. B., 203, 432
Fisher, J. S., 126, 365
Fjeldstat, J. E., 225
Francis, J. R. D., vi, 222, 325
Frenkiel, F. N., vi, 211
Fukushima, H., 434, 435, 436, 437

Gade, H. G., 306, 307, 308
Gameson, A. L. H., 280, 282
Garvine, R. W., 138, 139, 328, 329, 330,
  331, 332, 333
Gibbs, R. J., 379, 380
Gilligan, R. M., 214, 218, 264, 265, 309
Glenne, B., 413, 415, 416

Goldstein, S., 29
Gole, G. V., 428
Gordon, C. M., 356, 357, 358
Grace, S. F., 92, 268, 269
Grant, H. L., 401

Hachey, H. B., 322
Haefner, P. A., 325
Hansen, D. V., 3, 4, 125, 126, 127, 197,
    199, 201, 203, 207, 265, 365, 408, 410
Harleman, D. R. F., 93, 112, 240, 348, 382
Harris, J. W., 367, 368
Heisenberg, W., 34, 36
Herlinveaux, R. H., 398, 419, 421
Hetling, L. J., 352, 353, 354
Higuchi, H., 438, 439
Hinze, J. O., 29
Howard, C. S., 326
Howe, M. R., 256, 357
Hughes, P., 258, 292, 293
Hughlings, N. C., 372, 373
Hunt, J. N., 282, 283

Ichiye, T., 34, 36, 371, 372, 373, 374
Inglis, C. C., 283
Ippen, A. T., vi, 93, 348, 382
Iselin, C. O'D., 220, 225
Iwai, S., 439
Izatt, J. B., 191

Jacobsen, J. P., 123, 201, 211, 214, 218,
    307
Jaeger, J. C., 26, 221, 223
Jakhellyn, A., 144
Jarman, R. T., 268
Jeglic, J. M., 347
Jones, G. L., 222, 223, 225
Jones, M. L., 372, 373, 374
Joseph, von J., 240

Kashiwamura, M., 435, 436
Keen, D. J., 380, 381
Kendrick, M. P., 266
Kent, R. E., 240, 364
Ketchum, B. H., 158, 160, 162, 227, 230,
    231, 235, 341, 343, 380, 381, 405
Keulegan, G. H., 166, 167
King, H. L., 277
Kirby, R., 270, 272, 273, 274

Kraus, E. B., 222
Krone, R. B., 271, 272, 414
Kullenberg, G., vi, 314, 320, 321
Kuo, A. Y., 360

Lamar, W. L., 366
Lamb, H., vi, 53, 80
Lauff, G. H., vi
Lindsay, R. B., 27
Lofquist, K., 166

Macagno, E. O., 166
Marriage, L. D., 411
Masch, F. D., 379
Mason, W. D., 344
McAlister, W. B., 385, 388
McPherson, M. B., 240
Meade, R. H., 326, 327
Melchoir, P. J., 95
Monk, J. D., 138, 139, 328, 331, 332, 333
Montgomery, R. B., 58, 218
Morgan, G. W., 112, 375
Munk, W. H., 210, 213, 218, 225
Murty, T. S., 139

Neal, V. T., 411
Nelson, B. W., vi
Neu, H. A., 322, 323
Neumann, G., 151
Newcombe, C. L., 349
Ng, A. C., 334
Nichols, M., 357, 358, 359, 360

O'Brien, M. P., 409
O'Connell, R. L., 352, 353, 354
O'Connor, D. J., 235, 248, 250, 337, 339,
    340, 345, 365
Officer, C. B., vi
Okubo, A., 240, 245
Olson, F. C. W., 34, 36, 372, 373
Orlob, G. T., 240, 407, 417
Otto, L., 310

Page, L., vi, 40
Palmer, M. D., 191
Parker, W. R., 270, 272, 273, 274
Pence, G. D., 347
Peterson, D. H., 417
Phelps, E. B., 250

Pickard, G. L., 385, 386, 387, 389, 390, 391, 393, 394, 395, 396, 397, 418
Pierson, W. J., 151
Pietsch, W. H., 344
Poincaré, H., 82
Poor, G., 357, 358, 359, 360
Preddy, W. S., 191, 278, 280, 281
Price, W. A., 266
Pritchard, D. W., vi, 1, 144, 162, 203, 349, 350, 351, 360, 361, 362, 363, 364
Proudman, J., vi, 53, 80, 82, 84, 93, 144, 255, 295
Pyatt, E. E., 367

Ramamoorthy, K., 422, 423, 424, 425, 426, 427
Ramming, H. G., 309
Rao, G. V., 139
Rattray, M., 3, 4, 126, 127, 172, 197, 199, 201, 203, 207, 265, 365, 410
Redfield, A. C., 90, 91, 92, 93, 381, 382, 383, 384, 388, 418, 419, 420
Reid, J. L., 139
Reid, R. O., 378
Reynolds, O., 49
Rhodes, R. F., 375
Richardson, L. F., 33, 34, 36, 218
Riley, G. A., 383, 384
Rochford, D. J., 3, 429, 430, 431, 432, 433
Rodgers, K., 395, 396, 397
Rossby, C. G., 58, 218
Rouse, H., vi, 58, 166
Ryther, J. H., 380

Saelen, O. H., 298, 299, 300, 301, 303
Saffman, P. G., 245
Santos, J. F., 406
Schijf, J. B., 112
Schlichting, H., 58
Schonfeld, J. C., 112
Schubel, J. R., 351
Selleck, R. E., 413, 415, 416
Sendner, H., 240
Shankar, N. J., 379
Sharaf El Din, S. H., 262, 264
Sheppard, P. A., vi, 211
Simmons, H. B., 3, 336, 366, 367, 375, 414
Sobel, M. J., 347
Stallard, R. F., 334

Stelzenmuller, W. B., 370
Stewart, R. W., 144, 364
Stoker, J. J., 96
Stommel, H., vi, 34, 36, 97, 102, 106, 162, 163, 166, 171, 178, 182, 185, 191, 222, 229, 230, 235, 267, 322, 324, 374, 377, 405
Stoner, J. D., 406, 407
Streeter, V. L., vi
Sutton, O. G., 29
Sverdrup, H. V., 313
Swecker, M. N., 411
Szekielda, K. H., 139, 348

Tabata, S., 384, 385
Takahashi, S., 435, 437
Takano, K., 144, 148, 149, 433, 434
Talbot, G. A., 314, 315, 316, 317, 318, 320
Talbot, J. W., vi, 314, 315, 316, 317, 318, 320
Taylor, G. I., 29, 33, 36, 82, 93, 201, 213, 218, 269, 271, 294, 309
Taylor, P. A., 282
Taylor, R. B., 369
Thakar, V. S., 428
Thomann, R. V., vi, 235, 240, 250, 346, 347, 355, 417
Tilley, L. J., 407
Trites, R. W., 322, 391, 393, 394
Tully, J. P., 222, 402, 403, 404, 405, 421
Turner, J. S., 222
de Turville, C. M., 268

Vaccaro, R. F., 343
Vreugdenhill, C. B., 166

Wada, A., 437, 438
Waldichuk, M., 397, 400, 401, 421
Ward, G. H., vi
Watson, E. E., 380
Webber, B., 280
von Weizäcker, C. F., 34, 36
West, J. R., 285, 286
Williams, D. J. A., 285, 286
Winter, D. F., 127, 172
Woodcock, A. H., 222
Wright, L. D., 375, 376, 377, 378

Yakuwa, I., 435, 436, 437

Subject Index

Advection, definition, 21
Alaska inlets, 385-388
Alberni inlet, 402-406
Alligator harbor, 371-374
Amazon river, 379-380
Amphidromic region, 84
Ariake bay, 438, 439
Arrested salt wedge, circulation, 105-112
   definition, 2
Austausch coefficient, 24
Australian estuaries, 429-433

Bab el Mandeb, 310-313
Baltimore harbor, 203, 350
Baroclinic force, 125
Barotropic force, 125
Bay of Fundy, 380-382
Bernoulli equation, 14
Bilharmonic equation, 146
Biochemical oxygen demand, definition,
   226, 246
   in Delaware estuary, 345-347
   in James estuary, 248
   in New York harbor, 337-341
Biscayne bay, 369-370
Black sea, 313-314
Blackwater estuary, 318-319
Bores, 96
Bosporus, 310-313
Botany bay, 431, 432-433
Bottom friction, in Bristol channel, 269
   in Irish sea, 295
   in Red Wharf bay, 288, 289, 292
   in Solent, 277
   in Thames estuary, 283
Bottom friction coefficient, 54
Bondary layer, 53
Bristol channel, 266-274
British Columbia inlets, 389-
   397
Bute inlet, 394-395

Chao Phya estuary, 428
Charleston harbor, 367
Chesapeake bay, 349-360
Chezy coefficient, 69
Chile inlets, 418
Choptank estuary, 356-358
Circular frequency, 74
Circulation, in Alberni inlet, 402-406
   in Black sea, 313-314
   in Columbia river estuary, 408-411
   in Duwamish river, 406-407
   in Frierfjord, 303-304
   in Hardangerfjord, 298-302
   in James estuary, 360-365
   in Knight inlet, 395-397
   in Long Island sound, 382-384
   in Mediterranean sea, 313-314
   in Mersey estuary, 260-266
   in Miramichi estuary, 324-325
   in Nordfjord, 302-303
   in Oslofjord, 306-308
   in Rappahannock estuary, 357-360
   in Silver bay, 385-388
   in Snohomish river, 407-408
   in Southampton estuary, 274-277
   in Vellar estuary, 422-427
   lateral effects, 144-154
   longitudinal effects, 116-125
Coefficient, apparent eddy diffusion, 30
   austausch, 24
   Chezy, 69
   circulation, 198
   correlation, 29
   Darcy resistance, 69
   decay, 230
   diffusion, 22, 24, 26, 28, 30, 193, 200
   dispersion, 177
   exchange, 24
   neighbor separation, 33
   tidal damping, 88
   tidal mixing, 180, 181, 186, 190

viscosity, 43, 48, 50, 86, 87, 123, 124
Columbia river estuary, 408-411
Connecticut river, 326-353
Continuity, conservative pollutant equation,
    18
  mass equation, 16
  salt equation, 18, 26
Controls, 62, 98
  in Mississippi river, 374-375
Correlation coefficient, 29
Cotidal lines, 84
Coupled equations, 245-250
Critical velocity, 64, 100
Currents, bottom friction, 55
  drift, 151
  geostrophic, 42

Darcy resistance coefficient, 69
Dardanelles, 310-313
Decay coefficient, in New York Harbor, 338
Delaware estuary, 344-349
Density, in Dardanelles, 311
  in Mississippi river, 376
  in Oslofjord, 308
Density gradient, 117
Diffusion, coefficient, 22
  definition, 20
  eddy, 21
  equation, 24
  kinematic coefficient, 24, 26, 28, 30, 31,
      193, 200
  molecular, 21
  neighbor separation, 33
  scale effects, 31-36
  tidal mixing, 178-191
Diffusion coefficient, in Blackwater estuary,
    319
  in Botany bay, 433
  in Columbia river estuary, 411
  in Fal estuary, 319
  in Irish sea, 295
  in James estuary, 364-365
  in Kennebec estuary, 326
  in Mersey estuary, 262
  in North sea, 319
  in Osaka bay, 438
  in Oslofjord, 306
  in Randers fjord, 308
  in Red Wharf bay, 293
  in Schultz's grund, 308

Diffusion induced circulation, 203-208
Discovery passage, 401-402
Dispersion, circulation, 197
  coefficient, 177
  conservative, 227-230
  definition, 175
  nonconservative, 230-235
  tidal diffusion, 186
Dispersion coefficient, in Ariake bay, 438
  in Columbia river estuary, 411
  in Hudson river, 339
  in Irish sea, 297
  in Mersey estuary, 259
  in Potomac estuary, 354
  in San Francisco bay, 415-417
  in Severn estuary, 267
  in Southampton estuary, 276
  in Tay estuary, 286
  in Thames estuary, 282
  in Yaquina estuary, 412
Dissolved oxygen, definition, 226, 246
  in Delaware estuary, 345-347
  in Duwamish river, 407
  in James estuary, 248
  in New York harbor, 337-341
  in Nordfjord, 303
  in Oslofjord, 307
  in Penobscot estuary, 325
  in Saanich inlet, 398
  in Snohomish river, 407-408
  in Vancouver island inlets, 397
Distortion rate, 45
Duwamish river, 406-407
Dye diffusion, in Ariake bay, 438-439
  in Blackwater estuary, 317-318
  in Chesapeake bay, 349-351
  in Fal estuary, 316-317
  in North sea, 319-320
  in Potomac estuary, 352-354
  in San Francisco bay, 414

Ekman spiral, 152
Elbe estuary, 309
Ems estuary, 310
Energy, hydraulic, 98
  kinetic, 79
  potential, 79
  total, 80
Energy flow, 80
Entrainment, 102, 167

Entrainment mixing, 165-174
Equipotential surface, 12
Equivalent downestuary transport, 156
Estuary, arrested salt wedge, 2
  classification, 1-5
  definition, 1
  fjord entrainment, 2
  longitudinal division, 3
  stratified, 2
  well mixed, 2
Eulerian equations, 9
Exchange coefficient, 24
Exchange ratio, 159
  in Bay of Fundy, 381
Expansion, 17

Fal estuary, 316-317
Fenholloway estuary, 370-371
Fickian diffusion equation, 24
Fjord entrainment flow, circulation, 102-105
  definition, 2
  mixing, 165-174
Flocculent layers, in Chao Phya estuary, 428
  in San Francisco bay, 414
  in Savannah river, 367-368
  in Severn estuary, 270-274
  in Thames estuary, 283-284
Flushing rate, 158
Flushing time, 20, 155, 157, 158, 159
  in Bay of Fundy, 380
  in Black sea, 314
  in Irish sea, 297
  in Mediterranean sea, 314
  in Mersey estuary, 259
Forces, baroclinic, 125
  barotropic, 125
  body, 13
  bottom friction, 53-58
  centrifugal, 143
  Coriolis, 40, 140
  density gradient, 38, 117
  geostrophic, 42
  gravitational attraction, 40
  interface slope, 37, 113
  internal friction, 42
  pressure gradient, 13
  surface slope, 37, 113, 117
  transverse, 139-144

viscosity, 42, 45
Frequency, 36
  circular, 74
Fresh water fraction, 155
Friction, bottom, 53-58
  interfacial, 106
  wind, 113
Frictional depth, 153
Frictional stresses, 50
Frierfjord, 303-306
Fronts, 132-139
  in Delaware bay, 132
  in Long Island sound, 133, 138-139, 328-333
Froude number, 68

Galveston bay, 378-379
Gaussian distribution, 28
George's river, 431
Georgia strait, 418-421
Gibraltar, 310-313
Grays harbor, 411

Hardangerfjord, 298-302
Hawkesbury river, 431-432
Helland-Hansen relation, 142
Hooghly estuary, 428
Hudson river, 335-341
Hydraulic jump, 62, 96
  internal, 101
Hydraulic radius, 69

Interface slope, 113
Irish sea, 294-297
Iron, in Merrimack river, 334
Irrotational motion, 11
Ishikari river, 434-437
Isobaric surface, 141

James estuary, 360-365
Japan rivers, 433-437
Juan de Fuca strait, 418-421

Kelvin waves, 83
Kennebec estuary, 325-326
Kinetic energy, 79
Knight inlet, 391-394, 395-397
Knudsen relations, 20-21

Lagrangian equations, 9

Laplace equation, 17
Level of no motion, 141
  in James estuary, 362
  in Rappahannock estuary, 358
Liverpool bay, 264
Long Island sound, 328-333, 382-384
Longitudinal salinity gradient zone, 3, 429-
      430

Macleay river, 431-432
Manning roughness factor, 69
Mass continuity equation, 16, 177
Mass transport, 145
Mediterranean sea, 313-314
Merrimack river, 334-335
Mersey estuary, 254-266
Miramichi estuary, 324-325
Mississippi river, 374-378
Mixing, in Alberni inlet, 402-406
  in Amazon river, 379-380
  in Columbia river estuary, 408-411
  in Duwamish river, 406-407
  in Frierfjord, 305-306
  in Grays harbor, 411-412
  in Hardangerfjord, 302
  in James estuary, 360-365
  in Mersey estuary, 260-266
  in Mississippi river, 375-378
  in Oslofjord, 306-308
  in Southampton estuary, 274-277
  in Thames estuary, 278-282
  in Vellar estuary, 422-427
Moments, 240
Mortality coefficient, in Raritan estuary,
      342

Navier Stokes equation, 45
Neighbor concentration, 33
Neighbor diffusivity, 33
Neighbor separation theory, 33
Neroutses inlet, 400
New York harbor, 335, 341
Nitrate, in Saanich inlet, 398-400
Nitrogen cycle, in Delaware estuary, 347-
      348
  in Potomac estuary, 355
Nordfjord, 303
North sea, 319-320
Northumberland channel, 400-402
Null zone, 417

Ochlockonee bay, 373-374
Osaka bay, 437-438
Oslofjord, 306-308
Overmixing, 162-165
  in St. John estuary, 324
Overtides, 77

Patuxent estuary, 355-356
Penobscot estuary, 325
Perfect fluid, 12
Period, 75
Port Ajiro, 433-434
Port Hacking, 430-431
Portland inlet, 389-392
Potential energy, 79
Potomac estuary, 352-355
Pressure, 13
Progressive wave, 78

Randers fjord, 307-309
Rappahannock estuary, 357-360
Raritan estuary, 341-344
Rayleigh number, estuarine, 126, 197
Reaction coefficients, in Potomac estuary,
      355
Reaeration coefficient, in New York harbor,
      338
Red Wharf bay, 287-294
Residence time, 417
Resonance, 81
Reversing falls, 322
Reynolds number, 68
Reynolds stresses, 49
  in Choptank estuary, 357
  in Kennebec estuary, 325
  in Mersey estuary, 262
  in Red Wharf bay, 288, 290, 292, 293
Richardson flux number, 209
Richardson number, 53, 209
  in Kennebec estuary, 326
  in Mersey estuary, 214, 262
  in Randers fjord, 214, 308
  in Schultz's grund, 214, 308
River flow, in Alaska inlets, 387
  in Amazon river, 379
  in Australian estuaries, 429
  in British Columbia inlets, 389
  in Columbia river estuary, 408
  in Connecticut river, 327
  in Duwamish river, 407

in Grays harbor, 411
in Hardangerfjord, 301
in Hudson river, 339
in Mersey estuary, 259, 261, 263
in Mississippi river, 374
in Potomac estuary, 352
in Raritan estuary, 341, 342
in San Francisco bay, 413
in St. John estuary, 324
in Tay estuary, 285, 286
in Vellar estuary, 423
in Yaquina estuary, 412
River runoff, 118
Rossby number, 68

Saanich inlet, 398-400
Salt continuity equation, 18, 26, 177, 191-192
Salt flux, by advection, 176
  by circulation, 176, 197
  by lateral circulation, 203
  by lateral velocity shear, 203
  by tidal diffusion, 176, 186
  by velocity shear, 176, 197
  definition, 18
  in Columbia river estuary, 410-411
  in Frierfjord, 305
  in James estuary, 361
  in Mersey estuary, 261, 263
  in Vellar estuary, 427
Salt wedge, circulation, 105-112
  definition, 2
  in Duwamish river, 407
  in Ishikari river, 434-437
  in Mississippi river, 375
Salinity, definition, 18
  in Alberni inlet, 402-405
  in Alligator harbor, 371-374
  in Bosporus, 311
  in Botany bay, 431
  in British Columbia inlets, 389-391
  in Connecticut river, 326-327
  in Frierfjord, 304
  in George's river, 431
  in Hardangerfjord, 299, 300
  in Hawkesbury river, 432
  in Hooghly estuary, 428
  in Irish sea, 296
  in Ishikari river, 436
  in James estuary, 362

in Knight inlet, 393
in Macleay river, 432
in Merrimack river, 332-335
in Mersey estuary, 260
in New York harbor, 335
in Ochlockonee bay, 374
in Oslofjord, 307
in Penobscot estuary, 325
in Port Hacking, 430
in Portland inlet, 392
in Randers fjord, 308
in Rappahannock estuary, 357-360
in Raritan estuary, 341
in Schultz's grund, 308
in Silver bay, 388
in Southampton estuary, 275
in St. Andrew bay, 374
in St. John estuary, 323
in Swan river, 433
in Vellar estuary, 423-427
  vertical distribution, 194, 195, 196, 197
San Francisco bay, 412-418
Savannah harbor, 366-367
Schultz's grund, 307-309
Sedimentation, in Chao Phya estuary, 428
  in Charleston harbor, 367
  in Mersey estuary, 266
  in Mississippi river, 375
  in New York harbor, 336
  in San Francisco bay, 414
  in Savannah harbor, 366
  in Thames estuary, 283, 284-285
Severn estuary, 266-274
Silica, in Merrimack river, 334
  in San Francisco bay, 416-418
Silver bay, 385-388
Similarity solutions, 126-127
Snohomish river, 407-408
Solent, 277-278
Southampton estuary, 274-277
Standing wave, 77
St. Andrew bay, 374
St. John estuary, 322-324
St Johns estuary, 367-369
Stratification — circulation diagram 3-5, 199
Stratified estuary, circulation, 112-116
  definition, 2
  mixing, 191-201
Stratified flow, 116-125
Stream function, 145

Stream lines, 12
Stresses, frictional, 50
  Reynolds, 49
  wind, 113
Subcritical condition, 64, 96, 101, 104
Supercritical condition, 64, 96, 101
Surface slope, 113
Suspended sediments, in Chao Phya estuary,
    428
  in Chesapeake bay, 351-352
  in Rappahannock estuary, 357-360
Swan river, 431-433

Tay estuary, 285-286
Temperature, in Bute inlet, 395
  in Knight inlet, 393
  in Osaka bay, 438
  in Oslofjord, 307
  in Silver bay, 388
Thames estuary, 278-285
Thermocline, seasonal, 218-225
Tidal mixing, 178-191
Tidal prism, 157
  in Mersey estuary, 258
Tidal prism method, 158-160
Tidal waves, 70
  shallow water constituent, 74
Tides, cooscillating, 80
  damping coefficient, 88
  energy flow, 94
  energy loss, 94
  gravitational energy, 94
  in Alligator harbor, 371-374
  in Bay of Fundy, 381-382
  in Biscayne bay, 369-370
  in Bristol channel, 268-271
  in Columbia river estuary, 409
  independent, 80
  in Galveston bay, 378-379
  in Georgia strait, 418-421
  in Irish sea, 294-295
  in Juan de Fuca strait, 418-421
  in Long Island sound, 383-384
  in Northumberland channel, 400-402
  in Solent, 277-278
  in Thames estuary, 282-283
Transitions, 62
  internal, 97
Turbulence, in Choptank estuary, 356-358
  in Discovery passage, 401-402

  in Kennebec estuary, 325-326
  in Mersey estuary, 254-258
  in Patuxent estuary, 355-356
  in Red Wharf bay, 287-294
Turbulence spectra, 35-36
Turbulent diffusion, 26-31

Vancouver island inlets, 397-402
Vellar estuary, 422-427
Velocity, bottom, 131
  circulation, in Bab el Mandeb, 313
    in Bosporus, 312
    in Columbia river estuary, 410
    in Dardanelles, 312
    in Frierfjord, 304
    in Gibraltar, 313
    in Hardangerfjord, 301
    in Ishikari river, 436
    in James estuary, 361, 365
    in Knight inlet, 396-397
    in Mersey estuary, 260
    in Port Ajiro, 434
    in Silver bay, 388
    in Southampton estuary, 275
    in vellar estuary, 426-427
  critical, 64, 100
  deviation, 24
  instantaneous, 24
  longitudinal, 118, 120, 122, 195, 196
  mean, 24
  tidal, in Mersey estuary, 261
  tidal, in Mississippi river, 376-377
  tidal, in Red Wharf bay, 292, 293
  tidal, in St. John estuary, 323
  tidal wave, 70
  vertical, 128
    in Hardangerfjord, 301
    in James estuary, 363
    in Southampton estuary, 276
Velocity potential, 12
Velocity shear, 45
Vertical stability, 53, 209
Viscosity, coefficient, 43, 48
  definition, 42
  eddy, 43
  kinematic coefficient, 50, 86, 87, 123,
    124
  molecular, 43
Viscosity coefficient, in Kennebec estuary,
    326

in Mersey estuary, 262
in Randers fjord, 308
in Schultz's grund, 308

Waccasassa estuary, 370-371
Wavelength, 72
Wave number, 35, 88
Waves, Kelvin, 83
  long, 70
  progressive, 78

standing, 77
tidal, 70
Wave velocity, 72
Well mixed estuary, circulation, 116-
  125
  definition, 2
  mixing, 191-201

Yaquina estuary, 411-412
York estuary, 360